Normal and Abnormal Fetal Face Atlas

Jean-Marc Levaillant
Jean-Philippe Bault • Bernard Benoit
Gérard Couly

Normal and Abnormal Fetal Face Atlas

Ultrasonographic Features

Springer

Jean-Marc Levaillant
Center of women and fetal imaging
Créteil
France

Bernard Benoit
Princess Grace Hospital
Monaco
France

Jean-Philippe Bault
Center of fetal imaging Ambroise Paré
Les Mureaux
France

Gérard Couly
Department of maxillo-facial surgery
Necker Hospital
Paris
France

Translation from the French language edition 'La face foetale normale et pathologique: aspects échographiques' by Jean-Marc Levaillant, Jean Philippe Bault, Bernard Benoit, Gérard Couly © Sauramps Médical, Paris, 2013; ISBN: 978-2840238690

ISBN 978-3-319-43768-2 ISBN 978-3-319-43769-9 (eBook)
DOI 10.1007/978-3-319-43769-9

Library of Congress Control Number: 2017933711

Printed on acid-free paper

This Springer imprint is published by Springer Nature
The registered company is Springer International Publishing AG
The registered company address is: Gewerbestrasse 11, 6330 Cham, Switzerland

Foreword

In the field of obstetrics and fetal medicine, there are a few invariables in clinical use of ultrasounds. The operator must have a clear representation of the structures he or she is analyzing and the possible anomalies, which supposes knowledge of basic physiology, anatomy, and embryology. He or she must be able to use the spectrum of technical solutions available, know what to expect from them, and use them in the most up-to-date fashion, such as not generating images that are aesthetically pleasing but lack information. He or she should take a systematic approach, standardizing imaging via an analysis through the planes of symmetry; the idea is to be able to reproduce the examination and to avoid creating nice-looking but fallacious images. When it is pertinent, use of objective quantitative criteria should be privileged.

Communication should also be taken into consideration—a communication based on transmitting information rather than producing flashy images. It is important to understand the expectations of the various partners involved—geneticists, pediatricians, surgeons, and parents—and be able to respond to their specific questions.

Sharing experience is an important part of our practice. Teaching involves disseminating knowledge and showing targeted images; it also involves transmitting how to obtain these images via a multitude of approaches: words, screen captures, hands-on tutoring, and live demonstrations.

This list must also include (self-) assessment, which is at the heart of any responsible practice.

This atlas of normal and abnormal fetal face ultrasound imagery presented by Jean-Marc Levaillant, Jean-Philippe Bault, Bernard Benoit, and Gérard Couly fits this approach. It provides a clear, didactic approach with a review of the basic embryology and anatomy needed to understand the images, along with up-to-date clinical knowledge and the links between clinical practice and the images under both normal and various pathological situations. It also includes a collection of biometric curves. Beyond the images themselves, readers will appreciate the authors' explanations of the optimization and utilization of the various ultrasound imagery modes, as well as how to obtain the various shots and corresponding images.

In this respect, this atlas is another step in the development of the French School of Gynecological and Obstetrical Ultrasonography, which includes some major developments, including pioneering developments in 3D ultrasonography in collaboration with Philips Research Laboratories, followed by the first industrial developments with KretzTechnic; training in 2D and 3D ultrasound imagery at the Saint-Denis Ultrasound School, with trainers from all over France and let by the tireless and ubiquitous applied engineer Bernard Meyer; and journal publications on fetal face ultrasound imagery.

Daniel Rotten
Department of Gynecology and Obstetric
Delafontaine Hospital
Saint-Denis
France

Acknowledgements

We would like to thank our colleagues who contributed to the imagery found in this atlas:

Caroline Alby
Bettna Bessières
Christian Bisch
Joseph Bonan
Rabi Chaoui
Sophie Couderc
Laurent Guibaud
Soraya Kabar
Brigitte Leroy
Anne Elodie Millischer
Daniel Moeglin
Marc Molho
Edwin Quarello
Ahmed Sadji
Karima Sedikhi

Contents

1 Introduction Craniofacial Epigenesis in Vertebrates

The facial bones, as they appear in a fetal ultrasound image, result from a continuous process of development, regulated and sequenced by genetic and molecular embryology, which progressively constructs the face like epigenesis.

The face, like the entire head, is "prefigured" by prescribed regions that are genetically determined during the gastrula and neurula stages. Via typological deformations, the two-dimensional neural surface transforms symmetrically and acquires a third dimension.

When the dorsal region of the neural grove closes into a tube, the "fourth layer" of migrating neural crest cells invade the future embryonic cephalic pole and provide the cellular filling for the facial prominences.

The face develops thanks to the five founding prominences or swellings of the ectoderm, which grow volumetrically and then fuse by contact and are no longer recognizable as having existed once the face is totally formed. These prominences are:

- The frontonasal prominence, which is median, single, symmetrical and different from the other four as it is formed by two left and right halves.
- The maxillary prominences, right and left.
- The mandibular prominences, right and left (first branchial arches).

1.1 The Key Stages in Craniofacial Development

The prescribed regions of the face and brain are differentiated as early as the gastrula and neurula stages.

Gastrulation lays out the embryonic body plan for all vertebrates with bilateral symmetry via the formation of the notochord and the mesoderm interposed between the endoderm and the ectoderm (Fig. 1.1) (this beginning explanation of epigenesis or the phenomenology of development by stages replaced the eighteenth century concept of preformation that stemmed from the building of the first microscope, when Leeuwenhoek, its inventor, identified the spermatozoid in 1677).

The future regions of the face are situated in the rostral regions of the gastrula.

Neurulation is a theater of morphogenetic operations that will differentiate the prescribed regions of the head (Fig. 1.2). This is the stage during which the central nervous system, the facial mass and the neck truly begin their morphogenesis.

The neural plate is the two-dimensional cellular surface whose fate mapping demonstrates the closeness and unity of the regions of the brain, the neuro-sensorial receptors, and the face, genetically "printed" on the neural plate.

J.-M. Levaillant et al., *Normal and Abnormal Fetal Face Atlas*, DOI 10.1007/978-3-319-43769-9_1

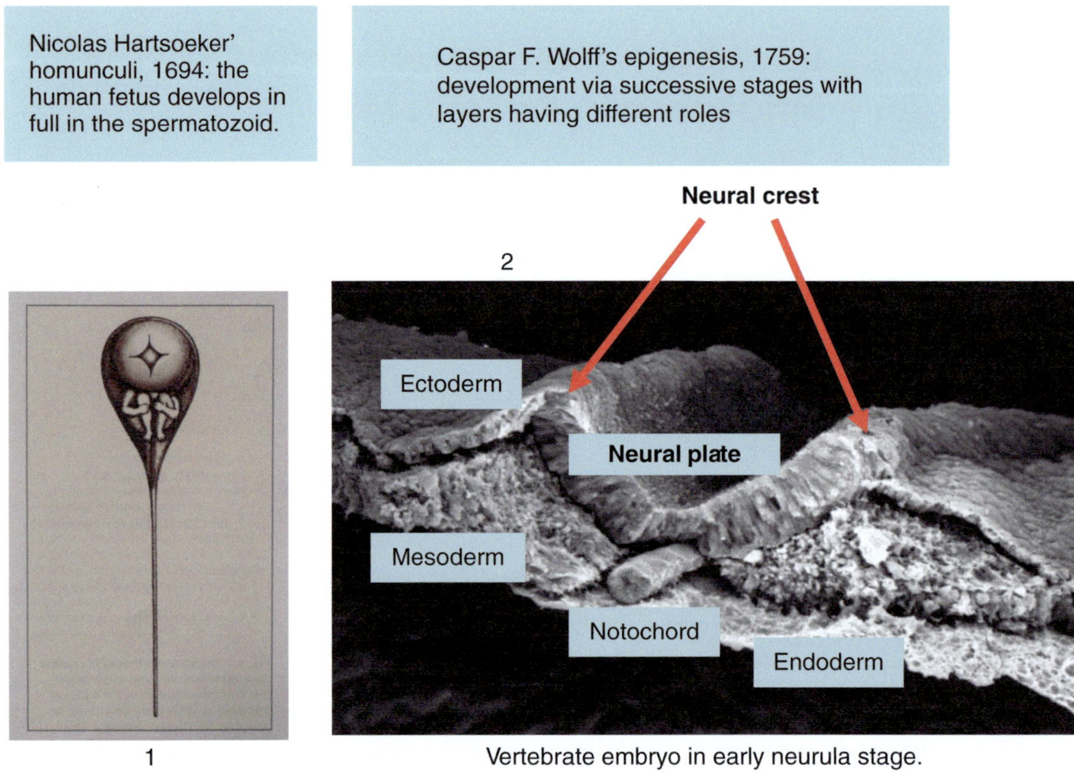

Fig. 1.1 Preformationism vs. epigenesis. *1* Homunculus. *2* End of gastrula phase, beginning of neurula stage

During neurulation, the topological transformation of the neural plate continues. The face is in the neurula and "grows" in front of it (Fig. 1.3).

The neural plate lengthens and folds, forming a groove with edges rising up and fusing into a tube (Figs. 1.4 and 1.5).

Closure of the neural tube's rostral opening represents an exceptional topological originality (Fig. 1.6): the anterior-most parts of the neural plate roll up forward, completing a half-circle rotation (180°), flipping these regions into a ventral position (it is an elegant and original solution to the problem of closing a tube that is open in the front (Figs. 1.7 and 1.8).

The lateral edges of the neural tube come together and fuse in the middle. As a result, the adenohypophysis (Fig. 1.9), initially the anterior-most neuro-ectodermic region, moves under the diencephalon and thus becomes the posterior-most region, above the future stomodeum. The two olfactory placodes sprout nerves and the olfactory apparatus, initially located above and on either side of the telencephalon and the adenohypophysis, approach the median line due to the median fusing of the neural ridges and position themselves ventrally under the tele-dicephalon, or the anterior brain (Fig. 1.10).

This is how the future nasal cavities and the rhinencephalon are linked for the sense of smell.

Jointly, the retina surfaces on the neural plate "blister" outwards and becomes the optic cups and then the optic vesicles (see Fig. 1.7).

The dorsal closure of the neural tube by molecular fusion triggers the migration of cephalic neural crest cells (CNCC) **(see** Figs. 1.4, 1.11, 1.12, 1.13, 1.14 and 1.15).

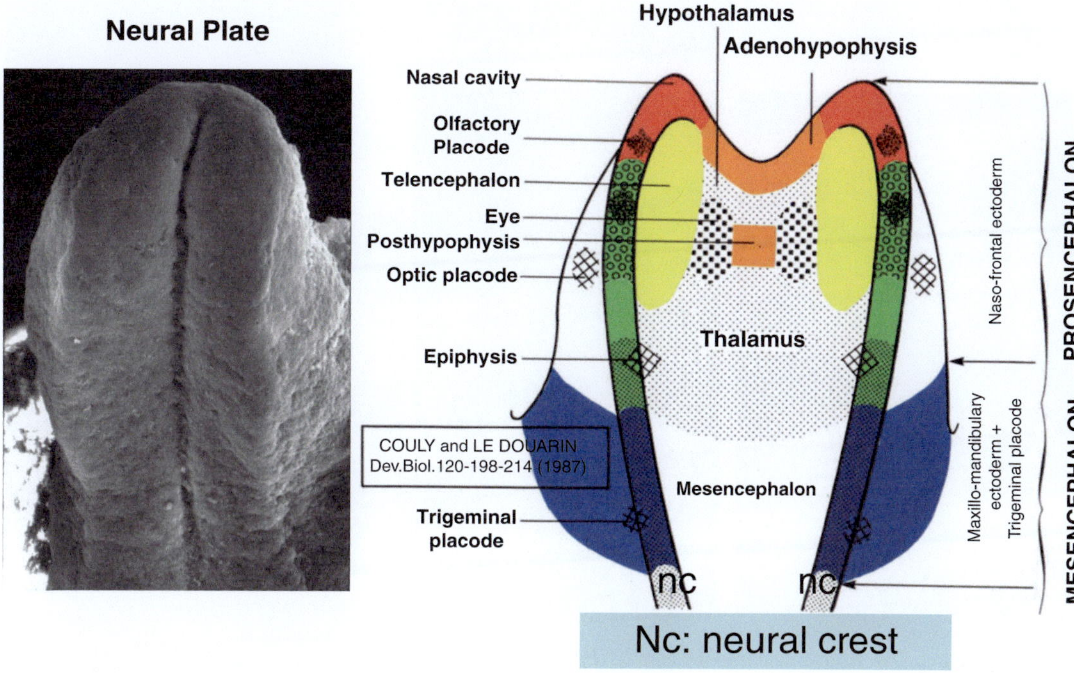

Fig. 1.2 Map of neural plate regions

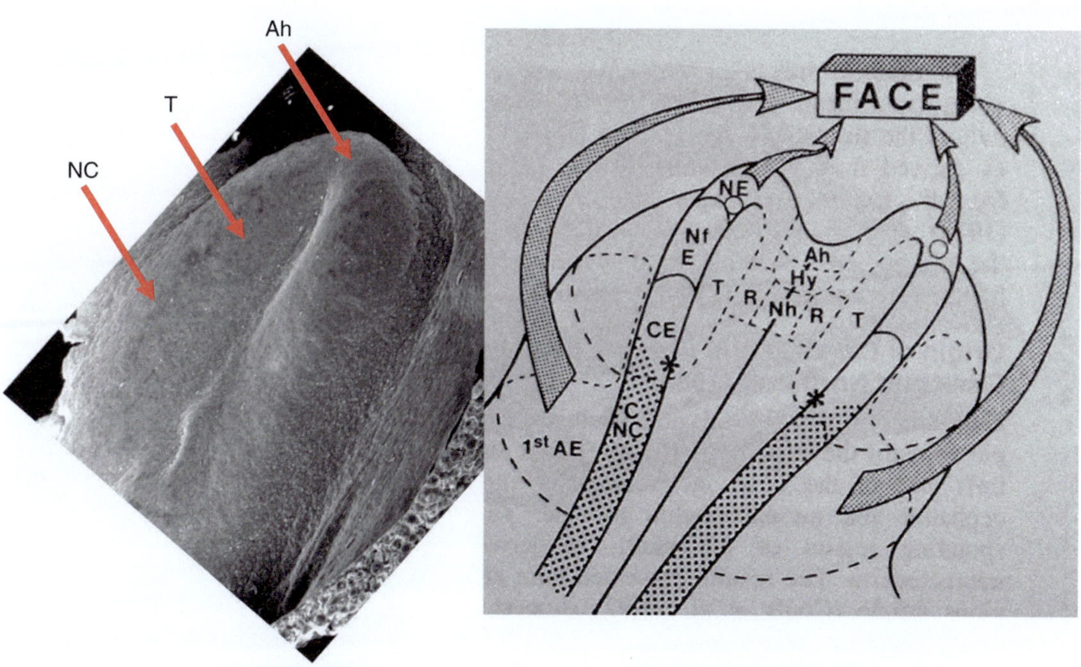

Fig. 1.3 The face grows in front of the neural plate

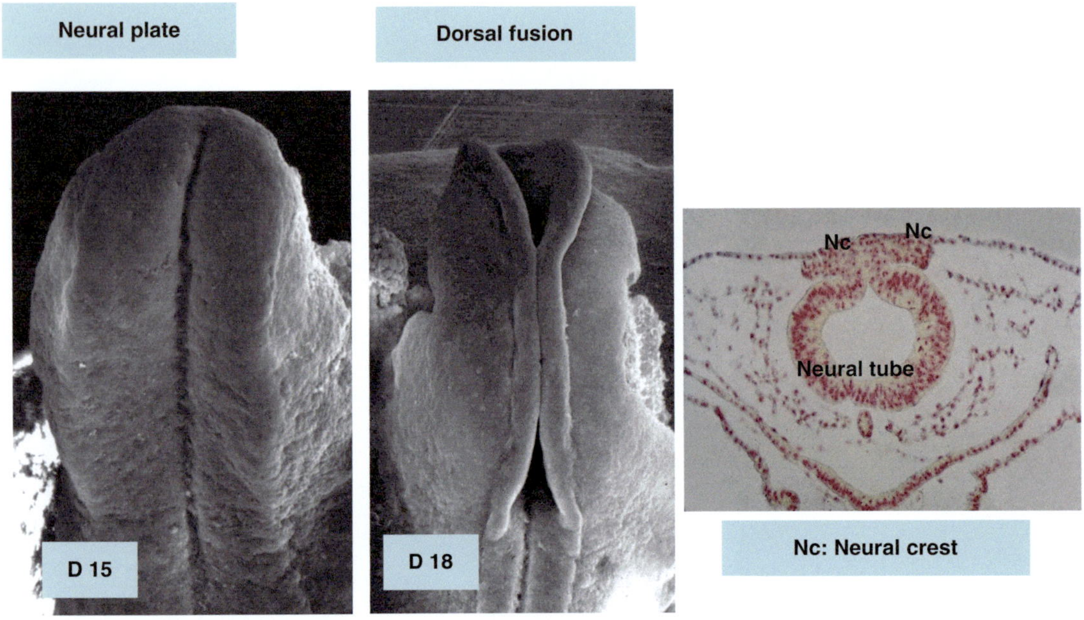

Fig. 1.4 Closure of the neural plate into the neural tube, with neural crest
D 15 – The neural plate
D 18 – Dorsal fusion
NC Neural crest
Neural tube

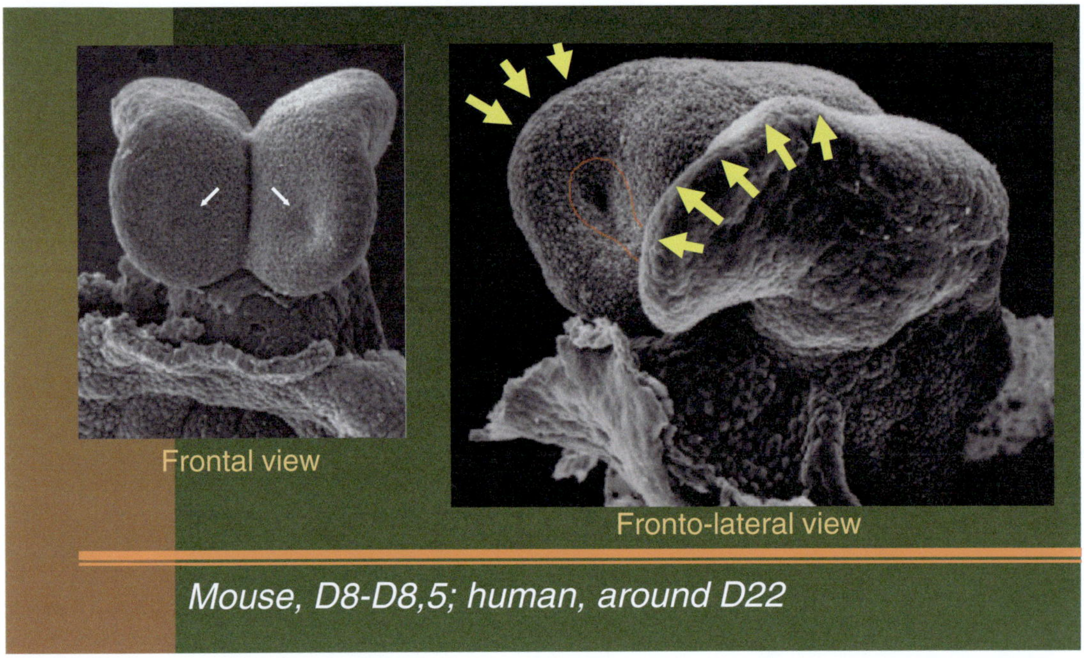

Fig. 1.5 Topogenesis of the anterior neural plate and beginning of the optic cup.
The optic cups invaginate. Elevated neural ridges move closer together on the dorsal side of the embryo
Frontal view
Fronto-lateral view
Mouse, D8-D8, 5; human, around D22

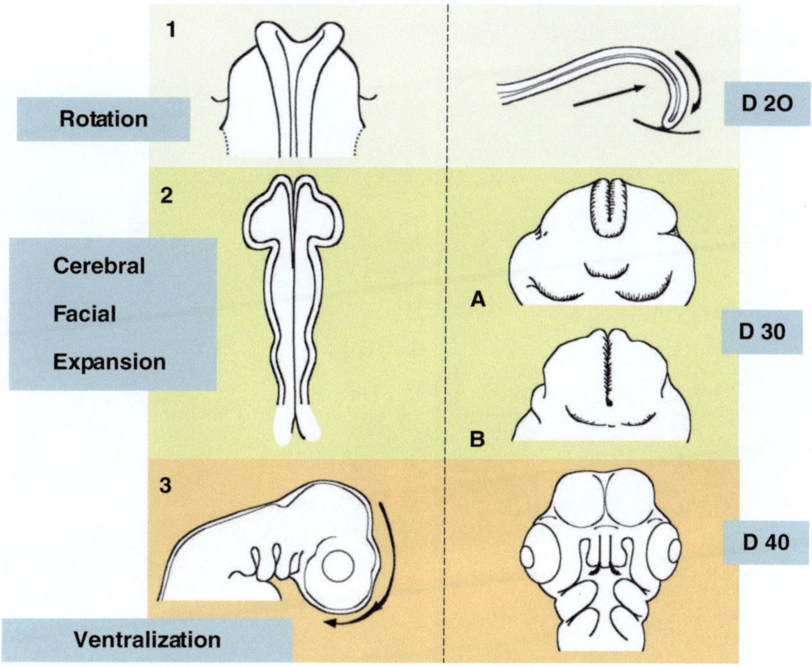

Fig. 1.6 Overall ventral rotational movement of the rostral extremity of the embryo
1. Rotation–D20
2. Facial cerebral expansion – D30
3. Ventralization – D 40

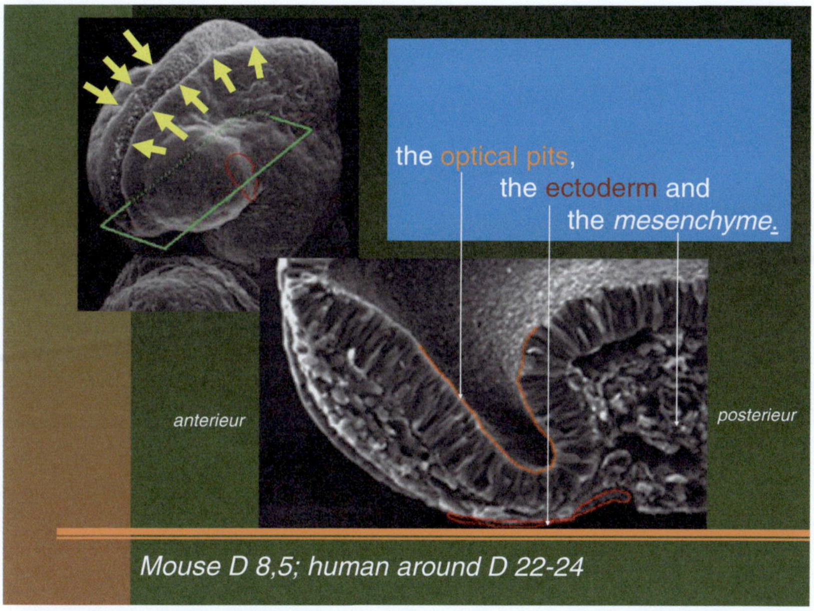

Fig. 1.7 Optical vesicles and the median closure of the rostral extremity
In this cut, one can observe the links between:
The optical buds, the ectoderm and the mesenchyme
Anterior
Posterior
Mouse D8, 5; Human around D22-24

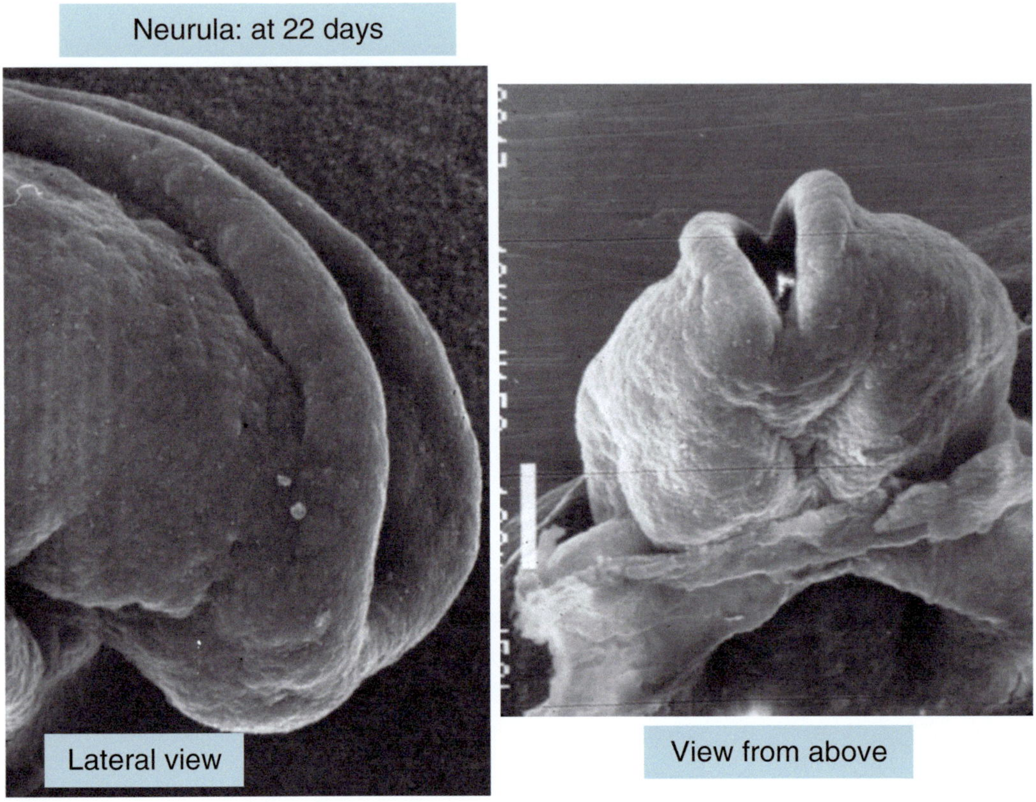

Fig. 1.8 Scanning microscope images of anterior neural topology
Neurula: Day 22
Lateral view
View from above

As the neural tube lengthens and rolls over 180° to achieve rostral and median closure, CCNC migrate dorso-ventrally, leaving the central nervous system and invading the cranial extremity of the embryo, between the ectoderm and the endoderm.

This very particular strategy determines the cellular volume of facial morphogenesis.

Thanks to their serious mitotic capacities, these cells will collect around the optic vesicles, the olfactory placodes and the branchial placodes. At this stage, the right and left olfactory placodes attract neural crest cells as a result of the expression **of mitogenetic molecules**, **growth factors such as FGF 8**, **transcription proteins**, **and the DLX5**, **SLX6 and SHH genes**, **which are strongly expressed in these placodes**.

The CNCC amass around these placodes, and thanks to their mitoses, ensure the development of the five prominences of the future face:

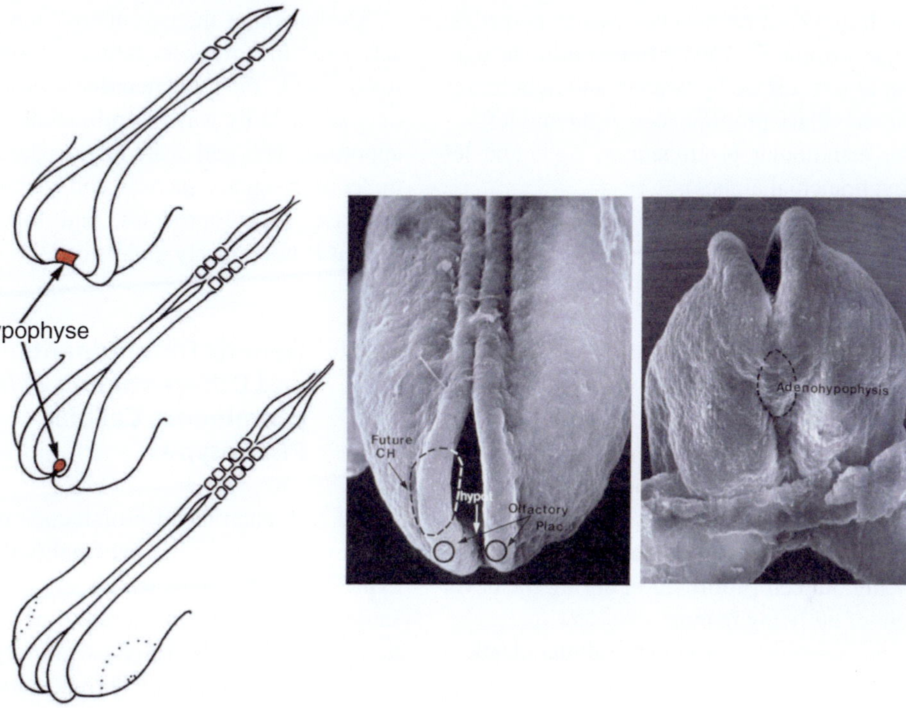

Fig. 1.9 Rostral neural topogenesis and final position of the adenohypophysis
The rostral dorso-ventral rotation outlining the pituitary gland in vertebrate embryo

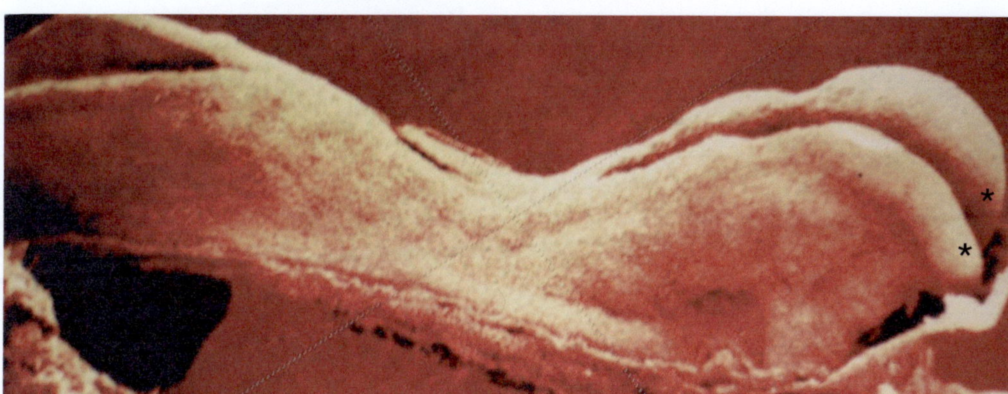

Fig. 1.10 Human neurula at 27 days (Dr. C. Nilsson)
Telencephalon and olfactory placodes (*)

- The frontonasal prominence, which is median, single, symmetrical and different from the other four as it is formed by two left and right halves.
- The maxillary prominences, right and left.
- The mandibular prominences, right and left (first branchial arches).

1.1.1 Cell Death

The fusion of the five face prominences after they come into contact with each other is only possible because of embryonic cell death, or apoptosis. This paradoxical developmental phenomenon was not recognized as fundamental at the time of its discovery in the 1960s (Gluckmann), but later became viewed as a process required for balancing out cell proliferation during the development of all living beings.

As an originator example, cellular death or morphogenetic necrosis of interdigital webbing contributes to separating the fingers and toes from each other. The first genes commanding cell death generate internal processes in each concerned cell, triggering a suicide, or cell sacrifice. These genes began to be identified in the early 1980s.

In every living cell, there exists a genetic death program, which when activated causes the cell to implode and disappear (this differs from cells that necrotize by explosion, trauma, or viral infection because they are surrounded by blood cells such as macrophages).

Cell death by apoptosis is quick, does not lead to inflammation, lesions, or scarring, which is why it went unnoticed for so long.

The first genes found to trigger cell death were CED-3 and CED-4 (Horwitz). Another identified gene—CED-9—antagonizes and represses these genes and their effects.

From the most primitive creatures, such as worms, through to mankind, these genes have remained functional, representing the same molecular process for millions of years. This system of cellular death and survival were preserved intact. The molecular trigger for cell suicide is linked to the activity of protease and caspase.

Cell death or apoptosis is essential to the formation of all facial anlage: **of the double symmetry of the upper lip and of the fusion of the basic face prominences**.

The fusion of the two internal nasal pits characterizes the median neural symmetry. This fusion with the two maxillary prominences is only possible by way of **embryonic cell** death or apoptosis. This cell death implies the existence of molecular signals such as collagenase, caspase, protease, metalloprotease, and folic acid. For now, it is not entirely understood.

1.1.2 Genetic Determinants and Differentiation of Intra-prominence Cellular Phenotypes

In the end, each facial prominence is characterized by its own genetic personality thanks to the expression of genes whose effects are represented by cellular proliferation and differentiation that ultimately develops into organs (bone, cartilage, muscles, teeth, dermis, nerves, vessels, etc.). Each prominence is thus responsible for the specific tri-dimensional construction of a specific skeletal and tissue territory of the facial mass, and of its anatomic and functional organization.

If a development gene is defective in a prominence, the latter cannot undergo satisfactory morphogenesis, which then impacts it final qualitative and/or quantitative development, by notably disturbing the fusion of this prominence with the adjacent ones, resulting in the various morphogenetic accidents that can be observed, such as clefts, hypoplasia, inclusion of embryonic relics, fistulas, and ectodermic cysts. Yet, these accidents will always conserve a developmental logic as well as facial and oral topographic coherence.

1.2 Cellular Phenotypes Originating from the Neural Crest (Fig. 1.15)

Classification of the normal and dysmorphic face (Fig. 1.16) — ***The six potential clefts around the mouth*** (Fig. 1.17).

Taking into account this genetic embryology, which leads to the **modular construction of the face**, it is possible to propose a modular topographical classification of facial and cervical

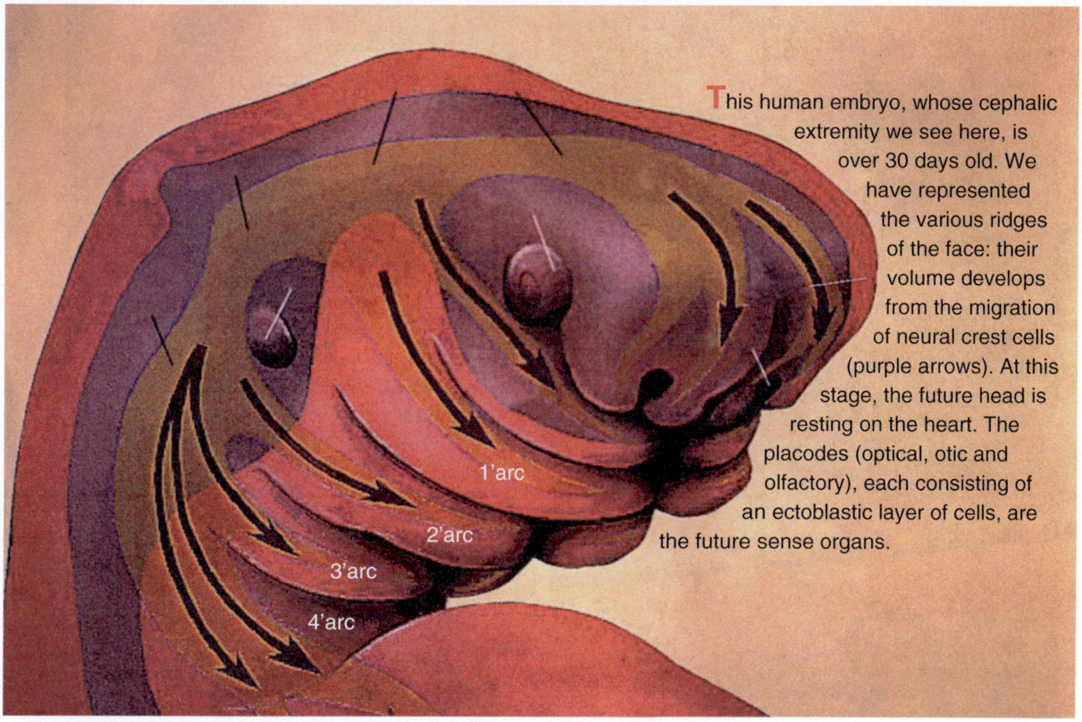

Fig. 1.11 Neural crest cells migrating and the development of facial prominences

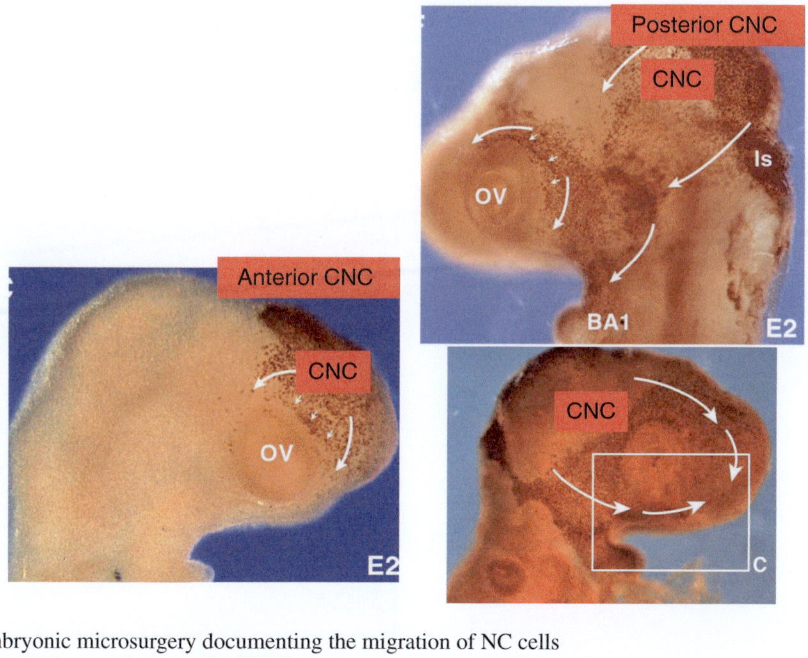

Fig. 1.12 Embryonic microsurgery documenting the migration of NC cells
Migration of cephalic neural crest cells (CNC)
Anterior CNC
CNC
Posterior CNC

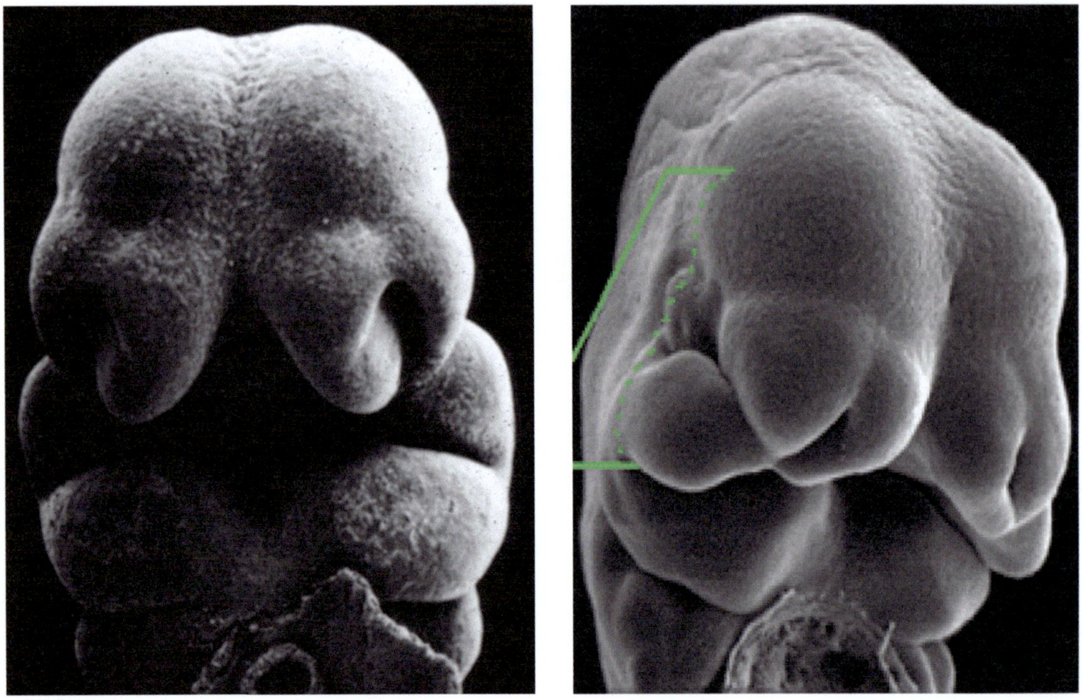

Fig. 1.13 Development of face prominences D31-36

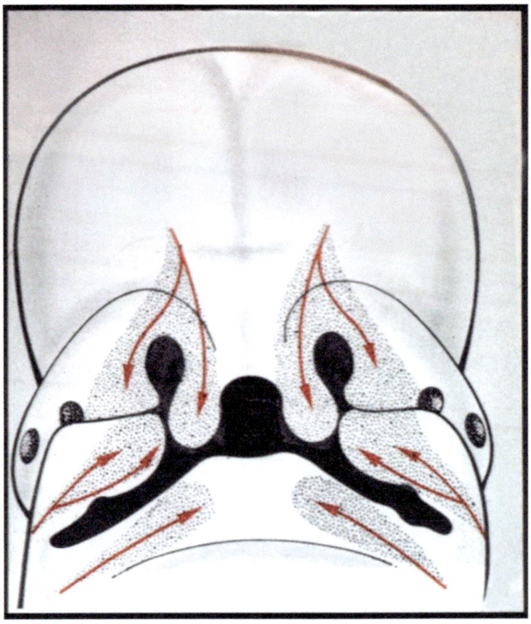

Fig. 1.14 Human embryo face at around day 42 Neural crest cells

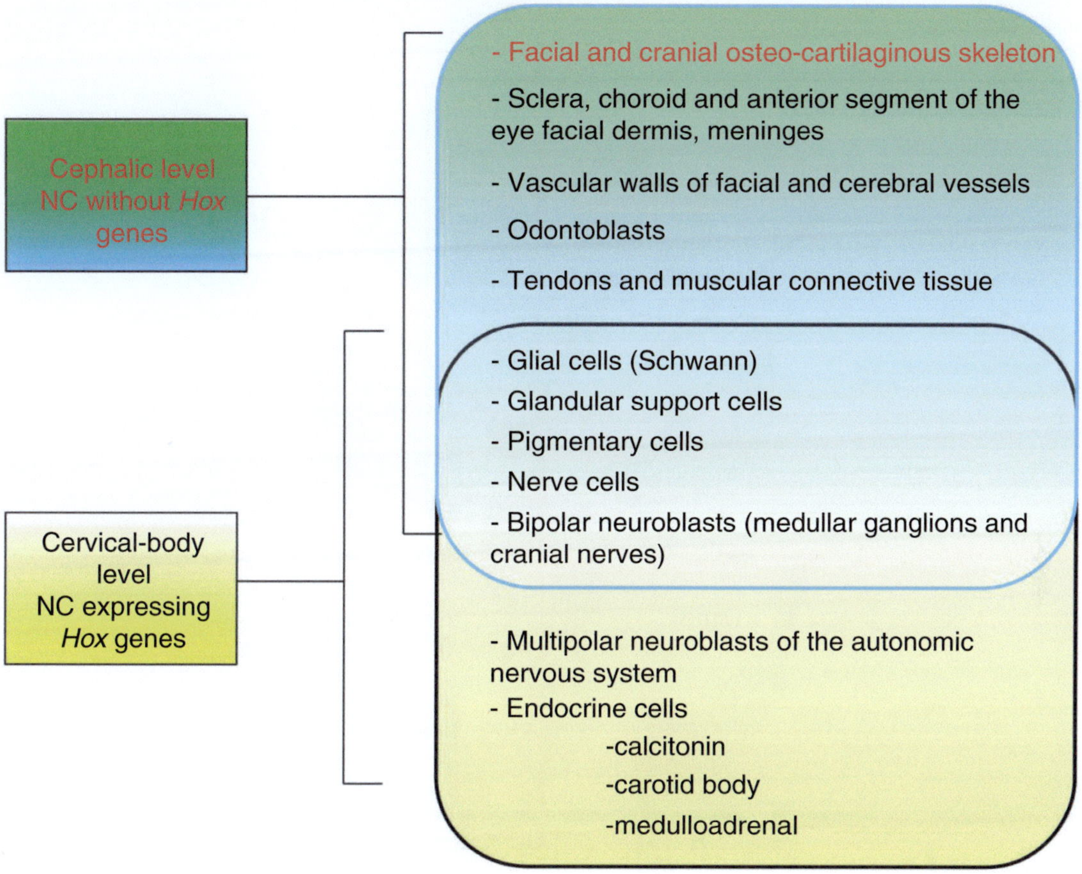

Fig. 1.15 Cellular phenotypes originating from the neural crest

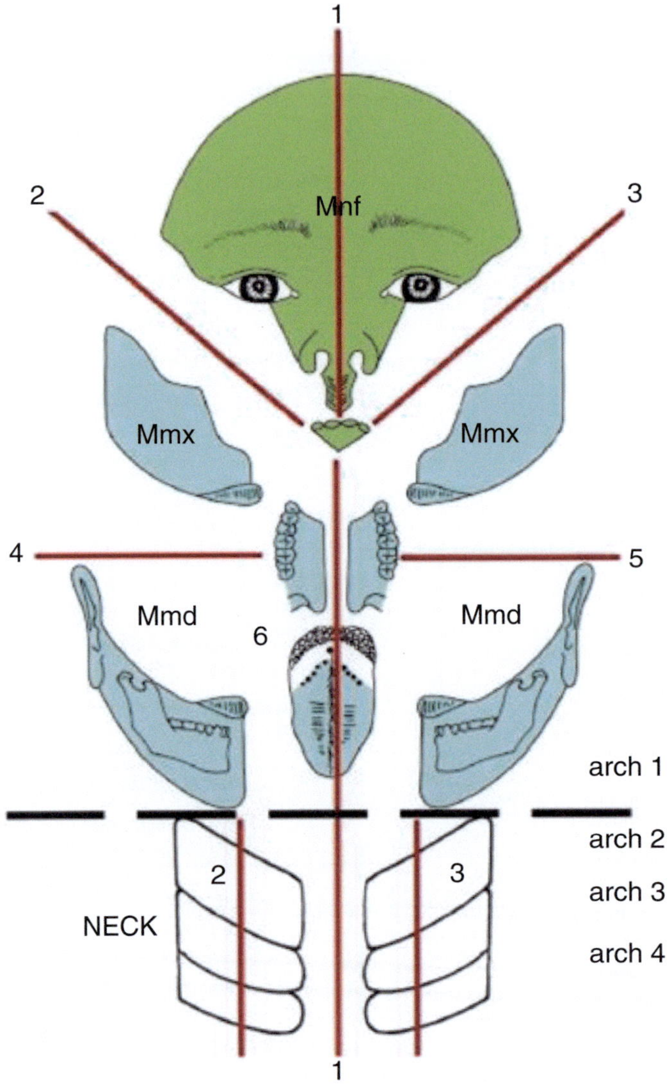

Fig. 1.16 Meridian and modular classification of morphological cephalic accidents

malformations and, as a result, provide keys for understanding them.

The median neural symmetry of the face results from the rostral closure of the neural tube (Fig. 1.18).

The frontal, single and median, prominence develops by the fusion of two halves, right and left, resulting from an original neural symmetry, the formation of which implies a process of fusion by molecular "median adhesion" by N.CAM (neural cell adhesive molecule) dependent on the expression of genes of the neural groove surface epithelium.

A total or partial defect in the median fusion of these two halves impacts the future nasofrontal prominence, which will present the frequent median defects—partial or total schism, encephalocele, fistula, cyst, etc.—that are observable with ultrasound.

This median facial cleft never extends beyond the adenohypophysis and the sella turcica. Embryology is a powerful predictive engine that is never at fault.

The maxillary and mandibular prominences have their own dysmorphisms, including intra-prominence clefts and hypoplasias. (**PL 1.19 and 1.20**).

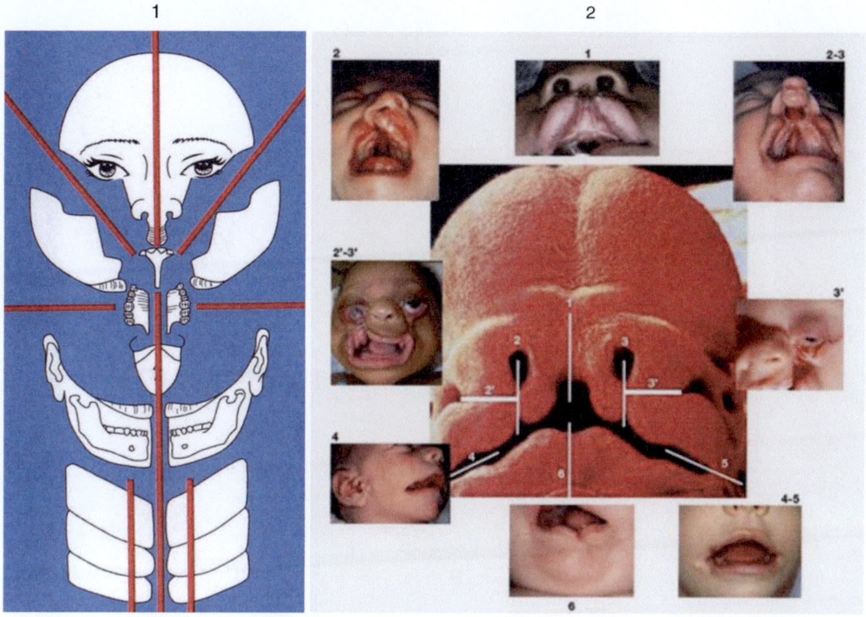

Fig. 1.17 The 6 clefts around the mouth

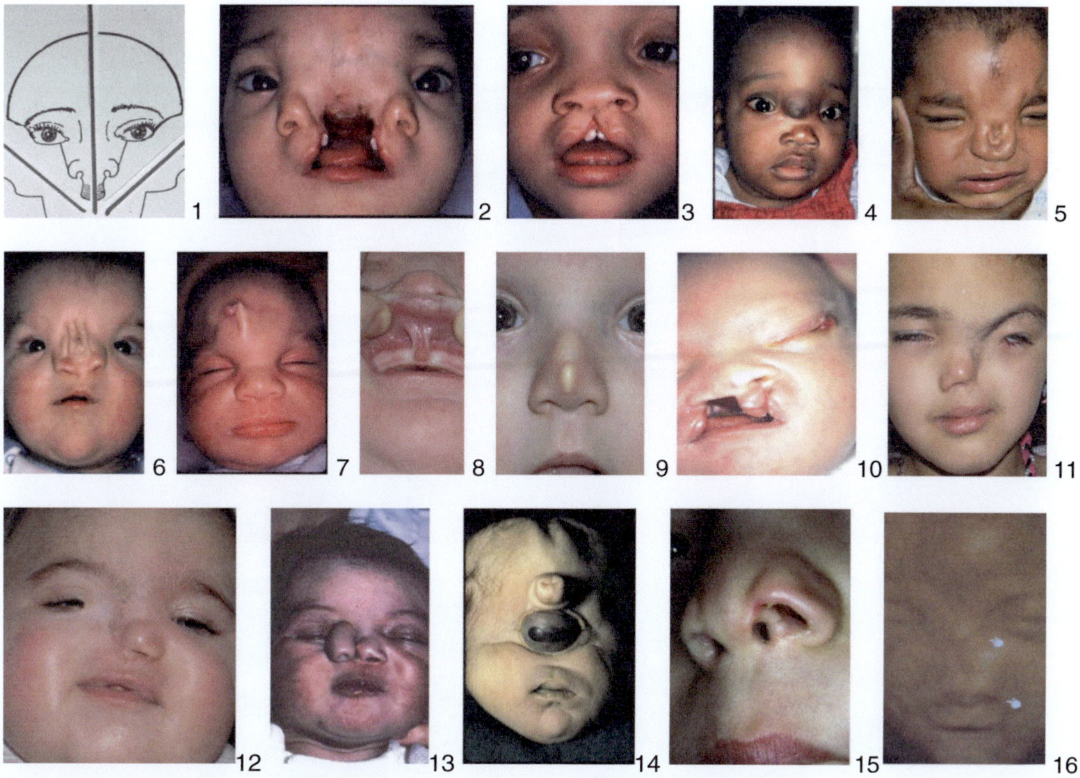

Fig. 1.18 Malformations of the nasofrontal-premaxillary prominence. From *1* to *9*: median clefts and frontonasal encephaloceles. *10*: bilateral labio-maxillary clefts and external orbital. *11*,*12*: hemi-arrhinia. *13*: nasal proboscis. *14*: cyclopia. *15*: nostril duplication. *16*: arrhinia (ultrasound)

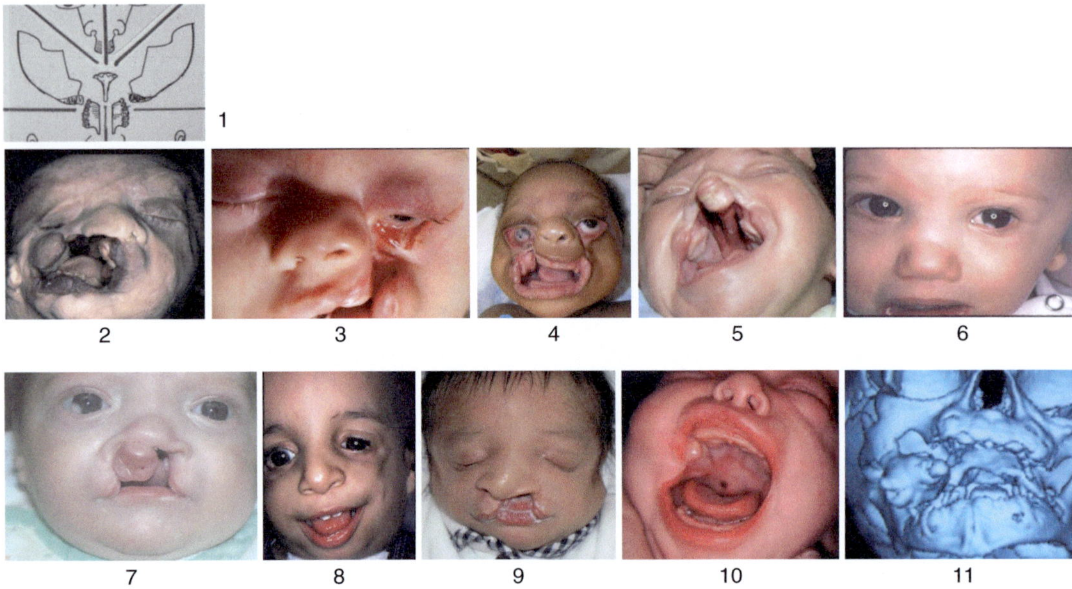

Fig. 1.19 Malformations of the maxillary prominences. *2*: aprosopia of the maxillary prominence. *3*, *4*: uni- and bilateral colobomas. *5*: labio-maxillary and bilateral velopalatine clefts. *6*, *7*: Goldenhar syndrome. *8*, *9*.: Franceschetti syndromes. *10*, *11*: homeotic duplication of the maxillary prominence

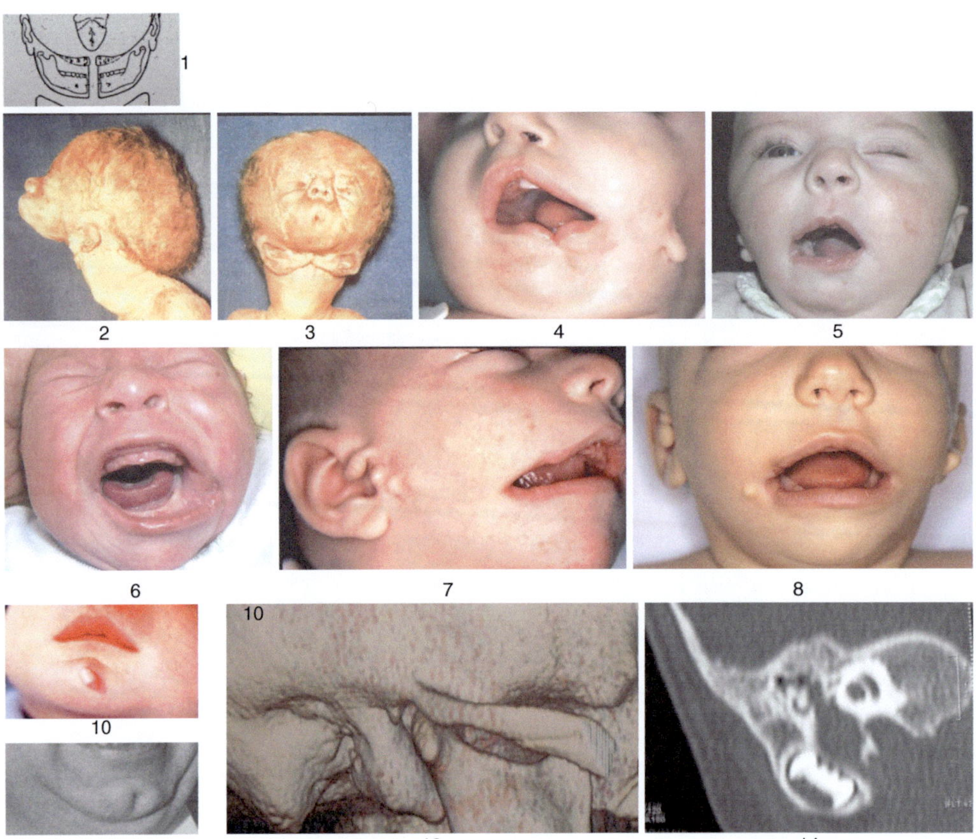

Fig. 1.20 Malformations of the mandibular prominences. *2*, *3*: otocephalia. *4*, *5*, *6*, *7*, *8*: First arch syndrome with macrostomia. *9*, *10*: fossa and fibrochondroma of the chin. *11*, *12*: homeotic duplication of the mandible

1.3 Summary

The final morphology of the fetal face results from a succession of stages punctuated by molecular signaling whose visible effects are successively topological deformations, cellular migrations, numerous cellular death or apoptosis, and then acquisition of cell phenotypes.

Defects in all or part of these molecular strategies in one or more of the facial prominences leads to the development of morphological accidents.

The facial mass and the brain and the skull, which make up the head, are inseparable as they both belong to the same neural compartment. (Figs. 1.21 and 1.22).

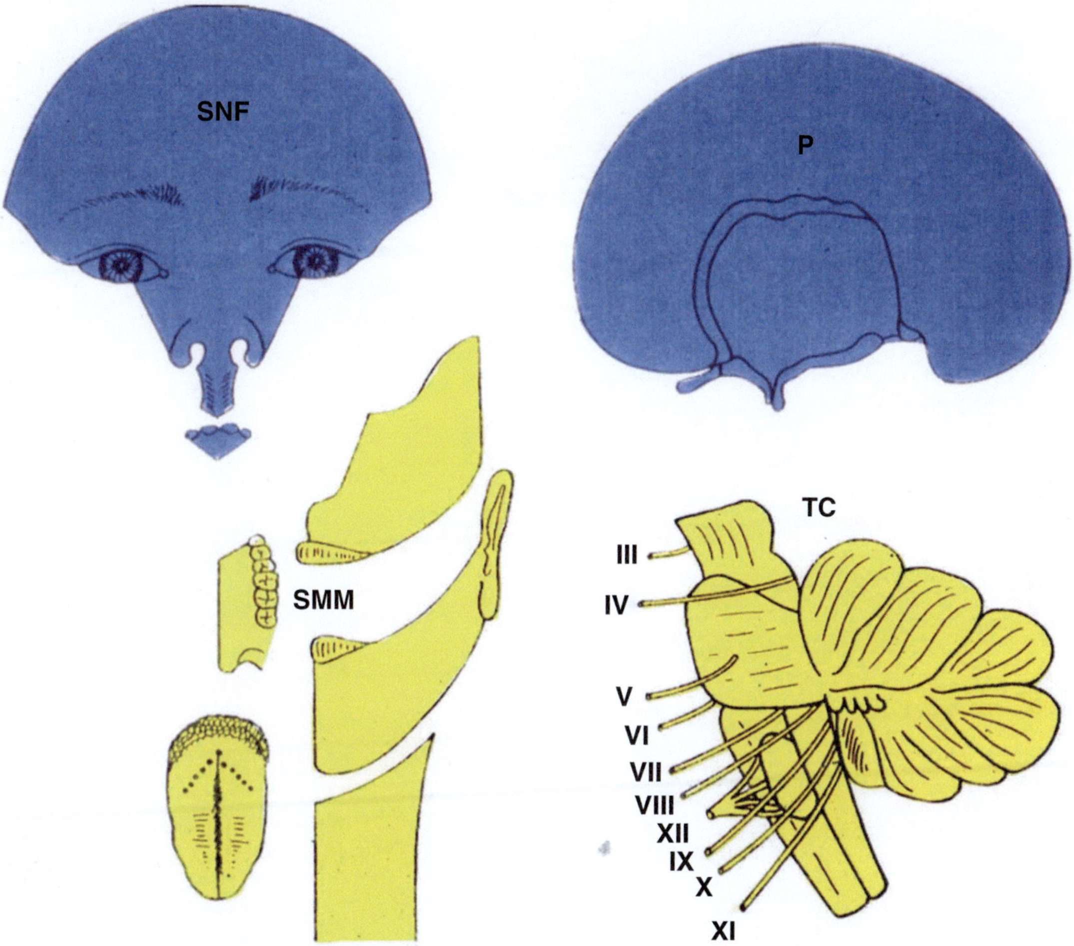

Fig. 1.21 Malformations associated with the face and the brain

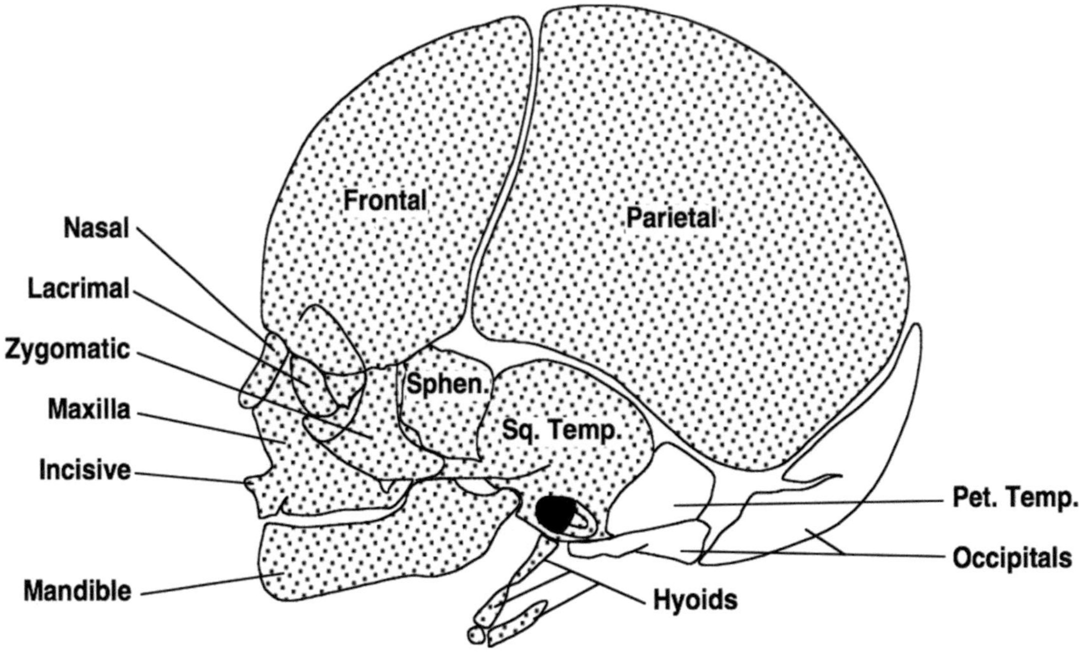

Fig. 1.22 Facial skeleton (*in grey*) is made up of neural crest cells

2 Normal Face

The ultrasound of the face is perhaps the most important moment in the examination of the fetus with regards to the impact for the parents who have come to "see their baby."

This exam procures both anxiety about the potential discovery of a malformation and pleasure in seeing their future child on the ultrasound screen.

This exam is also very important for the ultrasonographer for screening and diagnosis. Like a pediatrician who first of all examines a newborn's face, this view provides much information about potential pathologies. The same applies to the fetus, and yet, strangely, the "minimum" recommended exam minimalizes the examination of the face.

The chapters that follow will provide proof of the importance of this exam.

J.-M. Levaillant et al., *Normal and Abnormal Fetal Face Atlas*, DOI 10.1007/978-3-319-43769-9_2

2.1 What Is Included in a Basic Sonographic Survey and a Diagnostic Sonographic Survey?

Elements that must appear in the second trimester screening report:

- Continuity of the upper lip

Elements that must appear in the diagnostic report:

- Continuity of the upper lip
- Profile and nose bones with usual appearance
- Orbits with usual appearance
- Lenses with usual appearance

One should note that three-plane examination is essential in order to exam the face:

- The nose-mouth coronal plane
- The strict sagittal plane for the profile
- Stepped images on an axial plane, for the orbits, maxilla, tongue and mandible.

Fig. 2.1 Continuity of the upper lip

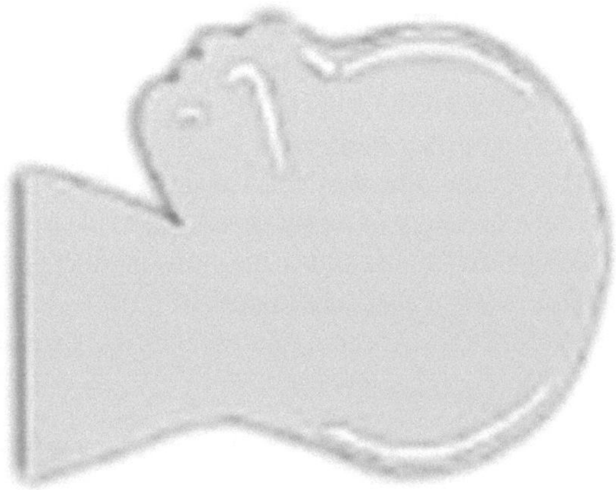

Fig. 2.2 Profile and continuity of the upper lip

2.2 Routine Trimester Examinations

2.2.1 First Trimester

- **Locate biparietal and back**
- **Profile: Herman image scoring sagittal view**
- **Analysis of profile: forehead, nose, upper lip and position of chin**

This is a quick analysis, which depends on the operator's subjectivity and enables the operator to raise any alarms based on various key points, notably using the Herman score of the sagittal image.

- Forehead: a smooth arch, not too domed, not too evasive
- Upper lip: straight, not looking like a comma
- Chin: aligned, individualized, not too far in advance, not too far behind

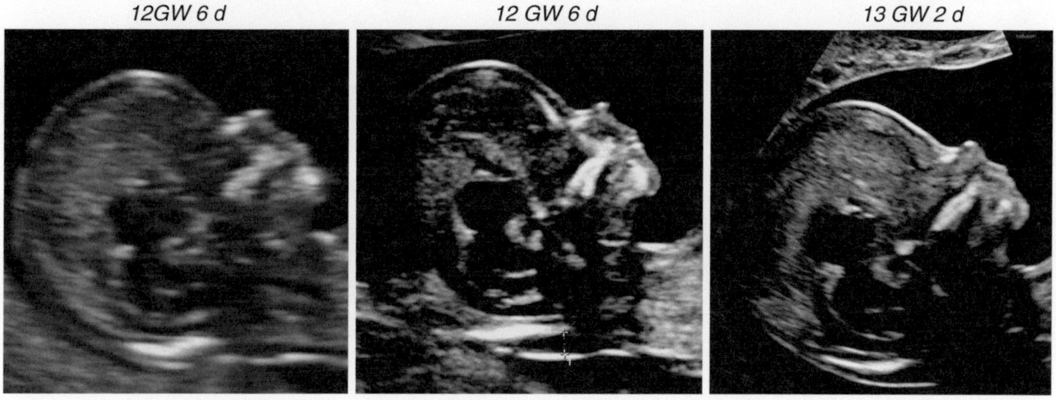

Fig. 2.3 Normal 2D profile

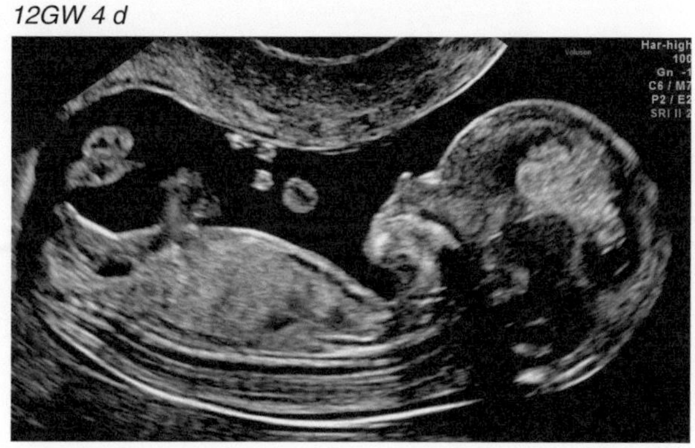

Fig. 2.4 Sagittal view, endovaginal probe

2.2.1.1 What 3D Can Do

Imaging volume from a strict sagittal view enables symmetrical restitution via the three planes—axial, sagittal, and coronal—which are orthogonal to one another:

- **A creation of structural symmetry, enabling one to see the strict sagittal view** of the nose bones.
- **A declination of the axial views**: frontal arch – orbital cut – maxillary arch – mandibular arch.
- **A "surface" coronal shot**: to analyze the upper lip and the nose.
- **A "bone" coronal shot**: to analyze the bone mass (metopic, actual bone junctions, maxilla).
- Reconstruction of the face in both surface and bone mode.
- Ear insertion: height and length

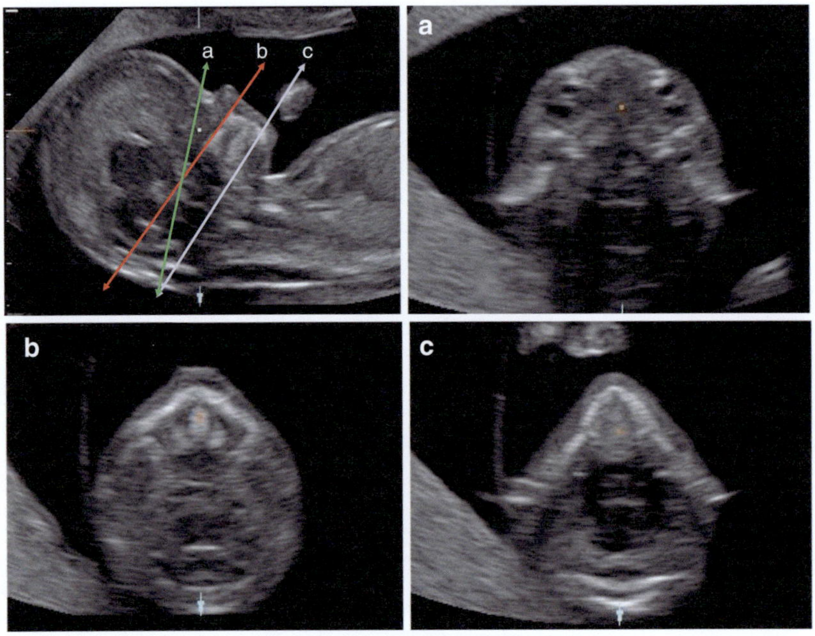

Fig. 2.5 Multiplane 3D images – different view levels
a. orbits and ears
b. maxilla and palate
c. mandible
d. orbits

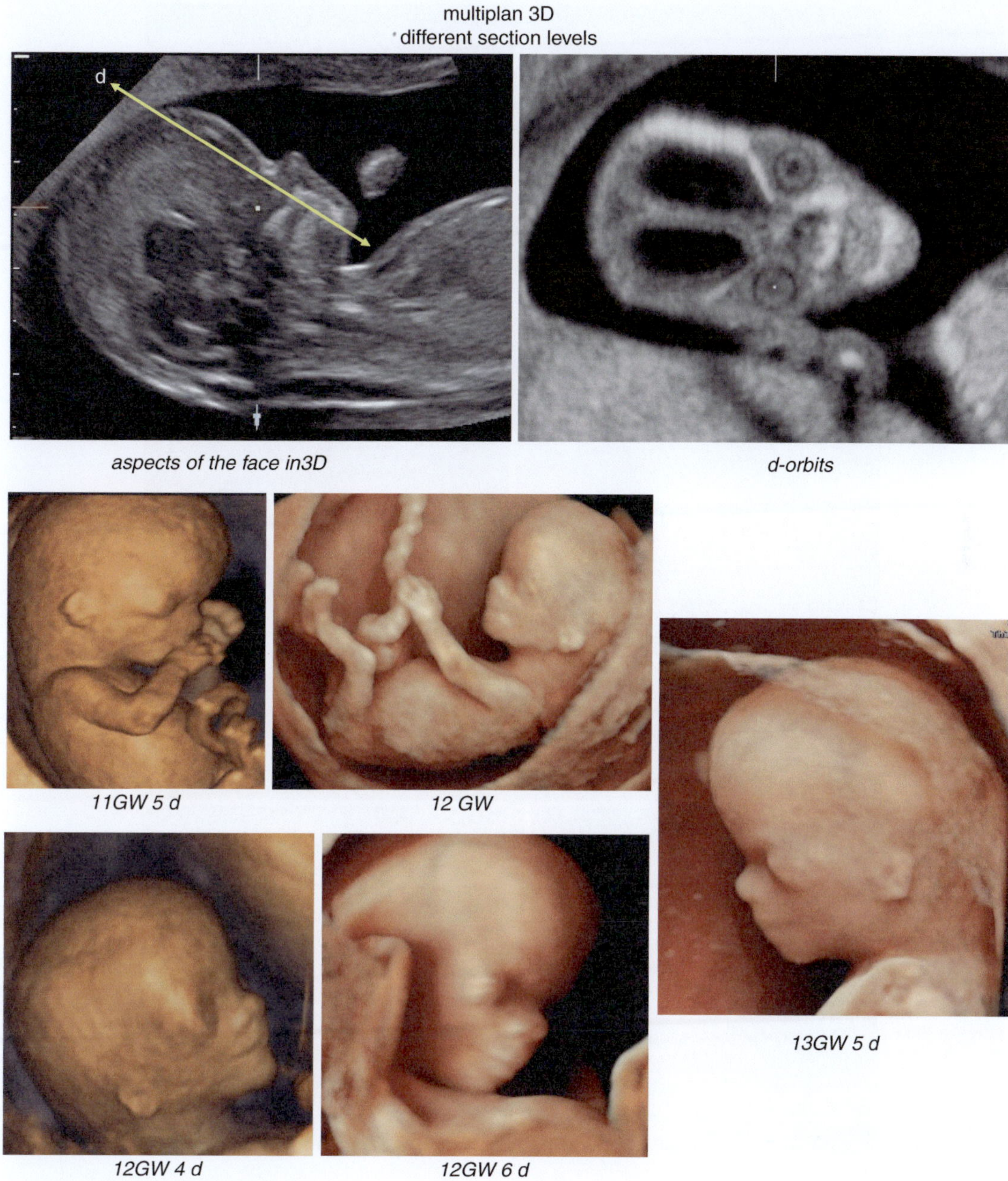

Fig. 2.6 Aspects of the face in 3D

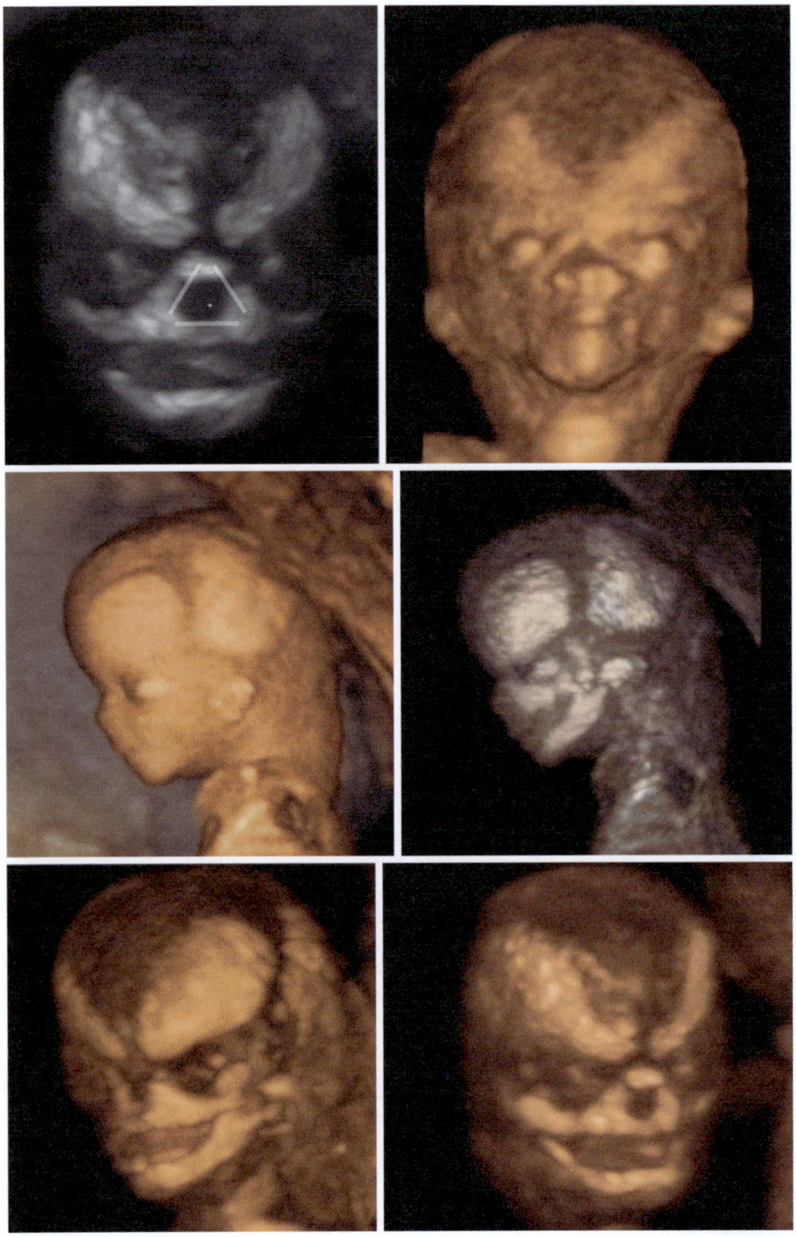

Fig. 2.7 Further images of the face
"Bone" mode, the retronasal bone triangle
At 12 SA the ears seem low because they are not yet in their final position
Face in "surface" mode
Face in "bone" mode
"Bone" mode – no nasal bones
"Bone" mode – nasal bones present

2.2.2 The Second and Third Trimesters

2.2.2.1 Proposed Screening Protocol

Example for a cephalic position

Starting point: biparietal diameter

1. Right from the start, locate the biparietal and the back with an oblique translation of the probe, at a distance from the biparietal diameter view, using a lateral sliding movement and then a rotation of 20–30° from its initial position.
2. Construction of the sagittal view with a 90° rotation of the probe around a fixed virtual point.
3. Construction of stepped axial views, from the forehead to the mandible with successive parallel views shot during a slow sliding movement rotating 90°, with the circular path ending with the nose-mouth view.
4. Arrival point: nose-mouth view.

2.2.2.2 Analysis of Sagittal and Parasagittal Views

The objective is to clear up the profile of a slightly bent head with the chin separated from the sternum.

This analysis makes it possible to check the various locations of facial anomalies:

- the forehead
- the angulation and the presence of the nasal bone
- the form of the upper lip
- the position of the chin

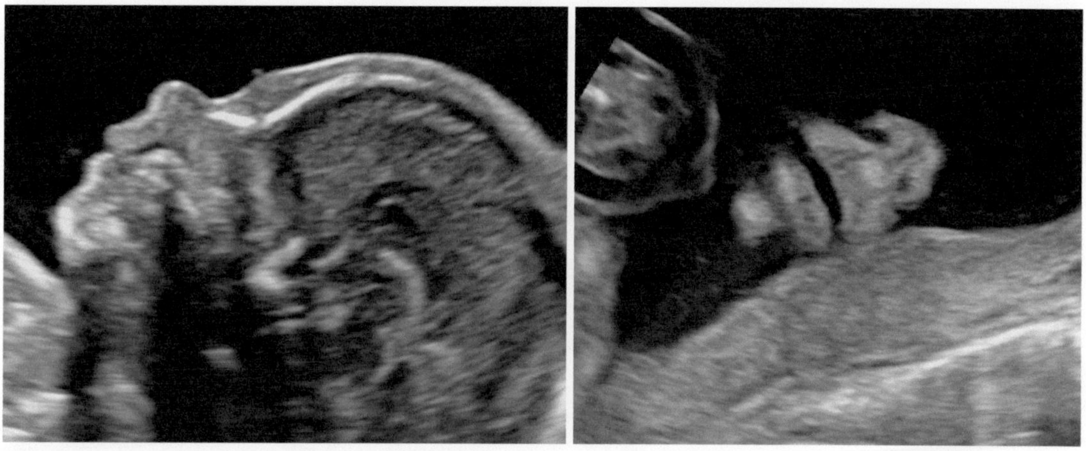

Fig. 2.8 2D profile and 2D nose-mouth view

2.2.2.3 Analysis of the Axial Views

The axial views enable the analysis of the face from top to bottom to view the orbits, the maxilla, the back edge of the palate, the tongue and the mandible.

- orbits and crystalline lenses
- maxilla
- posterior edge of the palate
- position of the tongue
- mandible

Other views are added to this paradigm:

- Ears

At least one ear could be analyzed if the examination is normal, and with frontal views, complete analysis of the various parts of the eye can be done.

The ears are round during the first trimester, slightly rimmed during the second trimester, but well rimmed during the third trimester.

- Eyes: full analysis of the various elements of the eye can be done on frontal views.

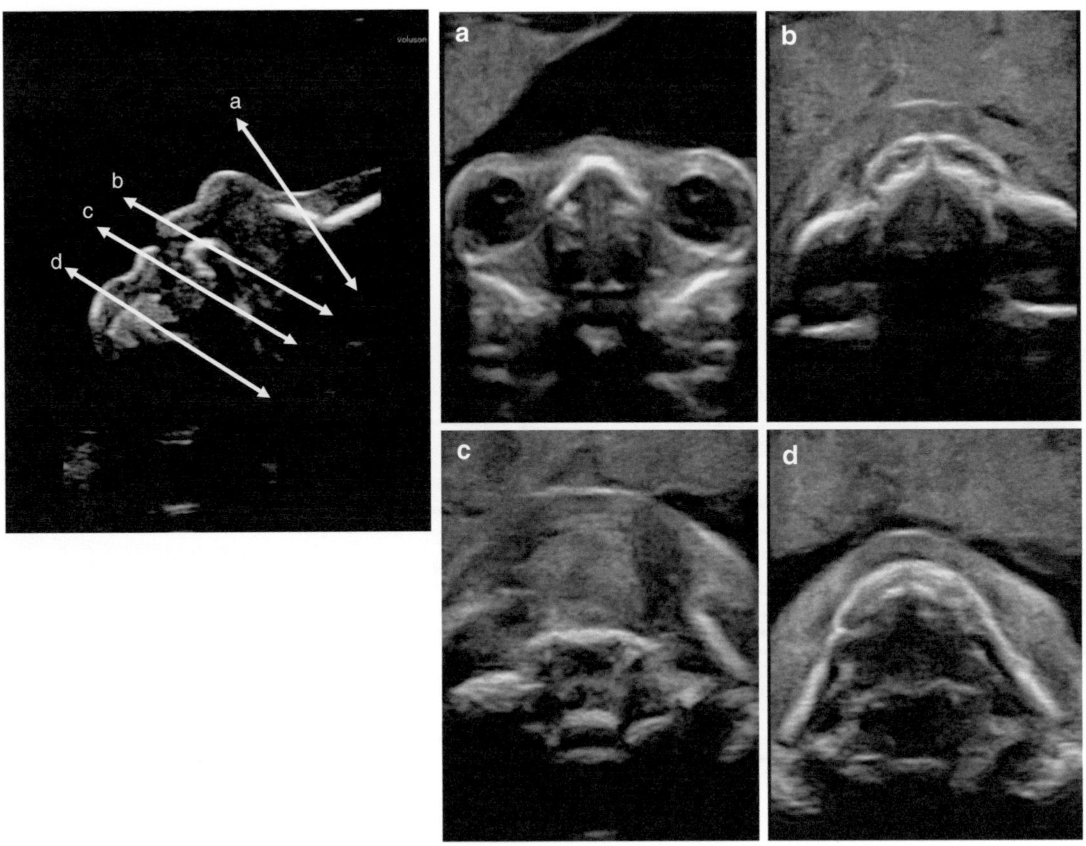

Fig. 2.9 Stepped views (high definition, 2D)
a-orbits
b-upper maxilla
c-posterior edge of bone palate
d-mandible

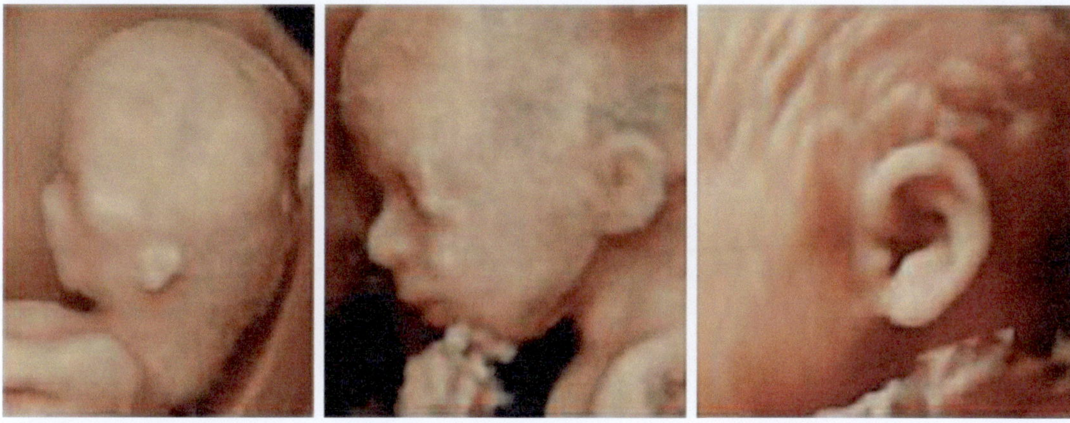

Fig. 2.10 Normal ears
12 GW
22 GW
32 GW

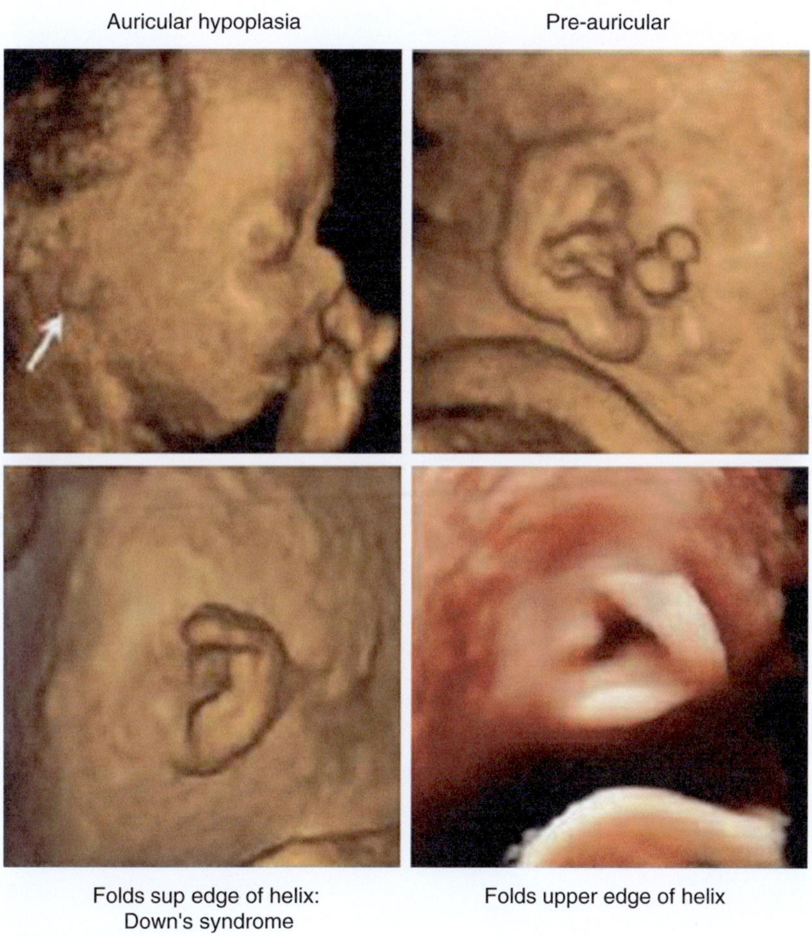

Fig. 2.11 Abnormal ears

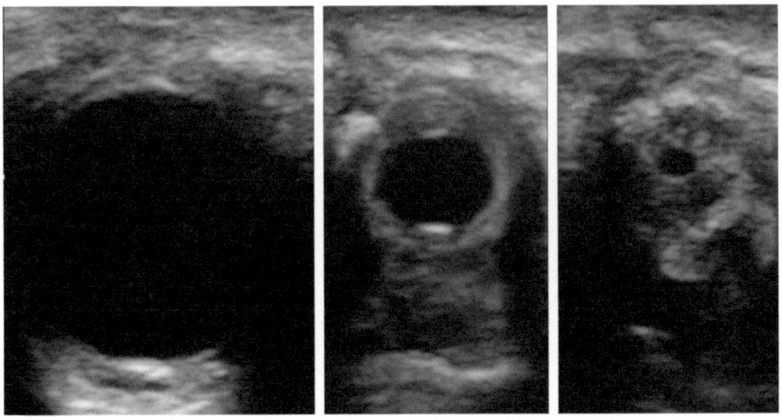

Fig. 2.12 Frontal view of the eye
Orbit, crystalline lens and pupil, 32 GW

2.2.2.4 Analysis of the Frontal View

- Nose-mouth view: This comes at the end of the circular path of analysis used on a face presumed normal in order to obtain the view required by the basic report.
- One must see the alignment of the nose, the lips and the tip of the chin.
- Absence of visualization of the chin could be an indirect sign of retroganthia.

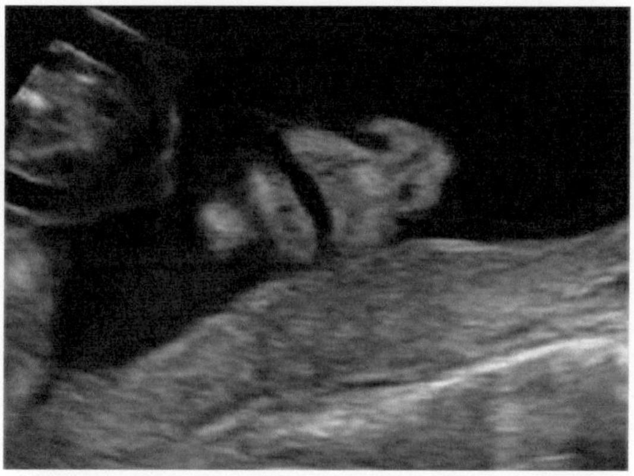

Fig. 2.13 2D nose-mouth view

Two remarks about the subjective analysis of the face during the second trimester:

- Sometimes the mid-level can show a pseudo-exophtalmy, which is physiological.
- At the chin level, the fetus sometimes appears to have retroganthia as a result of having a slightly bent head. This is why tools such as facial angle biometrics are so important.

2.2.3 Spotlight on the Third Trimester

Coronal, sagittal and axial views are sometimes difficult to capture, particularly when the examination is done at 32 GW, right before maternity leave begins, on a tired patient, or a fetus that is already low.

During the third term, capturing a perfect profile is particularly difficult, considering how the chin is bent onto the sternum, yet one should try to capture it by mobilizing the fetus's head using the abdominal hand to create a slight lifting movement in order to eliminate the possibility of real retroganthia, which would require a level-three team at birth and a pediatric team with ENT skills.

An important part of the routine is checking the nose-mouth view to confirm that there is no labial cleft. One should do this, not so much because such a cleft is a neo-natal emergency, but rather because it could be the sign that one should look for other syndromes.

2.2.3.1 What 3D Can Do

During the second and third trimesters, a 3D ultrasound enables one:

- To create symmetry of the views and the position of the tongue

- To work with the three planes—sagittal, axial and coronal, or the different views that are useful for ultrasound diagnosis of a normal face.

- To capture a restitution of the skeleton in "bone" mode, and of the face in "surface" mode

- When the captured images are of good quality, the volume restitution in "surface" mode renders an actual photographic map of the fetal face:
  - Eyelid inclination and openings
  - Eyelashes
  - Implantation of hair
  - Mouth and philtrum aspect

The technique for capturing 3D volume depends on which rendering one is looking for.

An aesthetic rendering of the face is possible from a capture that takes in three-quarters of the fetal face in an oblique sagittal view.

An analytical "anthropometric" rendering of the soft parts and the bone structures of the face requires a capture from a strict sagittal view.

An important moment in the examination: *KNOWING HOW TO PLACE THE PROBE FOR 2D AND 3D*.

Placing the ultrasound probe is essential for both the 2D view and for quality 3D capturing: the best example is that of visualizing the posterior edge of the palate.

Normal – Additional sagittal view – Normal – Face examination done

Abnormal – Additional sagittal and axial views – Normal – Abnormal – Send patient to referring ultrasonographer.

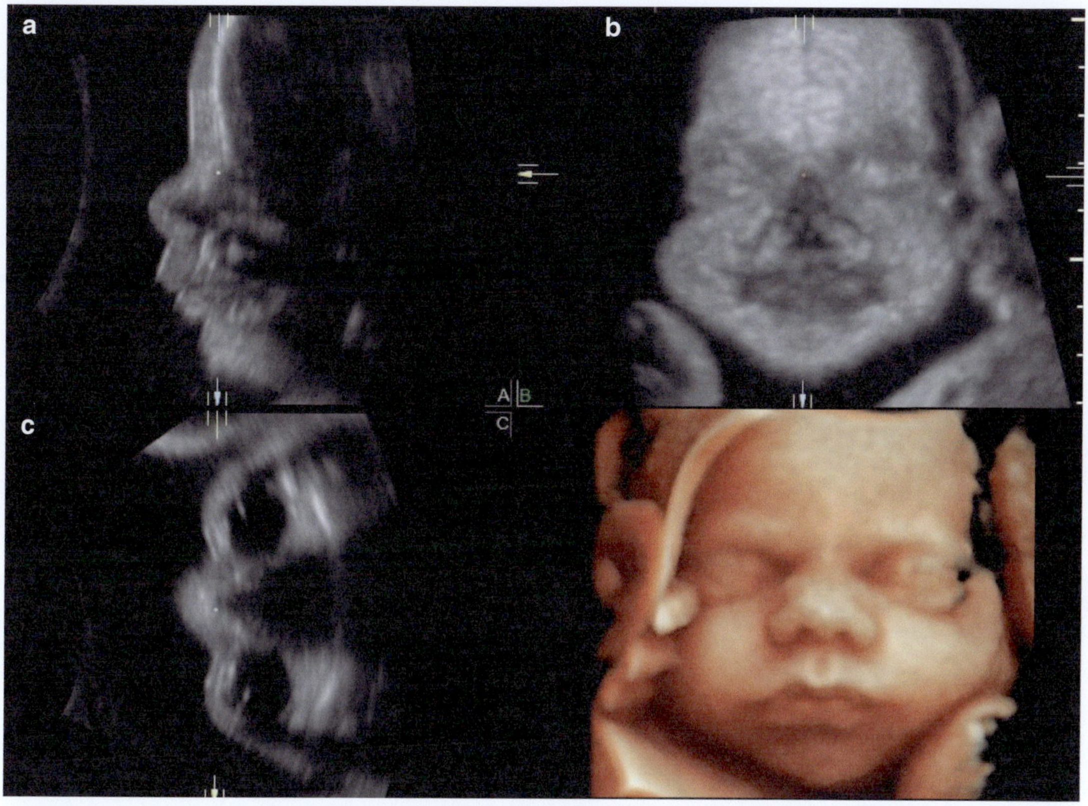

Capturing a strict profile in A, pinpointing the perfect position in B and C, and then 3D reconstruction

Fig. 2.14 Capturing a strict profile in (**a**), checking the perfect position in (**b**) and (**c**), and then 3D rendering

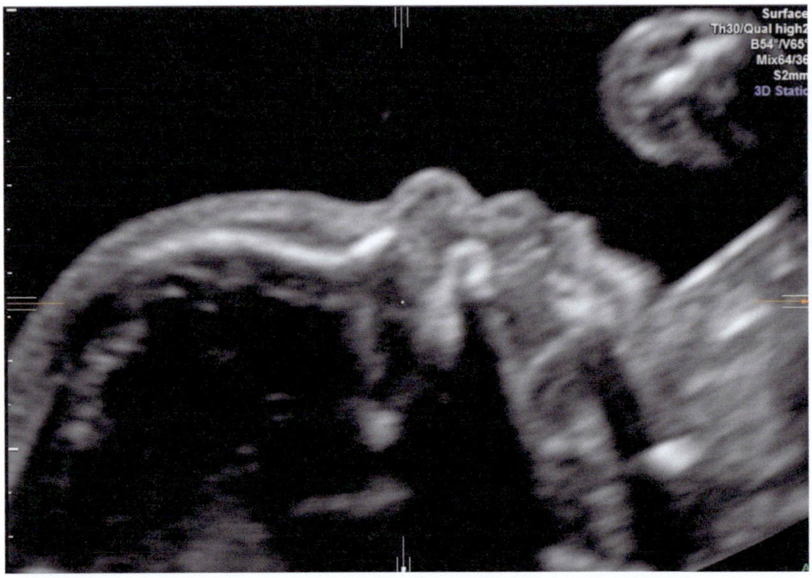

Fig. 2.15 Use of the 2-mm "thick view" with a mix of max mode and x-ray mode in order to accentuate the contrasts

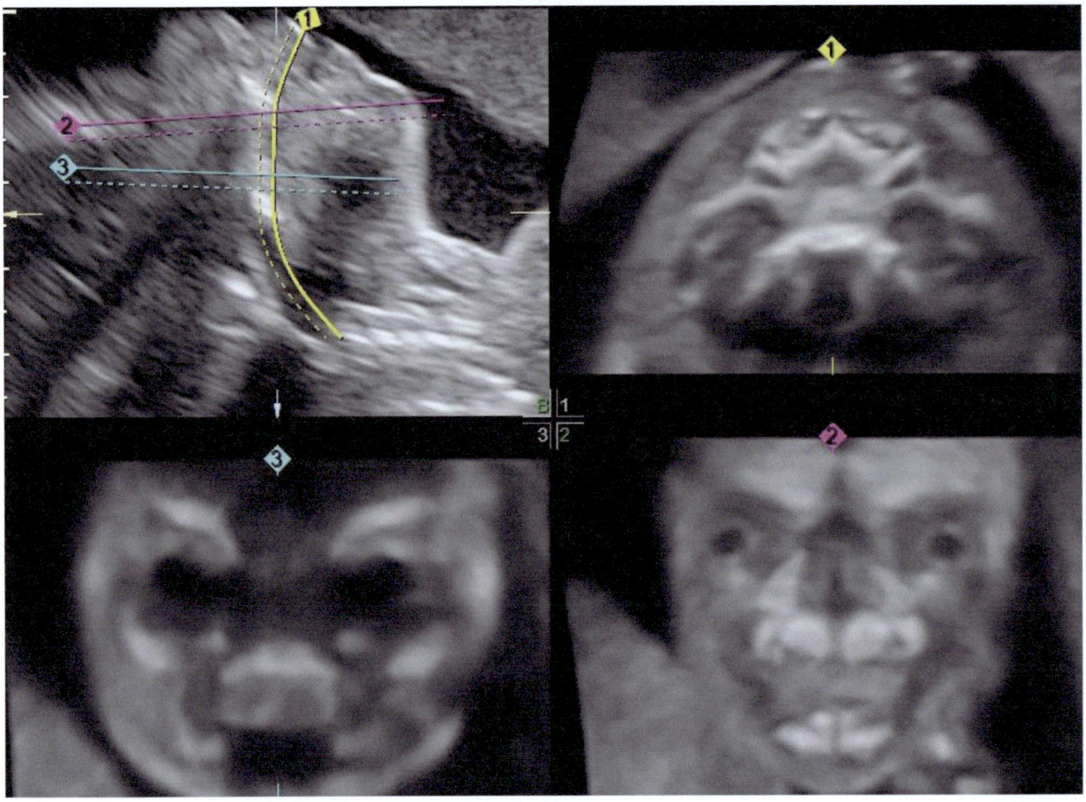

Fig. 2.16 Use of the omniview mode with the 2-mm "thick slice"
1. Sup. maxilla, palate
2. Frontal view with anterior portion of the mandible and the maxilla
3. Frontal view of palate

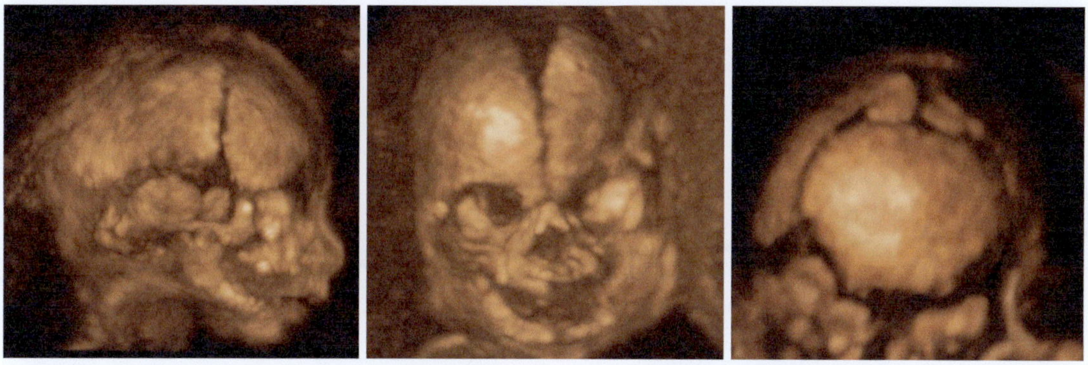

Fig. 2.17 It may be necessary to visualize the bone structures (with maximum mode)
Profile with sutures. Face: note separated nasal bones. Occipital bone, note the presence of two bones

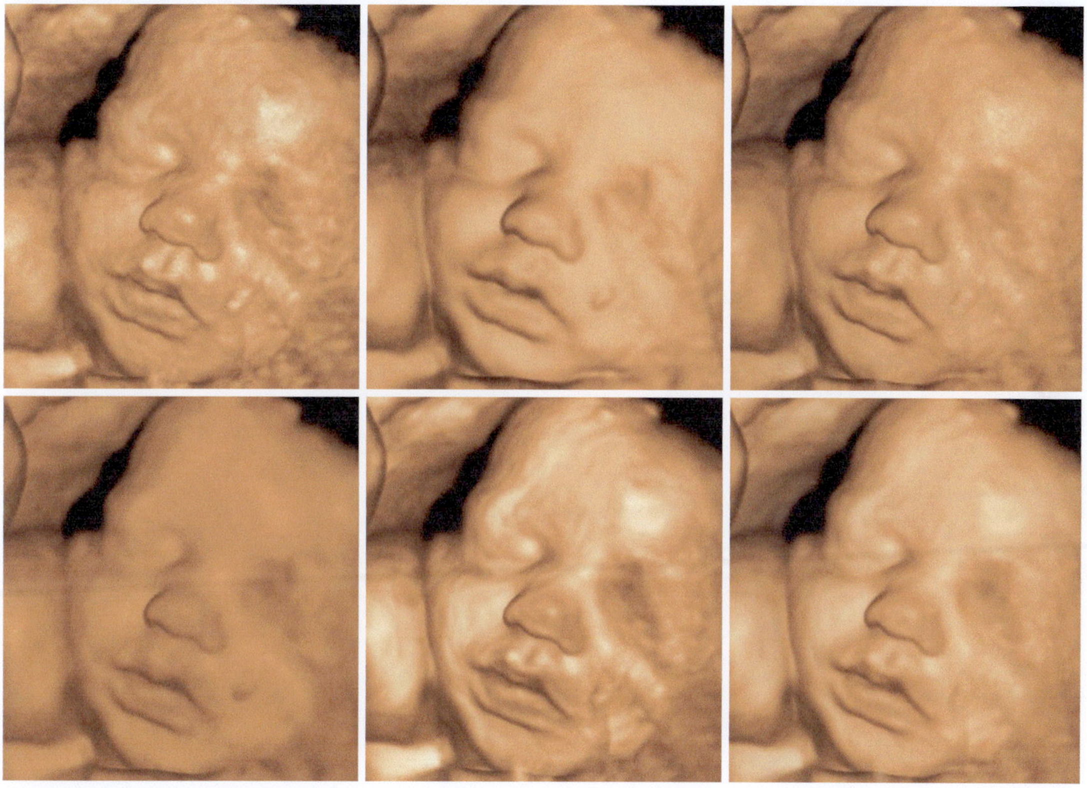

Fig. 2.18 Various surface modes
Surface texture
Gradient light
Mix of gradient light and texture
Surface texture
Enhanced surface
Mix enhanced and smooth

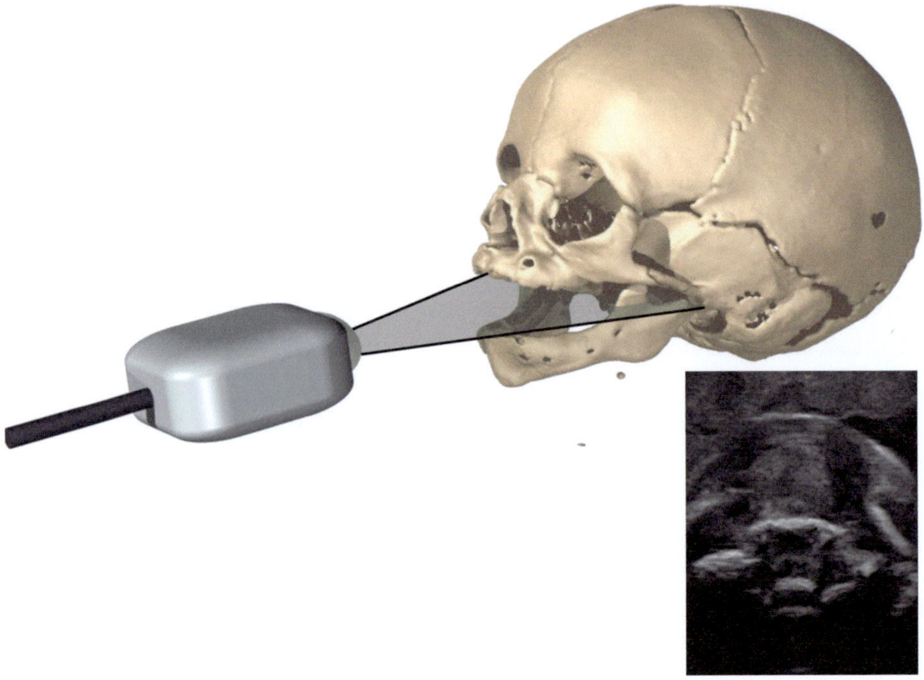

Fig. 2.19 Placement of the probe to look for the posterior edge of the bony palate

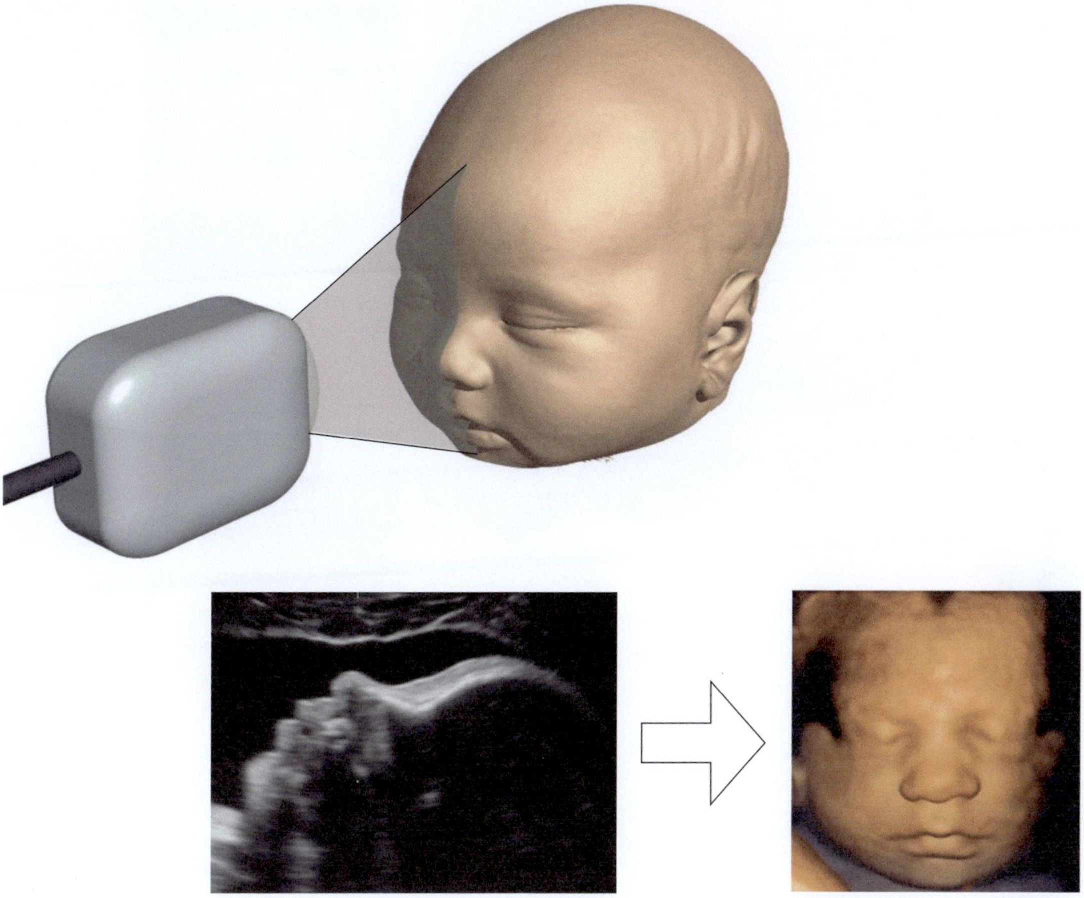

Fig. 2.20 Placement of the probe for a sagittal view
Starting 2D view
3D result

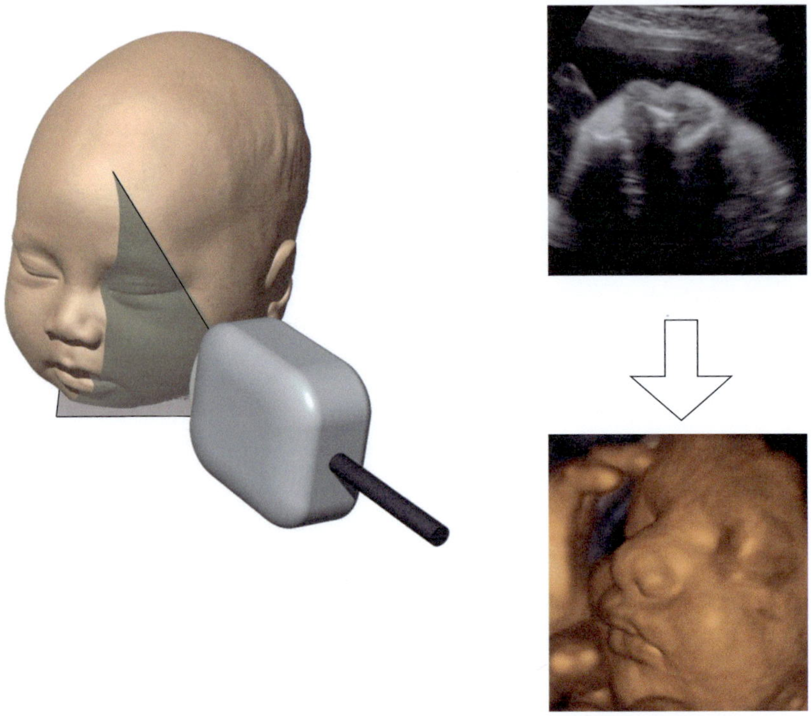

Fig. 2.21 Placement of the probe for an oblique view
Starting 2D view
3D result

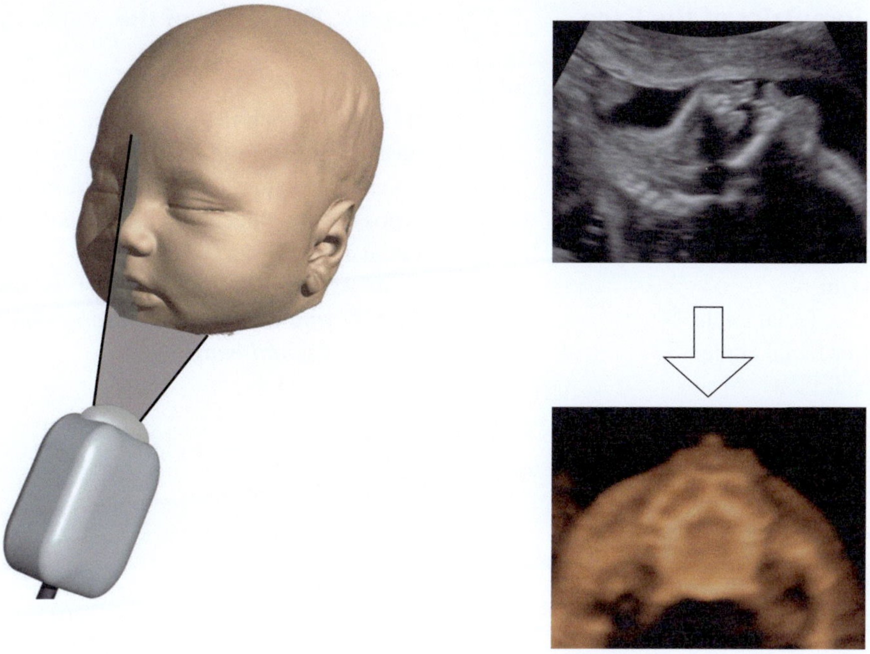

Fig. 2.22 Placement of the probe under the chin. Starting 2D view 3D result

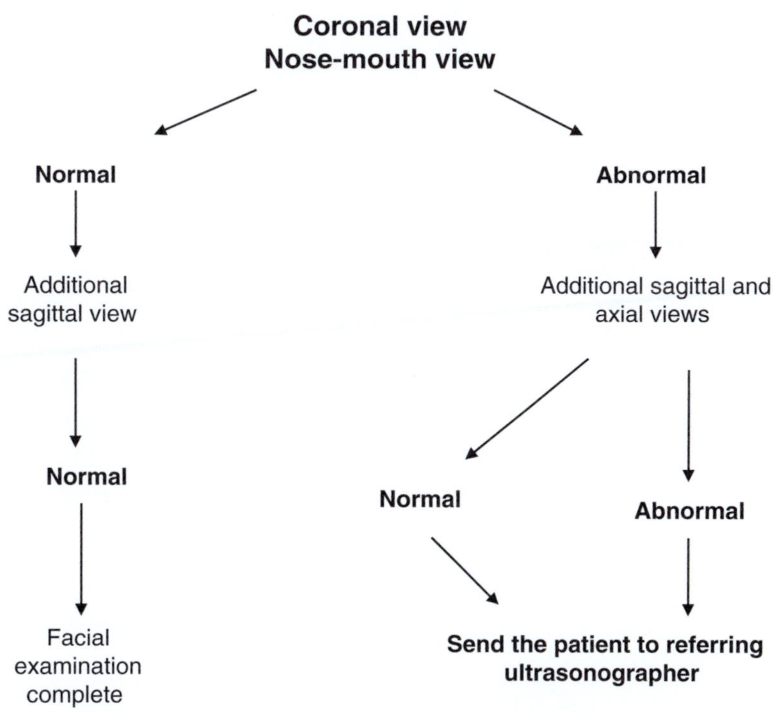

Fig. 2.23 Coronal view—nose-mouth view

2.3 Biometrics of the Face

Most of the biometric information useful for diagnosis is measurable:

- Strict sagittal view
  - prefrontal thickness
  - nasal bone
  - philtrum
  - superior facial angle
  - inferior facial angle
- Strict axial view
  - inter-orbit distance
  - maxilla width
  - mandibular width
- Coronal view
  - one the nose-mouth view, no additional biometrics

Table 2.1 22 GW 3.8 mm ± 0.6 mm; 32 GW: 5.93 mm ± 1.1 Levaillant N=638

GA (weeks)	Skin thickness compared to forehead (mm) average	SD
21	3.44	0.65
22	3.8	0.63
23	3.84	0.61
24	4.22	0.89
25–26	3.91	0.7
27–28	4.83	0.8
29–30	5.16	0.91
31–32	5.78	1.1
33	5.93	1.1

2.3.1 Thickness of the Forehead Skin

It is measured at the level of the synostosis of the nasal bone, from the external edge of the frontal bone to the external edge of the skin, and it makes it possible to objectify sub-cutaneous edemas and facial angle filling.

Measurements and views of the upper lip at 22 GW and 32 GW on the nose-mouth image and the sagittal image.

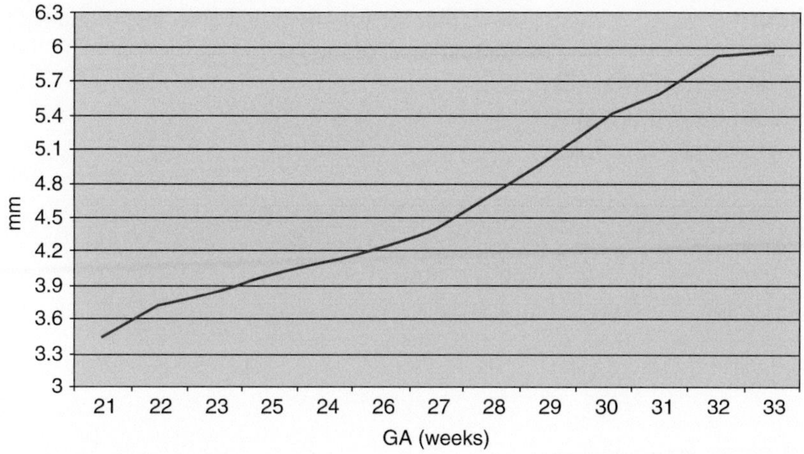

Fig. 2.24 Skin thickness

2.3.2 Length of Philtrum

This is measured from the nasal septum to the extremity of the white lip and makes it possible to differentiate normal-length upper lips from those that are very long, unbalancing aesthetic of the mid-level of the face.

Table 2.2 22 GW: 6.15 mm ± 0.7; 32 GW: 7.4 mm ± 1.37. (J.M. Levaillant. N=651)

GA (weeks)	Philtrum (mm) average	SD
21	6.06	0.99
22	6.15	0.79
23	6.44	1.03
24	6.47	0.99
25–26	6.7	1.02
27–28	7.34	1
29–30	7.02	1.36
31–32	7.4	1.31
33–34	7.65	1.29

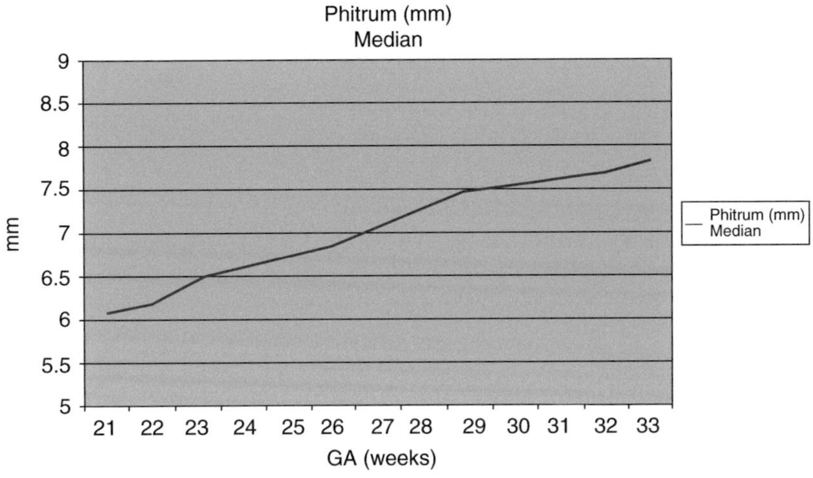

Fig. 2.25 Lenght of philtrum

2.3.3 The Nasal Bones

This is an important indication for trisomy 21 and other pathological syndromes. It is measured from the synostosis to the inferior extremity of the junction between the two nasal bones.

Table 2.3 Nasal bones

Gestation (weeks)	Mean (mm)	SD	−2SD	+2SD
14	4.183	0.431	3.321	5.045
16	5.213	1.062	3.089	7.337
18	6.308	0.654	5.000	7.616
20	7.621	0.953	5.715	9.527
22	8.239	1.102	6.035	10.443
24	9.362	1.300	6.762	11.962
26	9.744	1.277	7.190	12.298
28	10.72	1.459	7.803	13.639
30	11.348	1.513	8.322	14.374
32	11.580	1.795	7.990	15.170
34	12.285	2.372	7.541	17.029

(Guis et al. 1995)

Table 2.4 Fetal nasal bone length

Gestation (weeks)	Mean (mm)	SD (mm)
11–11.9	1.7	0.5
12–12.9	2.0	0.5
13–13.9	2.3	0.5
14–14.9	3.4	0.7
15–15.9	3.3	0.8
16–16.9	4.4	0.7
17–17.9	5.0	0.7
18–18.9	5.5	0.9
19–19.9	5.7	0.1
20–20.9	6.2	0.1

(Cuick et al. 2004)

Table 2.5 Nomogram of fetal nasal bone length

Gestation (weeks)	Mean (mm)	SD (mm)
11-11+6	1.69	0.26
12-12+6	2.11	0.37
13-13+6	2.34	0.39
14-14+6	2.94	0.48

(Sivri et al. 2006)

2.4 Facial Angles

Two facial angles are useful for the objective analysis of the face.

2.4.1 Upper Facial Angle

The upper facial angle is the angle formed between the line of the forehead that follows the synostosis of the nasal bones and the line of the nasal bones themselves.

Measurements at 22 and 32 GW

GA (weeks)	Average upper facial angle	SD
21–24	128.12	10.99
25–28	129.36	13.42
29–34	131.14	12.25

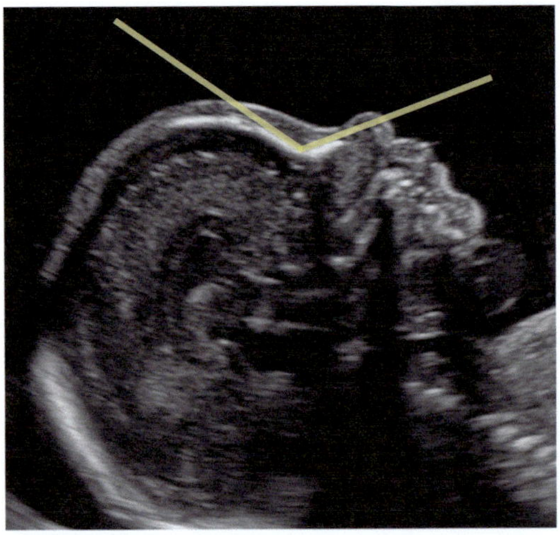

Measurement of facial angles

Fig. 2.26 Upper facial angle

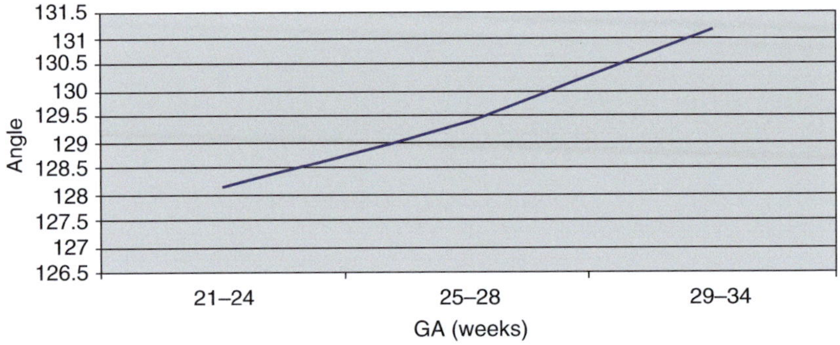

Fig. 2.27 Upper facial angle

2.4.2 Lower Facial Angle

The lower facial angle is the angle formed between the line of the forehead that follows the synostosis of the nasal bones and the line of the nasal bones themselves.

Measurements at 22 and 32 GW

GA (weeks)	Average lower facial angle	SD
21–24	65.1	7.79
25–27	67.2	8.43
28–34	69.06	7.96

JM Levaillant, N = 618

Fig. 2.28 Lower facial angle

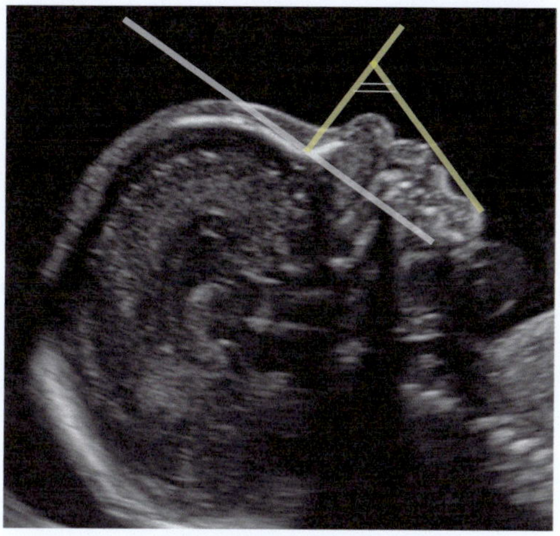

Measurement of facial angles

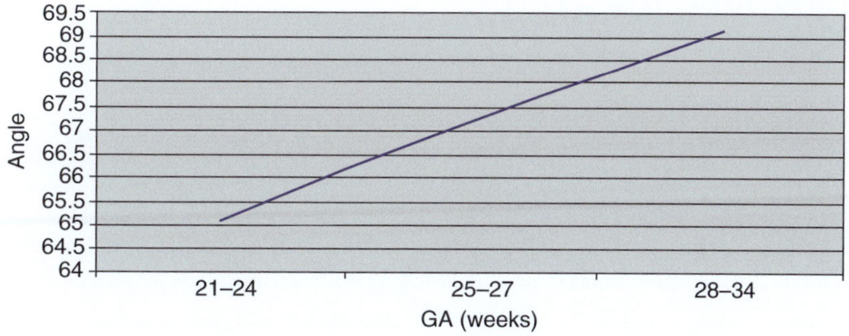

Fig. 2.29 Lower facial angle

2.4.3 Measurements of the Interorbital Distance

Interorbital distances—internal to internal, middle to middle, and external to external—are measured on the axial view.

Interorbital distance – a few points of reference:

Median distance: 22 GW: 25.2 mm ± 1.3 mm – 32 GW: 34.8 mm ±2.5 mm

Interorbital distance/biparietal: 22 SA: 0.47 ± 0.5 – 32 SA 0.42 ± 0.5

Measurements at 22 GW and at 32 GW for the three distances.:

GA (weeks)	Average median interorbital distance	SD
21	24.24	1.61
22	25.26	1.25
23	25.97	2.32
24	27.04	1.59
25–27	28, 86	1.99
28–29	30.91	2.43
31–32	34.15	2.35
33–34	34.49	2.14

JM Levaillant, N = 325

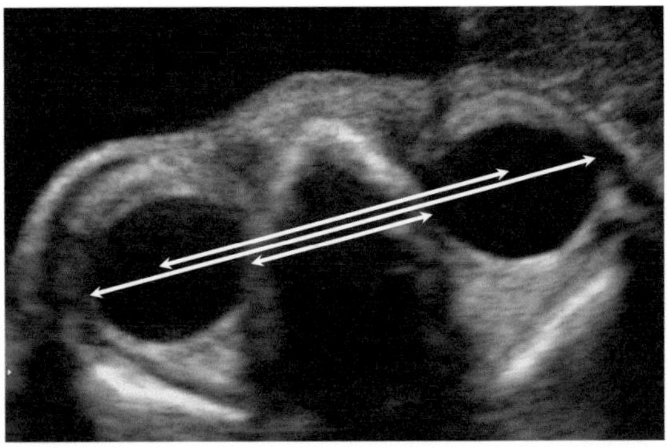

Fig. 2.30 Interorbital distance, via different methods: int./int., middle/middle, ext./ext

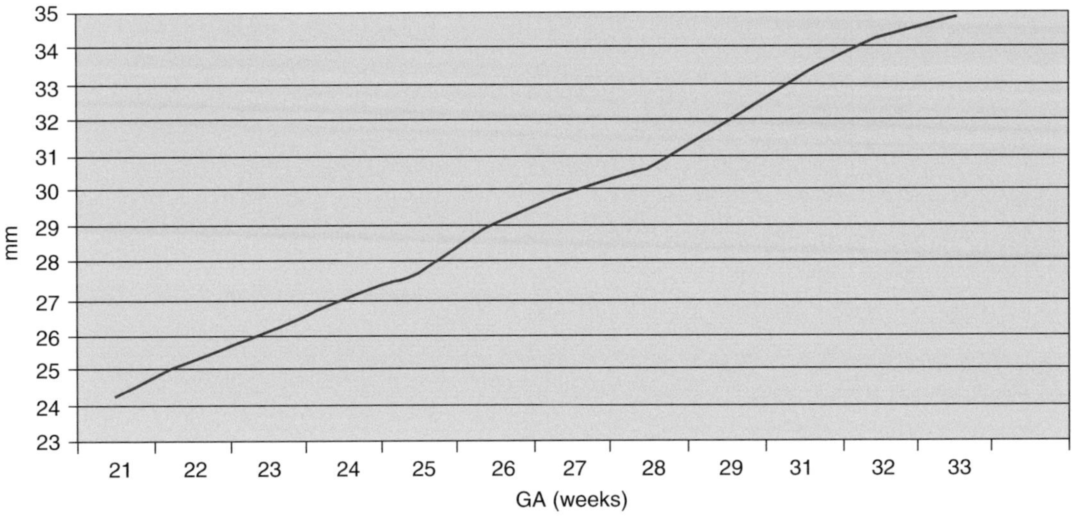

Fig. 2.31 Median interorbital distance

Table 2.6 Average internal interorbital distance

GA (weeks)	Average internal interorbital distance	SD
21	13.66	1.49
22	14.09	1.42
23	14.22	1.57
24	14.57	1.89
25–26	15.27	1.81
27–28	16.47	1.56
29–30	16.5	1.63
31–32	17.65	2.19
33–34	18.14	2.26

JM Levaillant N = 660

Table 2.7 Average External interorbital distance

GA (weeks)	Average external interorbital distance	SD
21	34.4	2.56
22	36.05	2.22
23	37.55	2.71
24	39.35	2.1
25–26	40.98	2.72
27–28	43.35	2.48
29–30	48.71	2.75
31–32	50.05	3
33–34	50.84	3.07

JM Levaillant N = 618

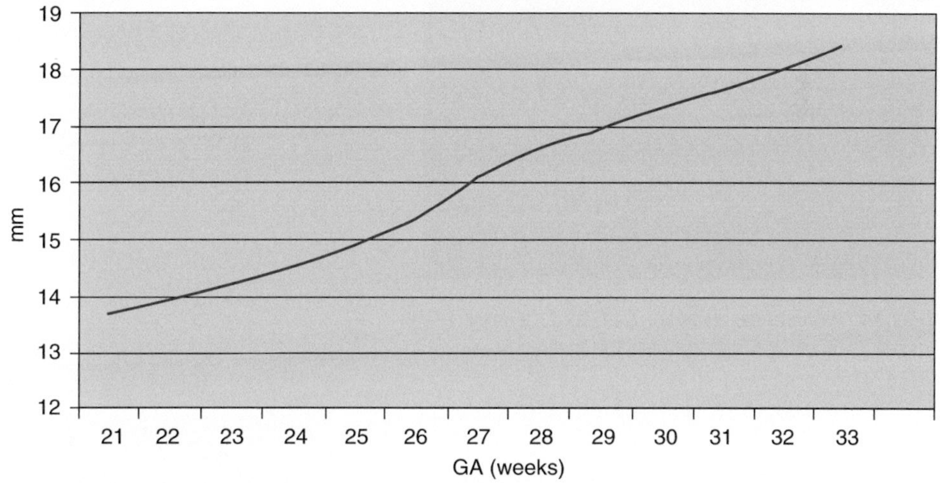

Fig. 2.32 Internal interorbital distance

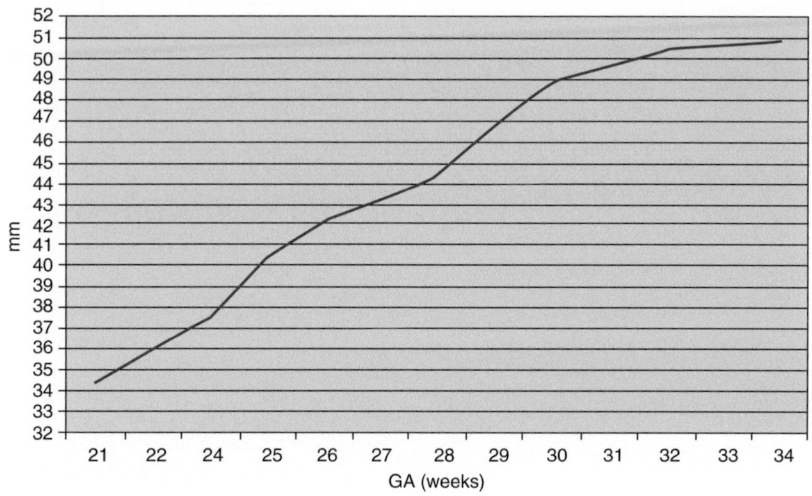

Fig. 2.33 External interorbital distance

2.4.4 Measurements and Views of the Mandibular and Maxillary Width

These measurements are made on axial views, from the external edge of the bone, 10 mm in, or more or less at the level of the canines.

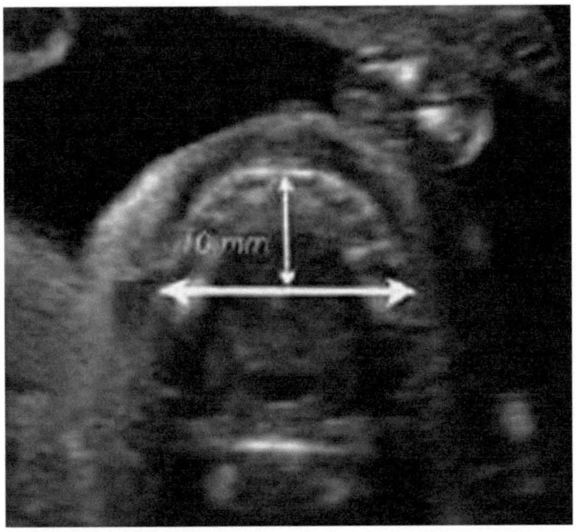

Fig. 2.34 Width of the mandible. 22 GW: 23.8 mm ± 1.9. 32 GW: 31.9 mm ± 4.7
Mandible/maxilla ratio: 1.01 ± 0.1

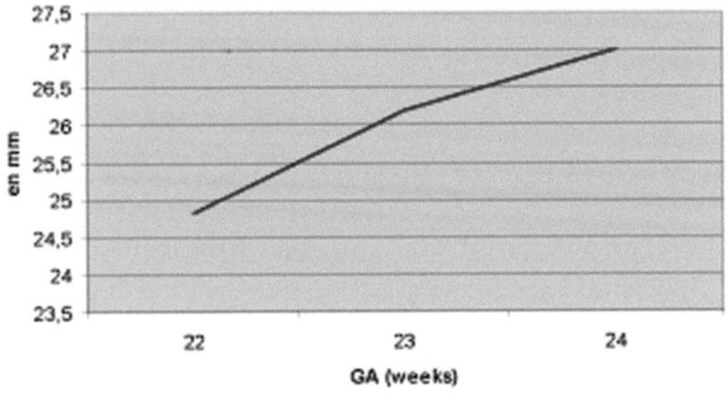

Fig. 2.35 Maxilla width

Measurements at 22 GW and at 32 GW

These profiles, all different, are either from normal fetuses or those with trisomy 21. **Can you tell the difference**? (*solution on next page*).

Table 2.8 Maxilla width (J.M. Levaillant. N=79)

GA (weeks)	Average maxillary width (mm)	SD
22	24.81	3.67
23	26.19	4.05
24	27	4.14

Table 2.9 Mandibula width (J.M. Levaillant. N=79)

GA (weeks)	Average mandibular width (mm)	SD
22	24.15	2.64
23	25.39	2.53
24	25.4	2.16

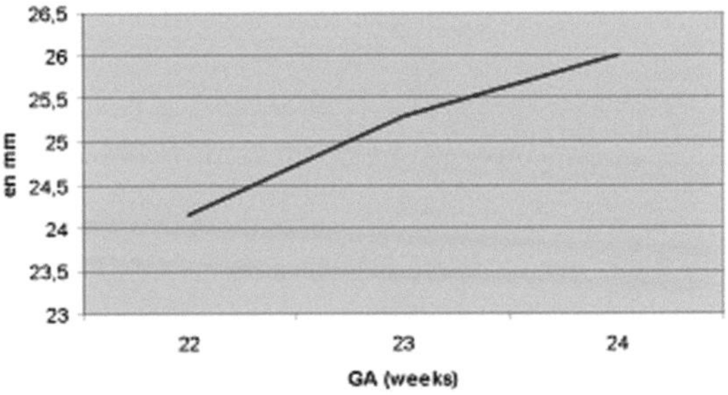

Fig. 2.36 Mandibular width

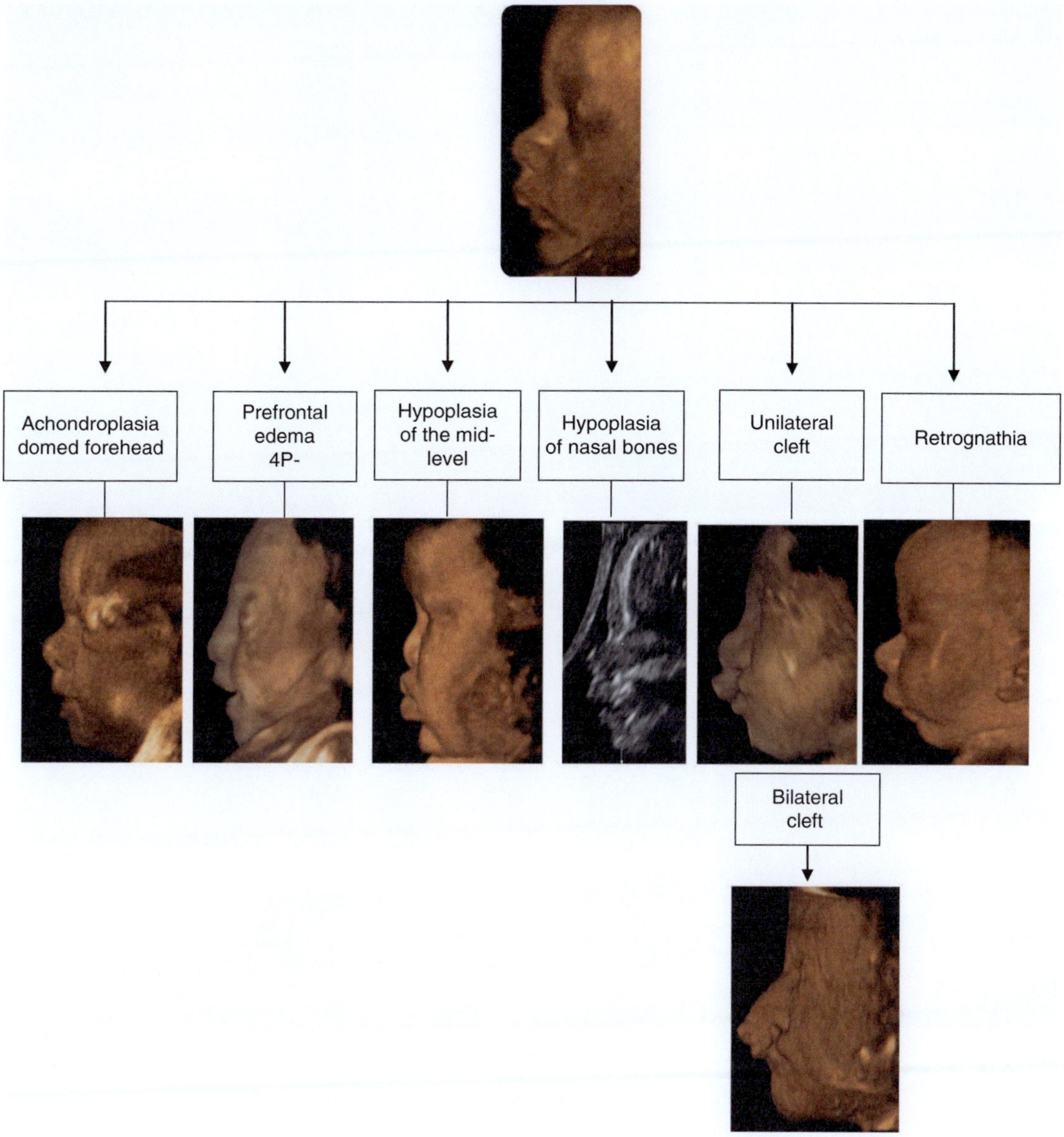

Fig. 2.37 Sagittal View – screening for rare, but serious, facial pathologies

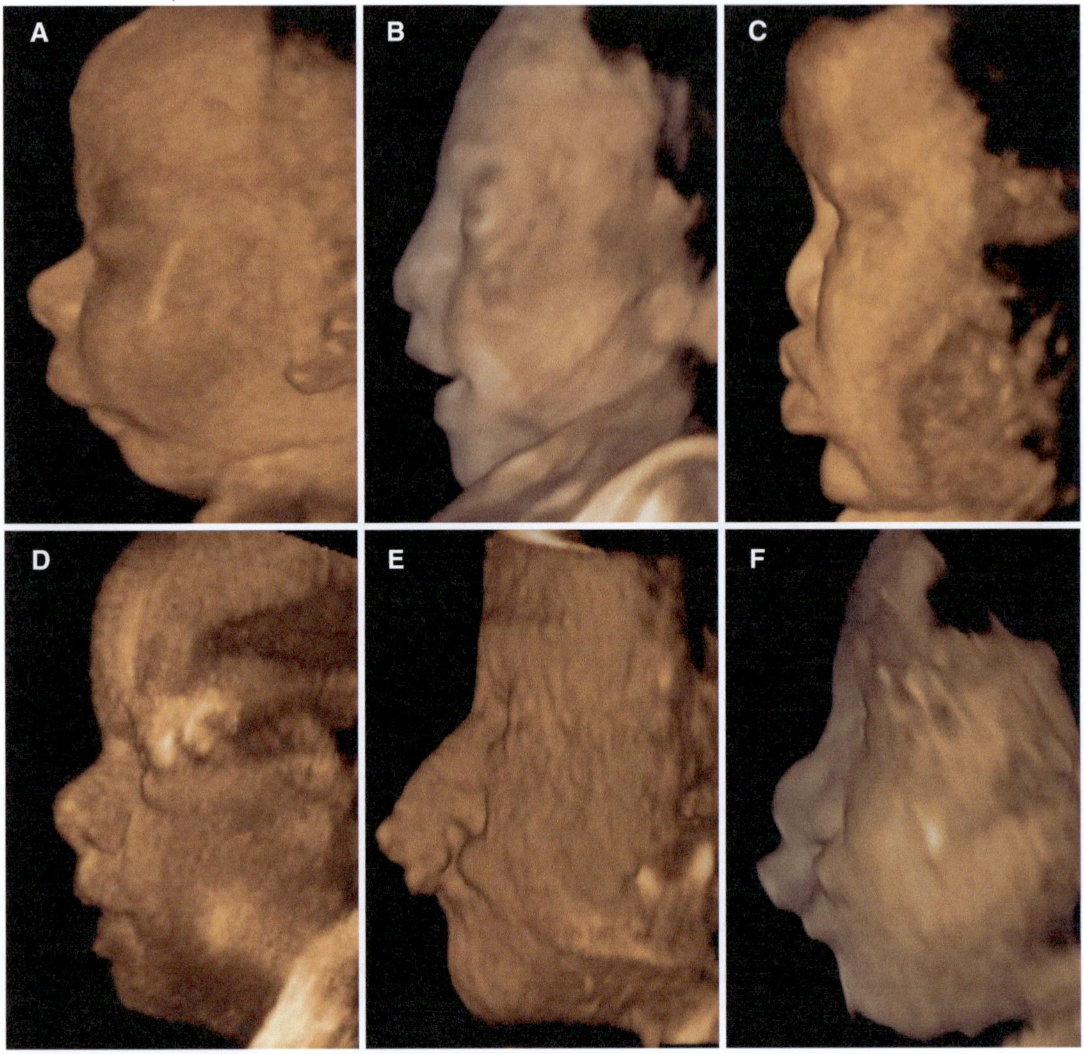

Fig. 2.38 A: Retrognathia; B: 4p-; C: Binder; D: Achondroplasia; E: Bilateral cleft; F: Unilateral cleft

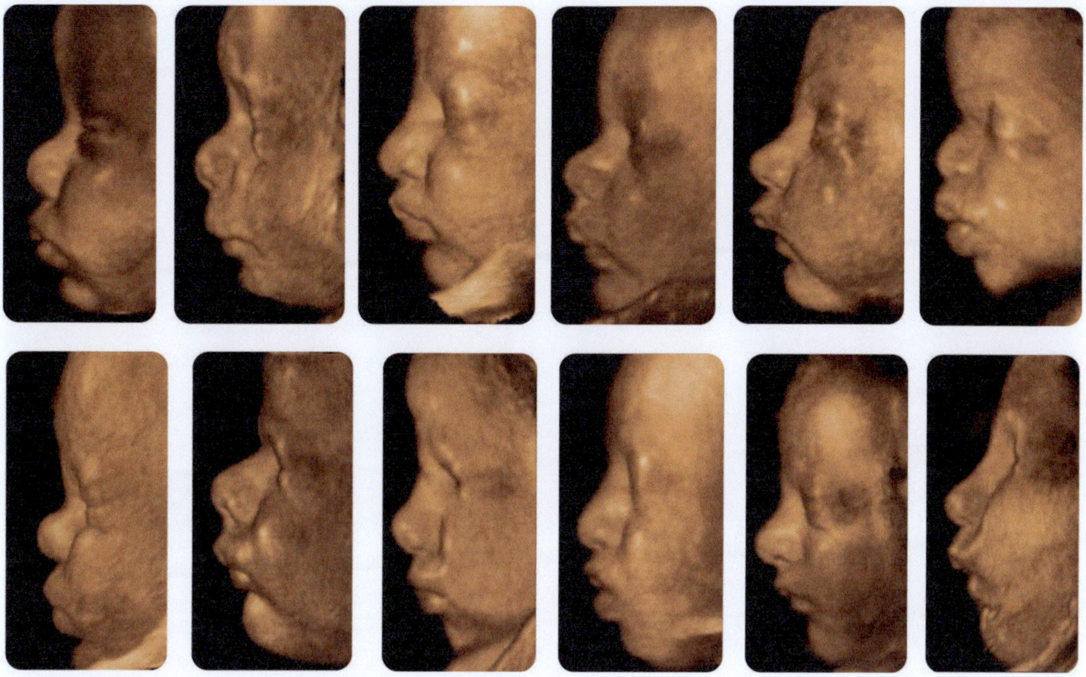

Fig. 2.39 Specific interest of 3D ultrasound images in the skin surface mode
Profiles of various fetuses at 32 GW. The black background effect is created by cutting the face volume in half

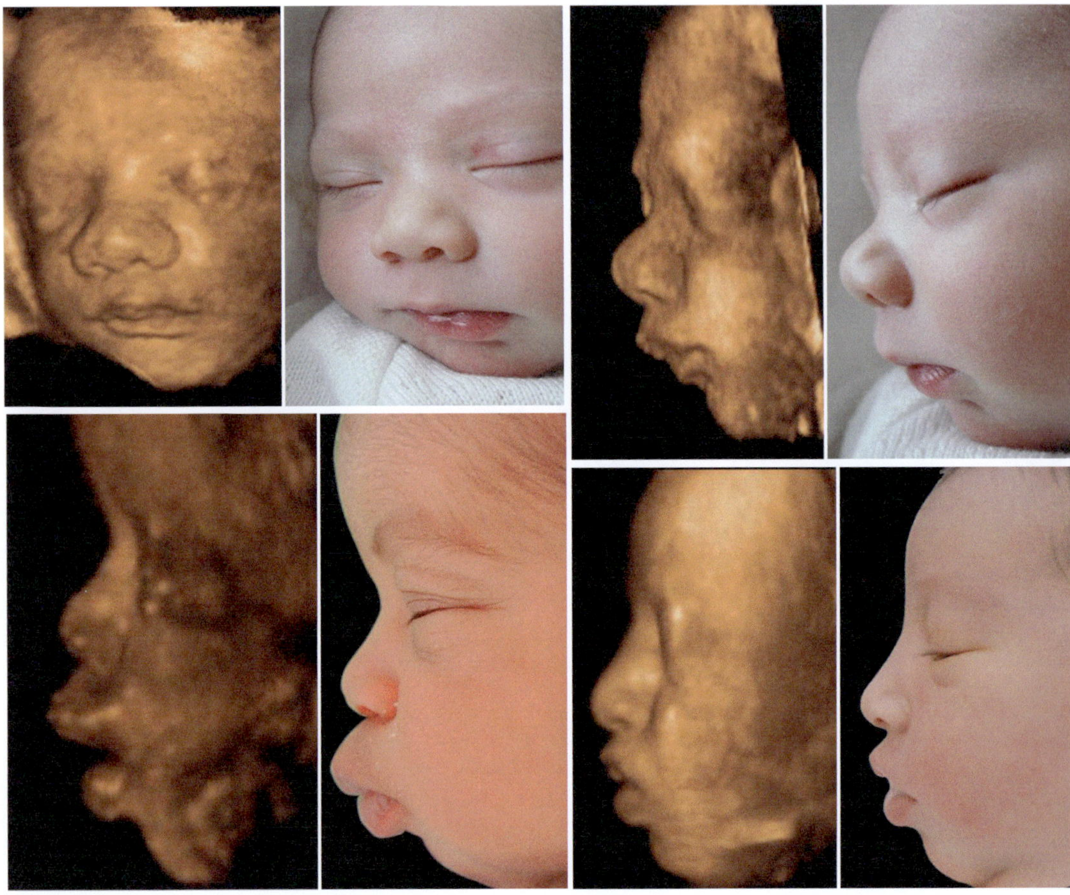

Fig. 2.40 Comparing profiles at 32 GW and birth

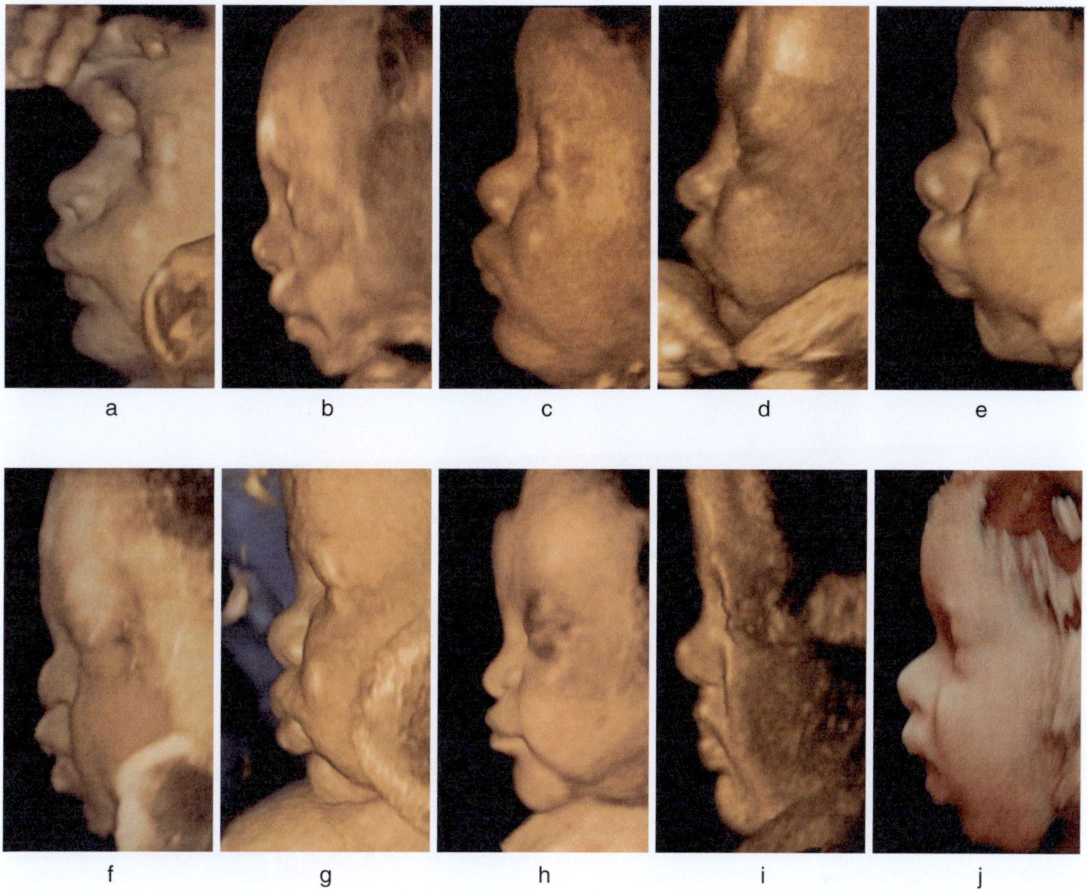

Fig. 2.41 Ten fetuses, five with trisomy 21. Which ones

2.4.4.1 Comments

As with 2D ultrasound, 3D imagery makes it possible to make a subjective diagnosis that must then be upheld by objective biometrics: that is the finality of this analytical approach.

The HD Live lighting technique highlights skin details, like images in dermatology.

Palpabral orientation, the sometimes reduced width of the mouth, the more or less visible nostrils and philtrum, are details that enable greater analytical precision of the skin surface.

Solution: *the Down's syndrome fetuses are b, d, f, h, j.*

2.5 Practical Review Work

- Table 2.10 is oriented by organ.
- Table 2.11 is for finding the most common markers for major chromosomic abnormalities.
- Table 2.12 classes the markers to look for the major deletions

These are not, of course, exhaustive tables, but practical guides to help ultrasonographers.

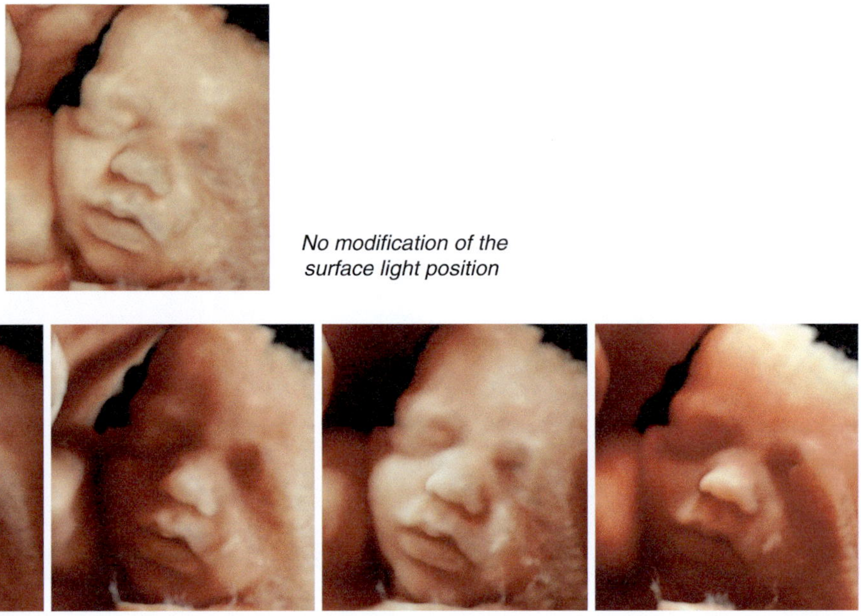

Fig. 2.42 HD Live surface mode
With not modification of the light source position
The same volume with different light source positions

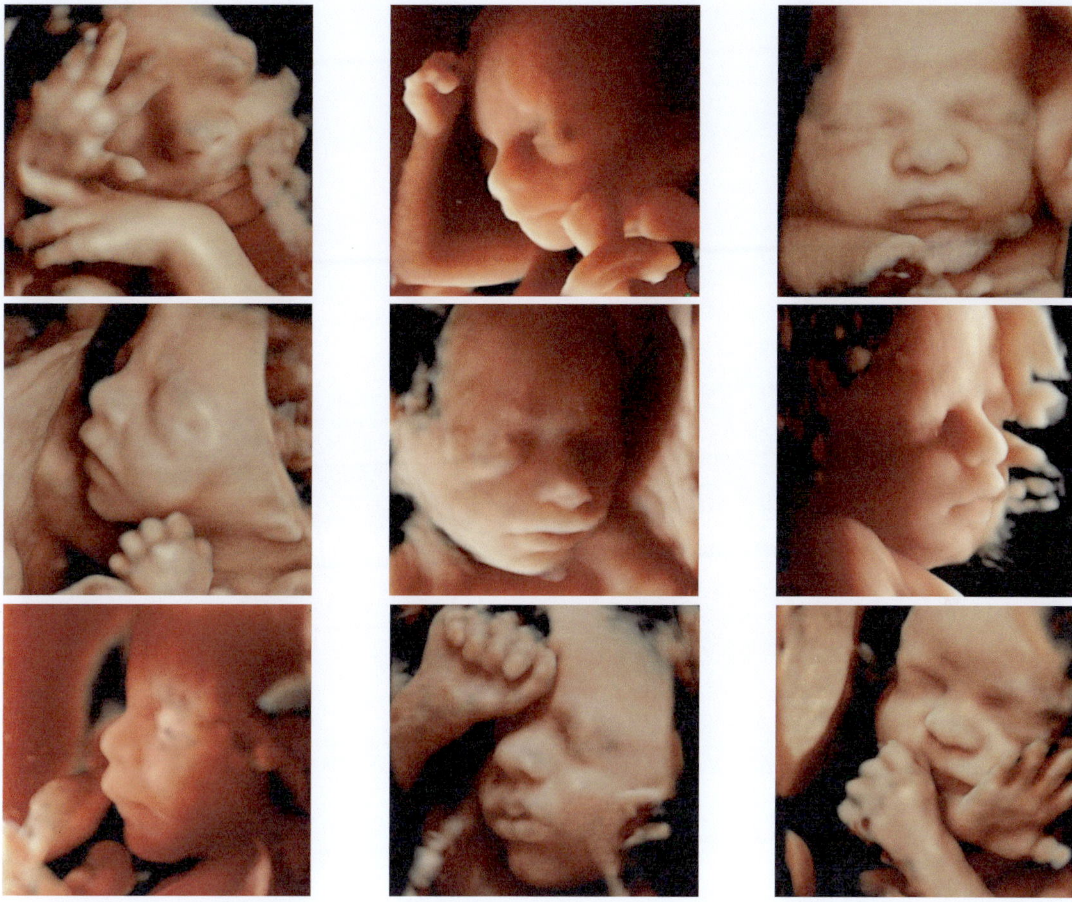

Fig. 2.43 Fetal faces in HDLive mode

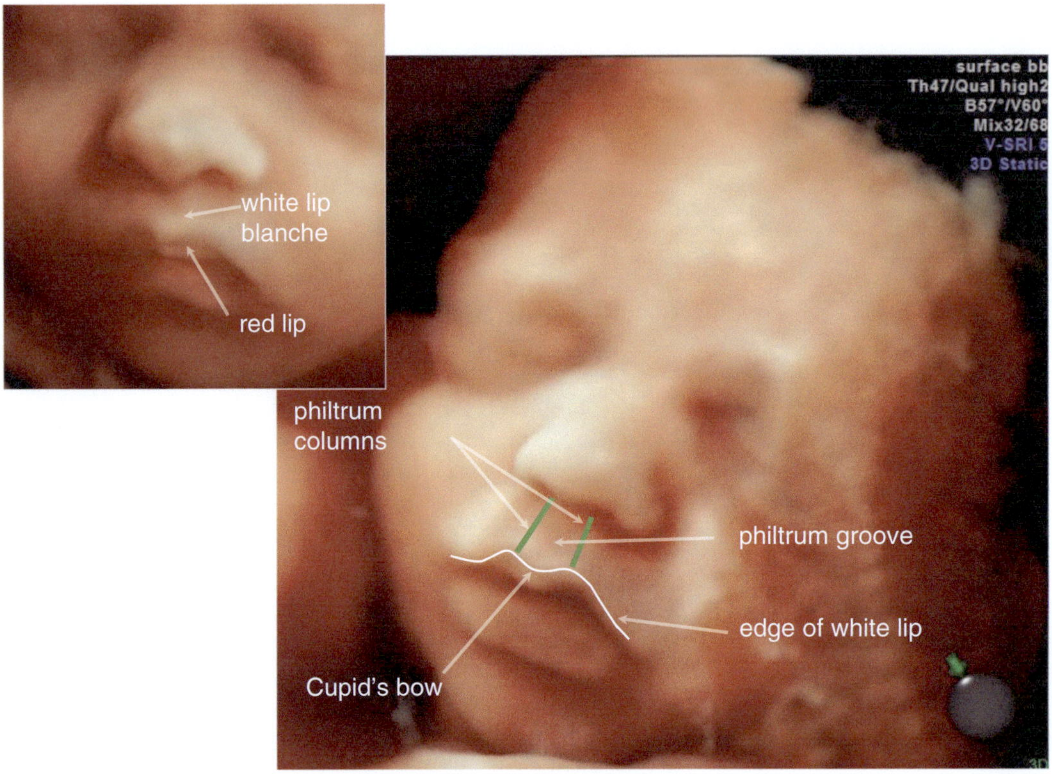
white lip
blanche
red lip
surface bb
Th47/Qual high2
B57°/V60°
Mix32/68
V-SRI 5
3D Static
philtrum
columns
philtrum groove
edge of white lip
Cupid's bow

Fig. 2.44

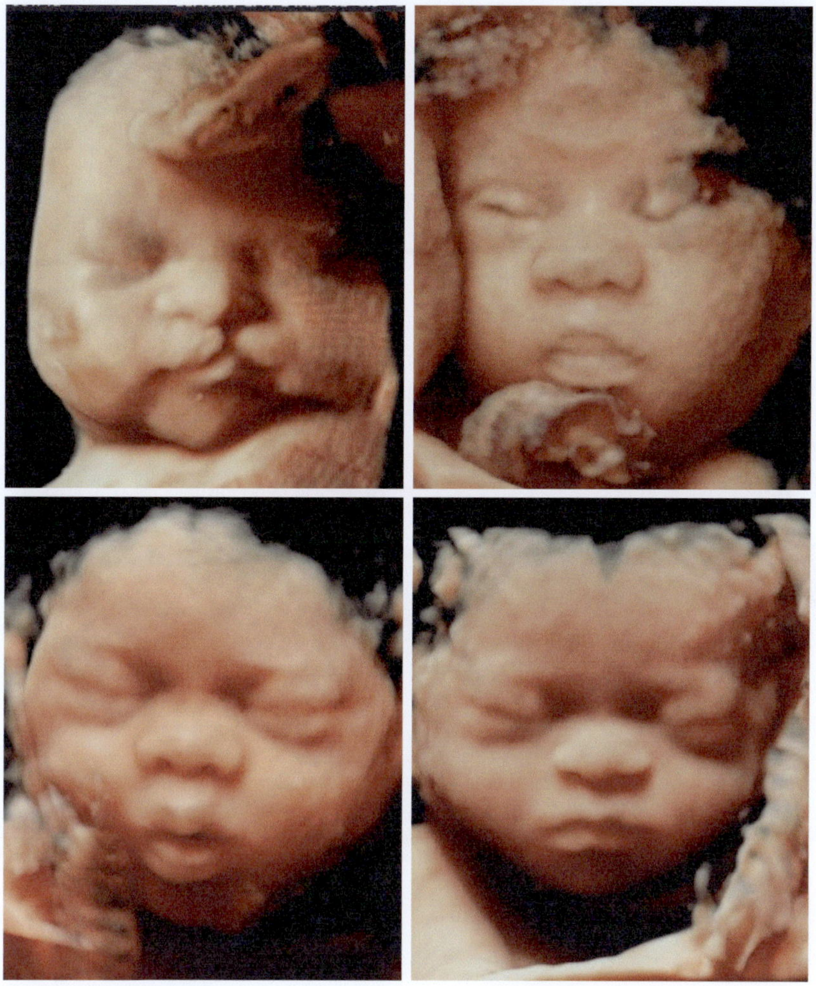

Fig. 2.45 4 trisomy 21 fetuses

Table 2.10 Prenatal screening of face abnormalities from ultrasound views

Views	Key organ visualized	Anomalies–diagnosis
Coronal view	Orbits and eyes	Anomaly of position Biometric distance Hypo or hypertelorism Size: microphtalmy or anophtalmy Cataract
	Eyelids and contours	Angle anomoly Presence of a daryocystocele
	Nasal bones	Morphological anomoly Disjointed bones Hypoplasia
	Philtrum	Long Smooth This with barely visible pillars
Sagittal view	Forehead	Very arched
	Nasal bones	Biometric hypoplasia
	Chin	Retrognathia Prognathia Larger or smaller facial angles
Axial view	Orbits	Visualisation of crystalline lens anomalies
	Tongue	Glossoptosis type position Macroglossia
	Mandible	"U" shaped

Table 2.11 Prenatal screening for face abnormalities from ultrasound views

	Trisomy 21	Trisomy 18	Trisomy 13	Trisomy 9
Forehead	Sometimes domed Prefrontal edema	Sloping forehead	Sloping forehead	Sloping forehead
Nose	Hypoplasia of nasal bones			Big, long
Philtrum				
Maxillary et palate		Narrow palate – palatal cleft	Labiomaxilliaray palatal cleft	
Mandible		Retro and micrognathia	Retro and micrognathia	Retro and micrognathia
Ears	Small and low	Small and low	Low	Low, prominent anthelix
Microcephaly		+	+	+
Tongue	Macroglossia			
Eyes	Palpebral clefts up and back		Hypertelorism – fentes Palpebral clefts down and back	Palpebral clefts up and back
Mouth	Small mouth			

Table 2.12 Microdeletions and facial dysmorphism

	Deletion 22 Digeorges	Deletion 4p16,3 Wolf Hirschhorn	Deletion 7q11 Williams	Deletion 17p13,3 Miller Dicker	Deletion 17p11,2 Smith-Magenis	Deletion 5p15 Cri du chat	Deletion 15q11 –q13 Prader Willi
Forehead	+	Prefrontal edema	Large forehead	High			Bifrontal narrowness
Nose	+	Greek helmet aspect	Bulbous nose tip	Anteverted nostrils – short nose	Sunken mid-level	Wide nose root	
Philtrum	+	Short	Long and smooth		Thick and long		
Maxillary et palate	Palatal or velar cleft					Ogival and narrow palate	Ogival palate
Mandible		Micrognathia	Retrognathia			Retrognathia	
Ears		Dysplasia and auricular appendage				Low ears	
Microcephaly		+				+	
Tongue		Hypertelorism				Hypertelorism	Lengthened cleft
Eyes		Falling corners	Thick lips		Falling mouth		Falling corners

3 Clefts and Pierre-Robin Syndrome

3.1 Clefts

3.1.1 Introduction

3.1.1.1 Why screen?

Facial clefts are found in 272 different syndromes, the frequency of cleft lips and palate is 0.5/1000, or 1/500, 41 % of which are associated with other organ anomalies, and only 6/10 being isolated.

3.1.1.2 Face and Brain: The Key Combinations of Pathologies to Know

When a facial cleft is discovered, complete analysis of the face associated with a morphological diagnostic study should be done, due to the frequent association between clefts and pathological syndromes, as was confirmed in this study by Gérard Couly, done at the Necker Hospital in Paris.

There is clearly a predominant association between the face and the brain (Table 3.1).

3.1.1.3 Facial Clefts and the Nose-Mouth Image During the Second and Third Trimesters, at 22 GW and 32 GW

The classic entry point for facial clefts is on the "nose-mouth" image, with analysis of the upper lip, the philtrum and below the septum.

A lip cleft is the starting point for labio-maxillary and/or palate clefts.

The second trimester report can no longer be written with such laconic sentences as "face unseen during this examination."

Now, during the second trimester of pregnancy, one should be able to effectively screen for facial clefts, at the very least cleft lip.

Table 3.1 Brain malformations associated with facial malformations

Facial malformations		Brain malformations		
Labio-maxillary and velo-palatal and Philtrum abnormalities	170	Unilateral 135	1 lissencephaly + (epilepsy) 1 agenesis of corpus callosum	14
		Bilateral 35	10 holoprosencephaly 1 encephalocele (Meckel) 1 cortical malformation	
Maxillo-madibular dysplasias	44	Francescetti S. Goldenhar S. 1st arch S.	1 agenesis of corpus callosum 5 posterior fossa anomalies (cyst, mega cisterna)	6
Pierre Robin syndrome	60		3 mega cisterna magna 1 occipital porencephaly	4

J.-M. Levaillant et al., *Normal and Abnormal Fetal Face Atlas*, DOI 10.1007/978-3-319-43769-9_3

3.1.2 Symptomatic Analysis of Clefts

3.1.2.1 Unilateral Cleft Lip

This is characterized by the bi-partitioning of the peri-buccal integuments and muscles and respects the gums and the dental alveoli in the upper maxilla.

It is possible to describe:

- The cleft side.
- The right and left processes.
- The skin and mucosa on either side of the cleft, the nostril bridge and nostril asymmetry.
- The seriousness of the defect on the median level and particularly at the philtrum level.

3.1.2.2 Bilateral Cleft Lip

This is characterized by the tri-partitioning of the peri-buccal integuments and muscles, and respects the gums and the dental alveoli in the upper maxilla.

It is possible to describe:

- The median process, under the nostrils
- The surface of the skin on the median process, and how much excess mucosa there is.
- Mucosa
- Skin
- The two clefts on either side of the median process and their possible asymmetry.
- For each side, describe as one would a unilateral labial cleft
  - The right and left processes
  - The skin surface and mucosa on either side of the cleft.

The sagittal and para-sagittal views enable one to the complete the examination with observation of related markers:

- Position of the tongue
- Orbits
- Bone structure: zygomatic arch, malar bone.
- Visualization of the soft palate.

The axial view, or more precisely the stepped axial views:

These will make it possible to see to what extent the cleft impacts the maxilla, the back side of the palate and the pterygoid apophyses, the tongue position and the mandibular arch.

3.1.2.3 What 3D Can Do

Multi-view sagittal and axial views, in TUI mode associated with thick VCI mode views.

The study of millimetric views enables the detailed analysis in three planes for possible deformations of bone structures.

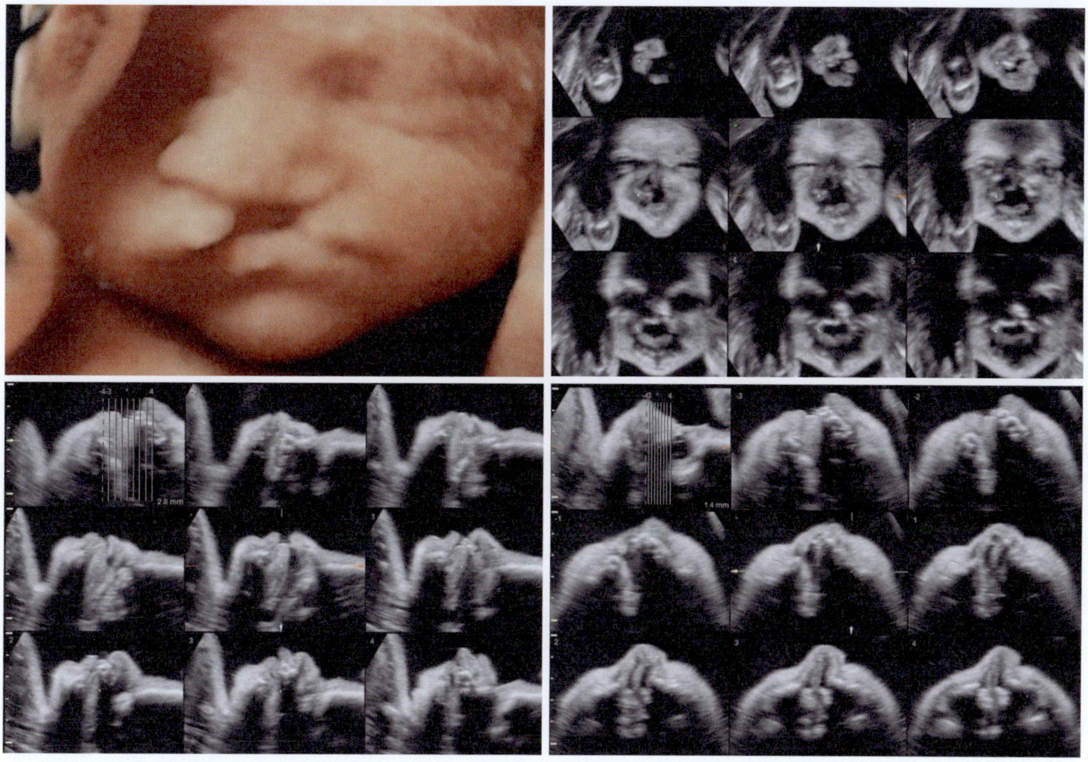

Fig. 3.1 Labioalveolar-palatal cleft. Overview in three planes

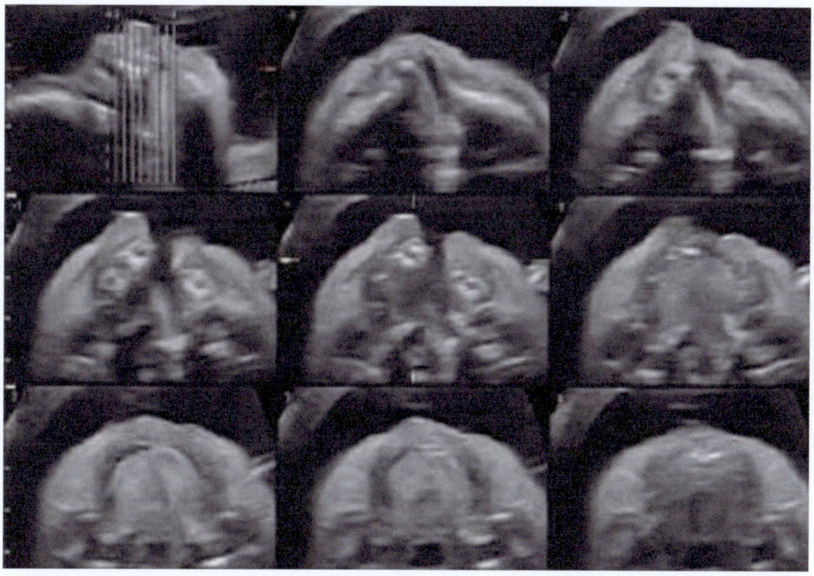

Fig. 3.2 Labioalveolar-palatal cleft. Axial view

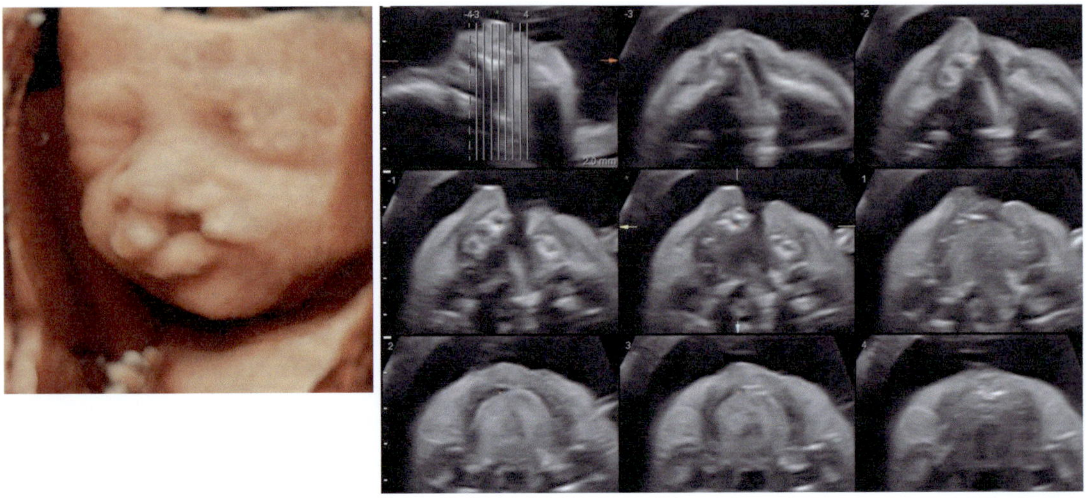

Fig. 3.3 Labioalveolar-palatal cleft. Axial view

3.1.3 Ultrasound Diagnosis of Clefts: Analytical Study

Once the diagnosis of a lip abnormality has been uncovered, the ultrasonographer must use a simple, strict diagnostic methodology:

- Description of the cleft lip
- Description of the cleft maxilla if there is one
- Description of the cleft palate if there is one.

3.1.3.1 Analytical Study of First Trimester Clefts

On the Herman score sagittal view, the upper lip could have a comma-shaped deformation, with slight eversion of its inferior portion, which should alert the ultrasonographer that there may be a cleft lip.

The nose-mouth view can then show the defect.

Maxillary and mandibular axial views are easy to capture and can show the bi-partitioning of the facial mass if there is a labio-maxillary unilateral cleft, or a tri-partitioning in the case of a bilateral labio-maxillary cleft.

An open-mouth sagittal view is rather easy to capture for an examination targeting the palate, and enables one to follow both the hard and soft palates. The image is almost easier to capture at this stage than during the second trimester, because there are fewer shadows linked to ossification.

What 3D Can Do

Skin surfacing objectifies the processes well.

The multi-planar technique and using thick views makes it possible to capture stepped axial views starting from the orbit on to of the mandibular cavities.

It is easier to look for associated anomalies during the first trimester. With trisomy 13, the associated hexadactylism can be seen in the same 3D volume, and for that reason, one must record the largest volume possible.

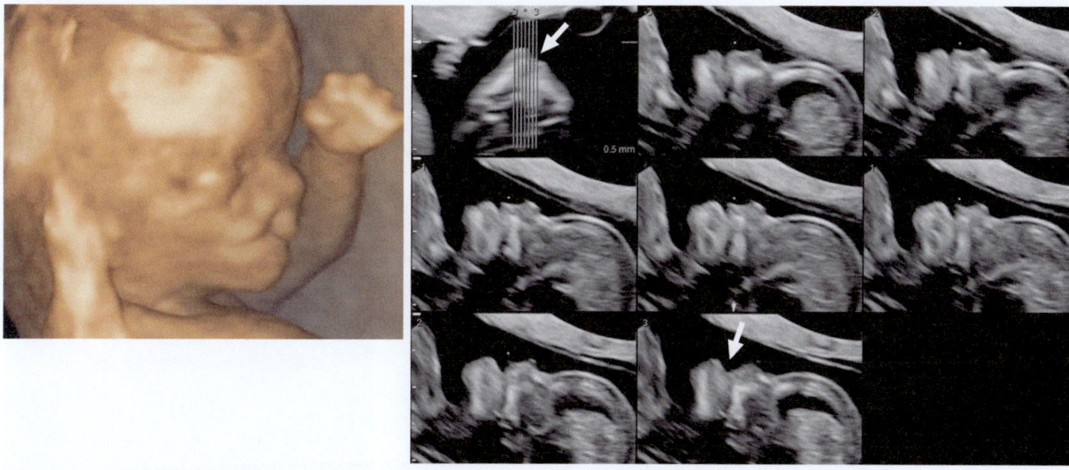

Fig. 3.4 Uniliteral labioalveolar cleft at 12 GW

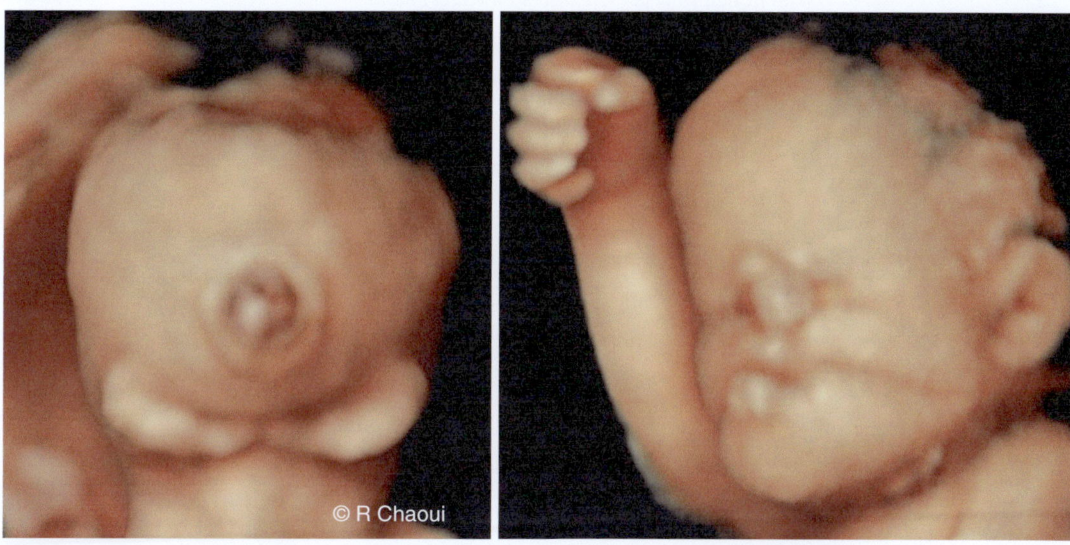

Otocephaly

Trisomy 13
facial dysmorphism and
hexadactylism

Fig. 3.5 Otocephaly

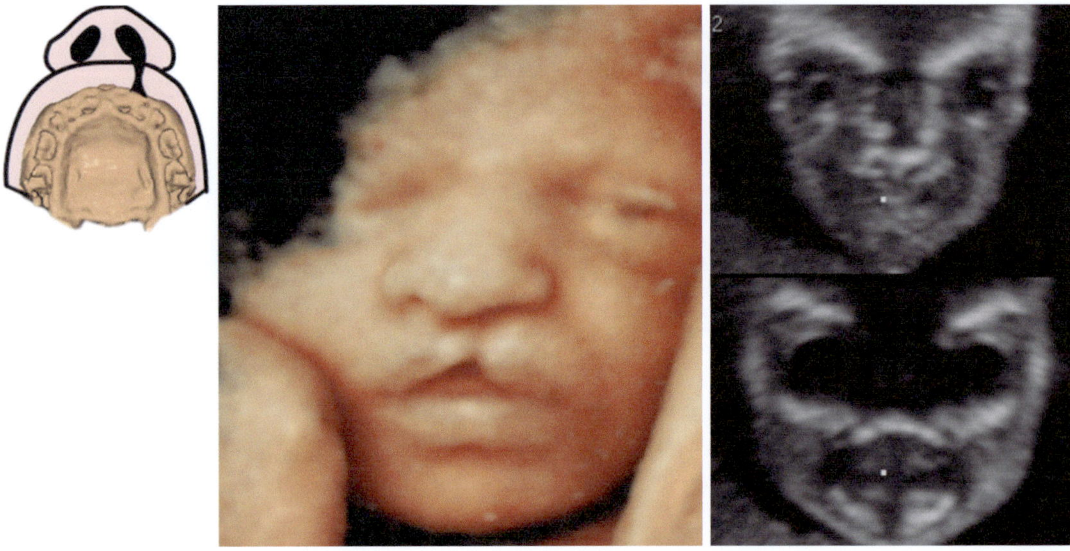

The frontal view demonstrates an intact palate

Fig. 3.6 Unilateral labial cleft

3.1.3.2 Analytical Study of Clefts from the 2nd Trimester

The most favorable time frame is between 20 and 25 GW.

The major types of clefts:

Unilateral cleft lip.

Unilateral labio-alveolar-palatal cleft.

Bilateral labio-alveolar cleft impacting the primary palate, with an intact secondary palate.

Median cleft: a particular and associated pathology.

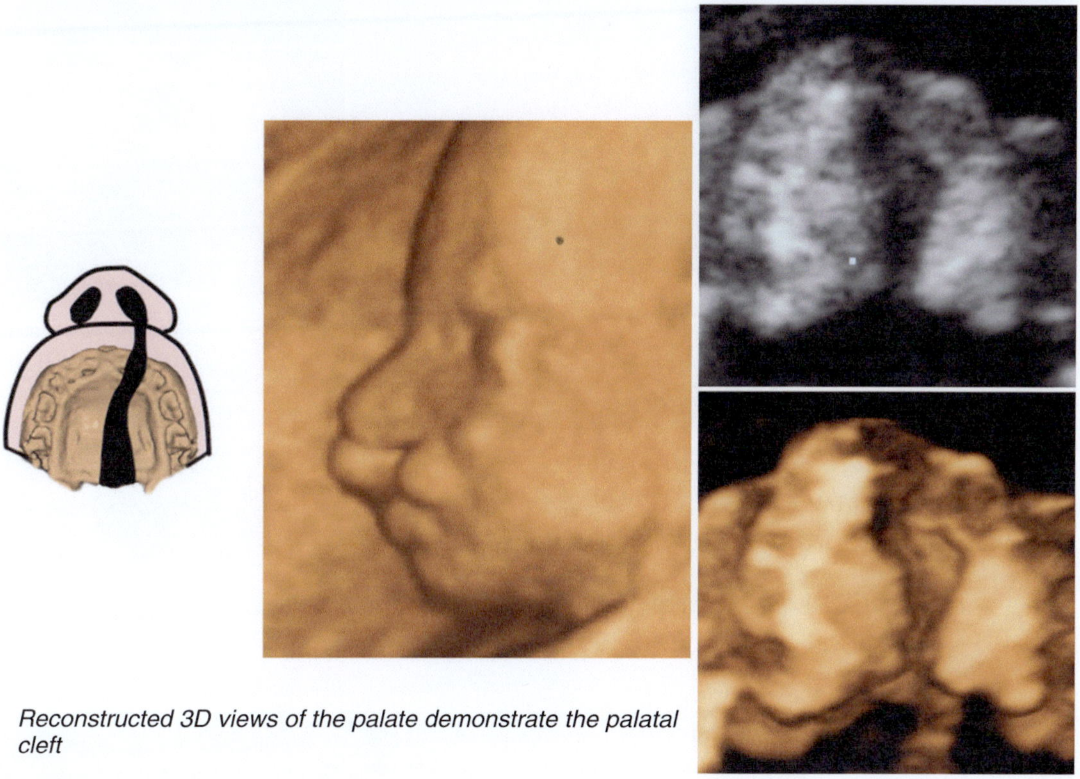
Reconstructed 3D views of the palate demonstrate the palatal cleft

Fig. 3.7 Labioalveolar and palatal cleft

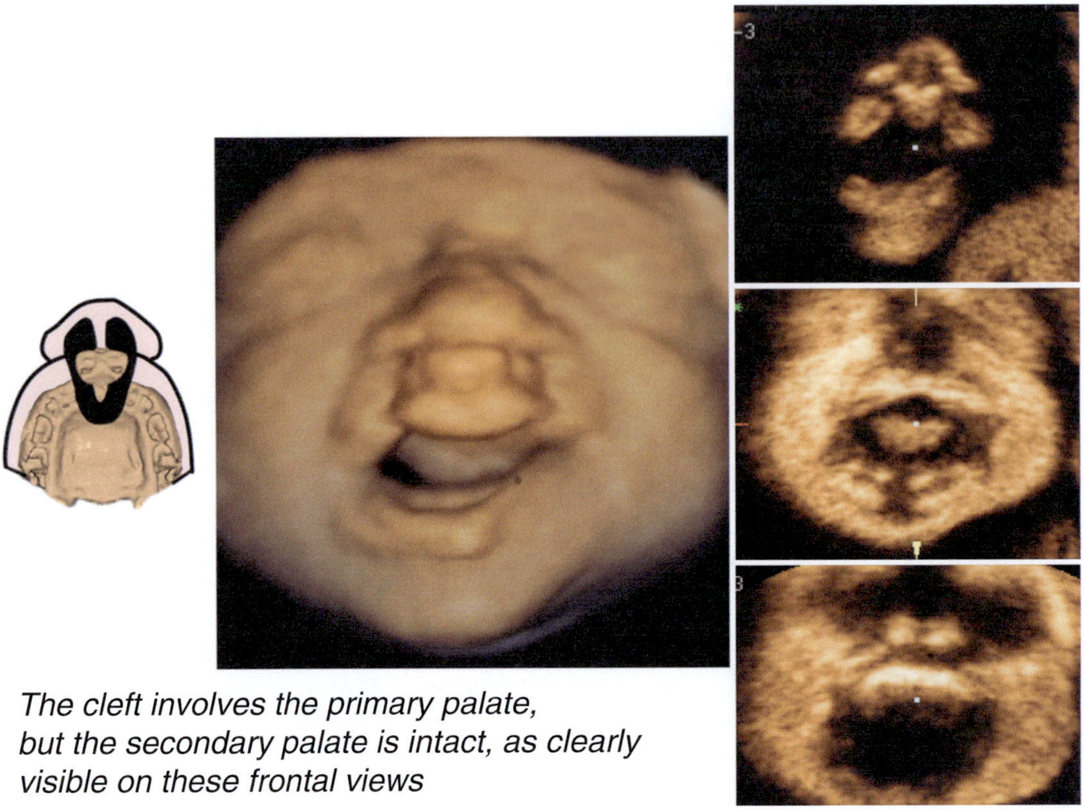

Fig. 3.8 Bilateral labioaveolar-palatal cleft. 3D reconstructed views of the palate

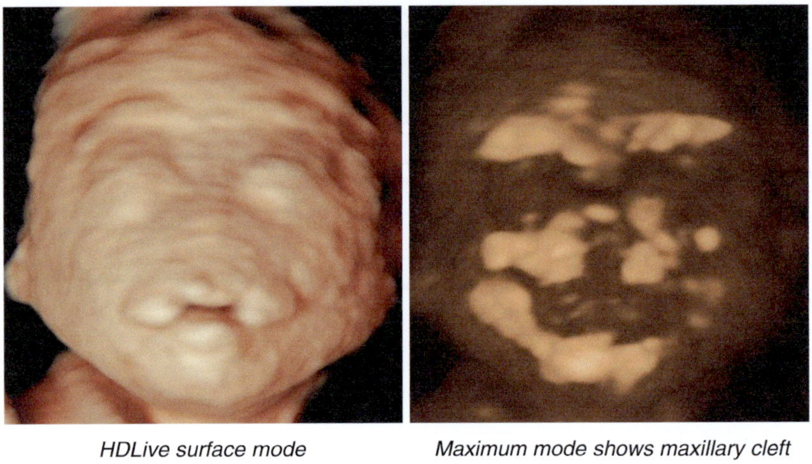

Fig. 3.9 17 GW median cleft lip – trisomy 13

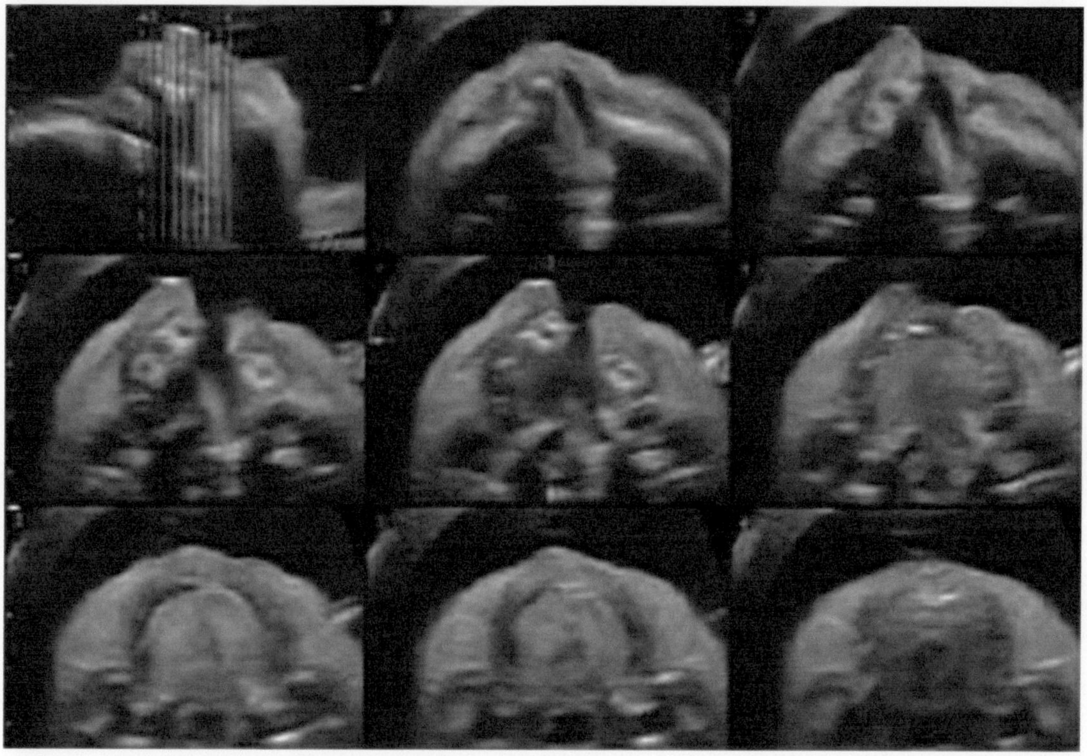

Fig. 3.10 Labioalveolar and palatal cleft, stepped axial views

3.1.3.3 Basic Principles for Obtaining 2D and 3D Views

Position of the head:

The position of the fetal head during image capture is important for precise rendering and for the diagnosis.

1. A denected position is used to capture and render images of the primary and secondary palates.
2. In 3D, one can approach the palate in three ways:
   - frontal mode
   - flipped mode
   - reverse mode

Sagittal, axial and coronal views:

By associating the various modes, such as thick 2-mm views, multiplanar mode and TUI mode, one can capture anatomical images of the various views in order to study the entire head in detail, from the integuments to the bone structure.

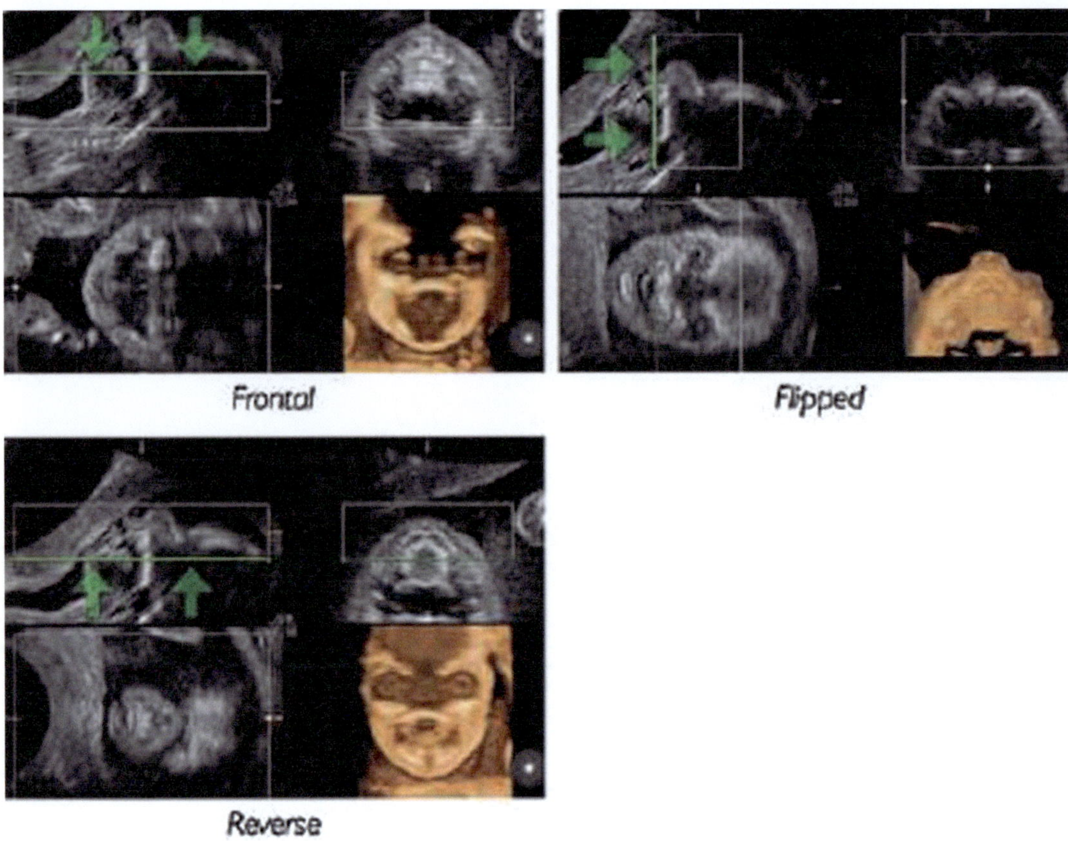

Fig. 3.11 Three ways to look at the palate from a volume image: frontal, flipped or reverse modes

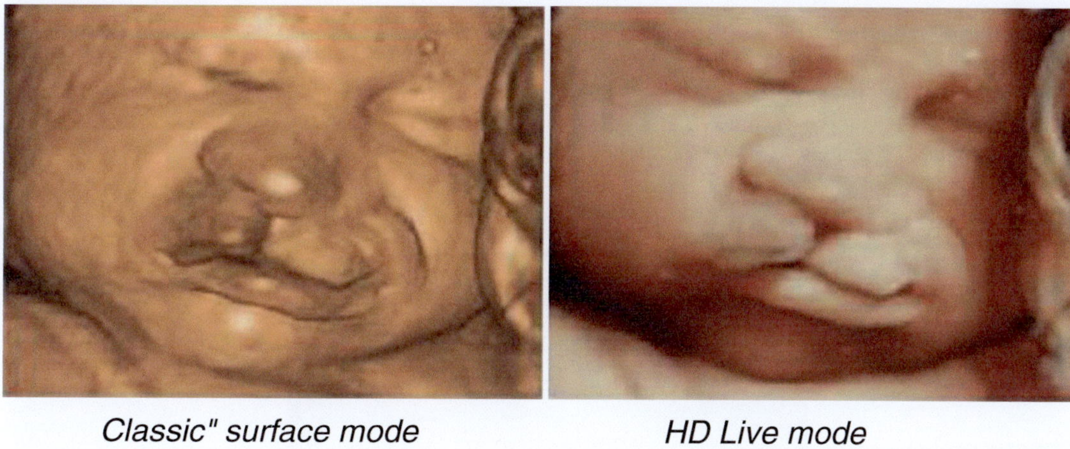

Fig. 3.12 Unilateral labioalveolar-palatal cleft

3.1.3.4 HD LIVE: Contribution to an Aesthetic Prognosis

Variable directional lighting software can enable more detailed analysis of the soft tissue. In the case of cleft lip, it is particularly interesting to differentiate the white and red lips and help the surgeon develop a prenatal estimation of the plastic repairs needed for the upper lip.

In this evaluation, it is important to look at two tissues:

- The lip skin surface on the cleft side
- The volume of the red lip on the same side, compared to the cleft's contralateral side.

3.1.4 Cleft Algorithm

An example of the cleft algorithm, in the case of a unilateral cleft lip, is shown in Figs. 3.13, 3.14, 3.15, and 3.16: unilateral labial cleft, bipartition of the primary palate, normal palate, and palatal cleft.

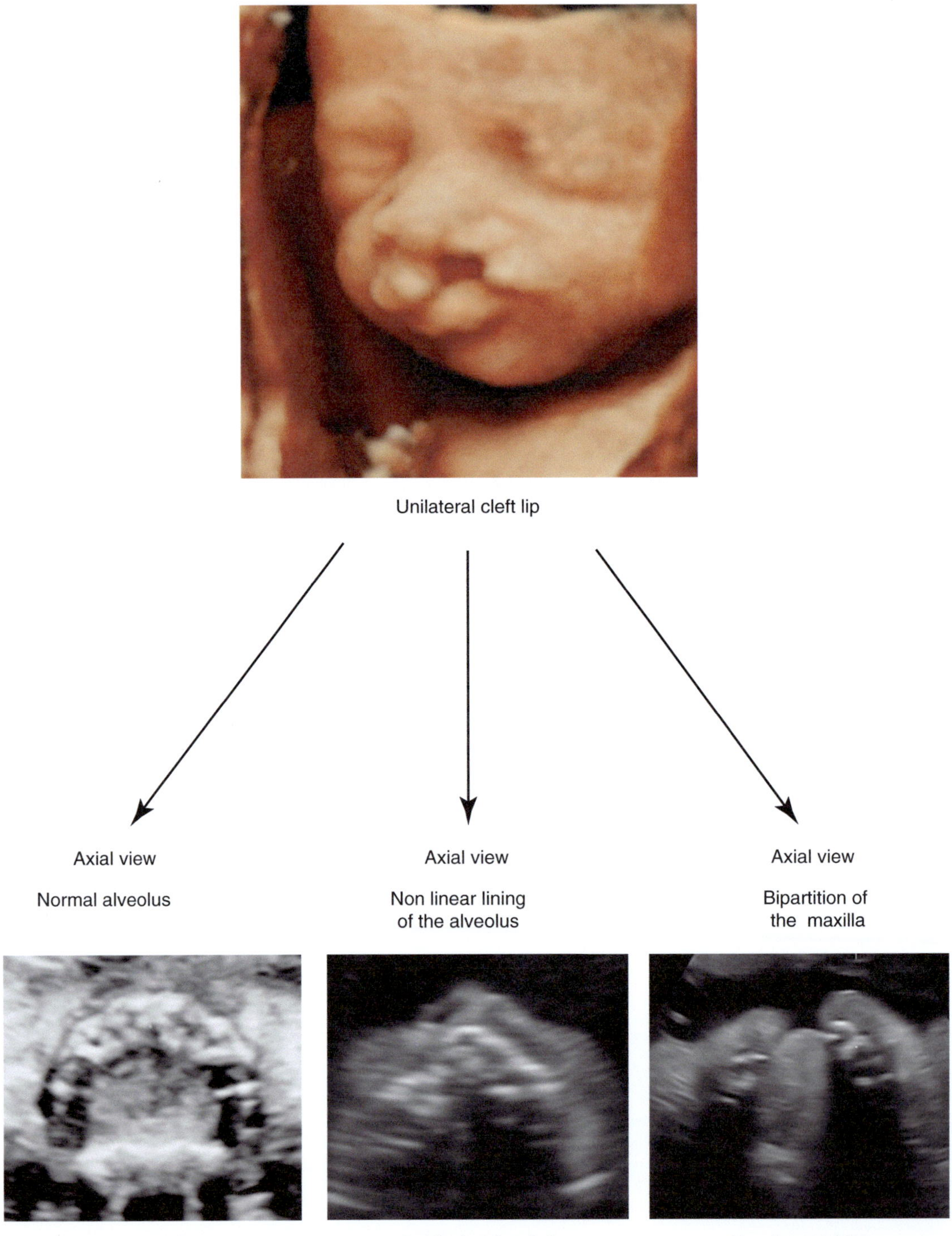

Fig. 3.13 Unilateral labio-alveolar-palatal cleft with tongue protrusion in the cleft. "Classic" surface mode. HD Live mode

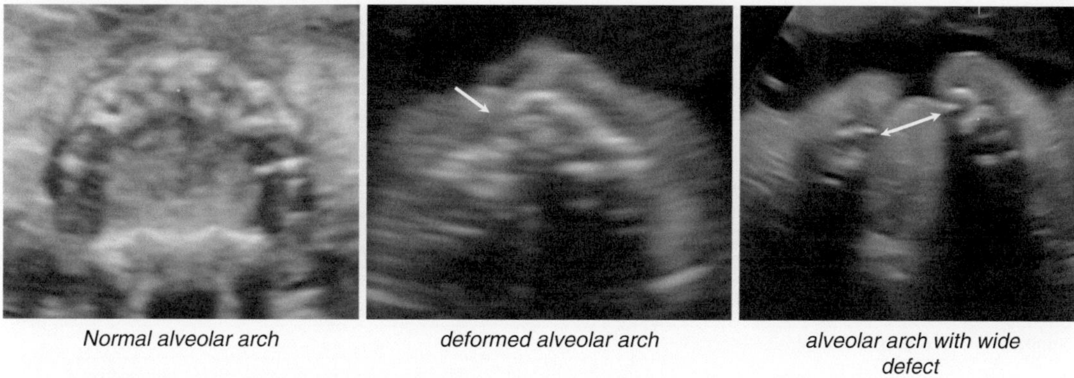

Fig. 3.14 Bipartition of the primary palate

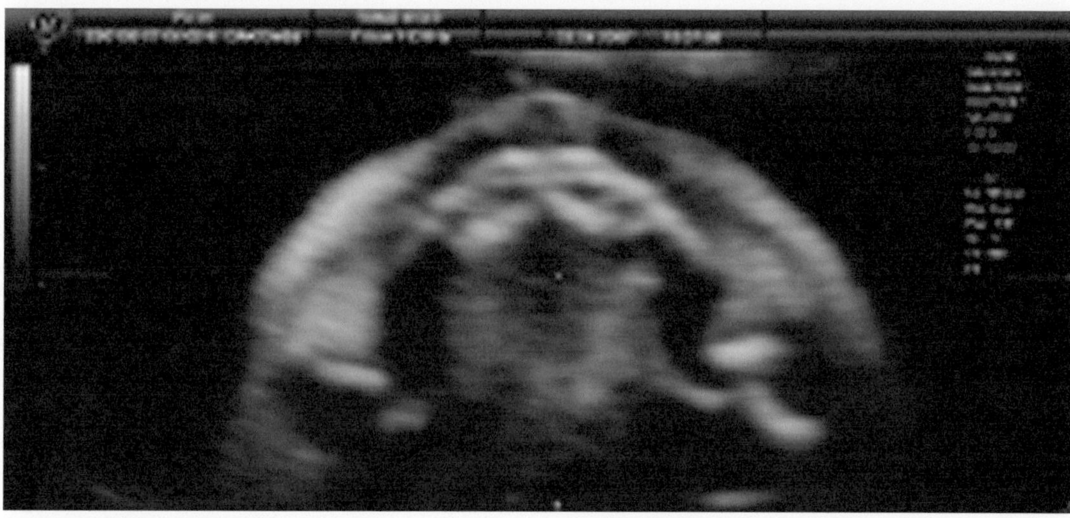

Fig. 3.15 Normal palate

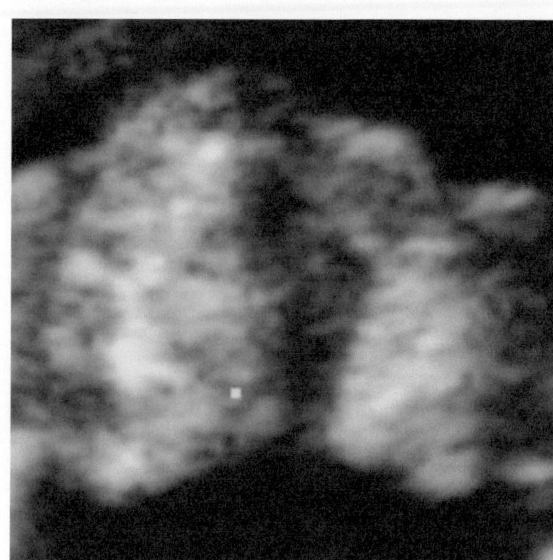

Fig. 3.16 Palatal cleft

3.1.5 Ultrasound Diagnosis of Labioalveolar-Palatal Clefts: Summary

Tables 3.2, 3.3, 3.4, 3.5, and 3.6 summarize what to look for and find on each of the image views in the different cases of facial clefts: Unilateral and bilateral cleft lip; unilateral and bilateral cleft alveolus; unilateral and bilateral cleft hard palate; posterior clefts, premaxillary agenesis and holoprosencephaly syndrome, respectively.

Table 3.2 Ultrasound of unilateral and bilateral cleft lip

Image view	Unilateral cleft	Bilateral cleft
Sagittal	**No direct image**	
	The indirect signs are: comma-shaped columella and loss of lip regularity.	The expected normal image is replaced by a budding growth of tissue resulting from pre-maxillary protrusion.
	A para-sagittal slice can visualize the cleft upper lip: a defect causing asymmetry in cupid's bow	There are two defects on the parasagittal image, dividing the facial mass into three.
	The nostrils can be seen, complete but asymmetrical.	
Axial	Bipartition of the lip with a solution for continuity.	The median process is very much in front of the two lateral lip lines, creating a division of the lip into three.
	Asymmetrical nostrils – the nostril is intact on the cleft side, flattened and widened.	Nostrils are symmetrical and complete.
Coronal	**This is an important view. The nose-mouth image required for the basic report.**	**This is an important view.**
	The unilateral cleft can be seen when the slice is perfect, without the interposition of a limb or the cord.	The division into three of the lip and the process "set" under the nostrils is clear.

Table 3.3 Ultrasound of unilateral and bilateral cleft alveolus

Image view	Unilateral cleft	Bilateral cleft
Sagittal or para-sagittal	The para-median zone at the level of the cleft lip shows an anechogenic defect and alveolar absence, more pronounced depending on the size of the alveolar cleft, which could go from simple misalignment of the alveolar arch to an anechogenic gap.	The pre-maxillary process is projected toward the front.
		Morphological analysis of the process shows the presence of soft tissue, associated with alveolar structures.
		The rupture between the remaining alveolar arch, to the left and right, appears like an anechogenic zone.
Axial	This is an important image. One visualized and quantifies the degree of rupture of the alveolar arch, from simple disturbed alveoli to a bi-partitioning of the facial mass.	Protrusion of the premaxillary and the prolabium.
		The pre-maxillary mass can be quantified: size, soft tissue, bone mass.
		The separation of the facial mass into three by full rupture of the pre maxillary respects the symmetry of the adjacent hemi-maxillaries and protruding pre-maxillary.
Coronal	There is a defect in alveolar continuity.	Continuity of the maxillary arch seen from the front is interrupted twice: to the left and right of the anterior process.
	To visualize, one must remain in the strict maxillary arch plane.	

Table 3.4 Ultrasound of a unilateral or bilateral cleft secondary or hard palate

Image view	Cleft palate with unilateral labio-maxillary cleft	Cleft palate with unilateral labio-maxillary cleft
Sagittal or para-sagittal	Successive slices show a defect in the secondary palate.	The bilateral cleft associated to a cleft palate is defined in the sagittal slice by the vomer bone, which is median, rectilinear and not deviated.
	It is median, asymmetrical, sinuous, often para-sagittal to the vomer bone, which can be a "trap" on the strict sagittal slice.	The para-sagittal slices are the ones that enable the visualization of defects in the hard palate.
Axial	The important image is the deviation of the vomer bone axis in relation to the line perpendicular to the maxillary.	The hyperechogenic median line, perpendicular to the maxillary axis, represents the vomer bone.
	The angle is around 10–20°	Not held back, it pushed the premaxillary toward the front, on the median axis.
		The defect is visible on both sides.
Coronal	The palatal arch is interrupted by the anechogenic defect.	Only the vomer can be seen, not deviated, in suspension, with lower bone support.
	The defect is situated opposite the cleft lip and maxillary.	From the two sides, one can visualize two little structures made of bone relics.
	The two palatal demi-arches are asymmetrical.	

Table 3.5 Ultrasound aspects of posterior clefts

View	Ultrasound aspects
Palate 10 mm sagittal and para-sagittal	The hard palate is seen on this view. It is subjectively short, inferior to 10 mm. This sign is not a certain sign, often looked for in inferior facial angle anomalies such as retrognanthia and glosoptosis.
Axial	On the axial view, the posterior edge is only visible if the slice goes through the pterygoid apophysis and the posterior palatal edge, which appears alone, in the form of a hyperechogenic line that is continuous and very slightly concave. If there is a cleft, the posterior edge forms an upside down V.
Coronal	Successive coronal view from front to back show a bi-partitioning of the bone mass behind the alveoli—a reference point for the primary palate. In the back, there is a rupture of the secondary palate, at 5–6 mm from the alveolar line. In 2D ultrasound, the successive views first show continuity of the line, then progressive flaring of the posterior edge.

Table 3.6 Ultrasound of median clefts. Pre-maxillary agenesis. Holoprosencephaly syndrome

View	Ultrasound
Sagittal or parasagittal	Flat nose, prominent lower lip.
Axial	No primary palate.
	In the back, visualization of secondary palate.
Coronal	Mucosa, labial and bone defect.
	Visible primary palate
	Nearly absent nose, flat
	X-shaped cleft lip, limited by the nose, the lower lip and two lateral defects.

Pre-maxillary agenesis. Holoprosencephaly syndrome

3.1.6 The Decision-Making Process After the Discovery of a Facial Cleft

The decision-making process following the discovery of a facial cleft is summarized in Fig. 3.17.

This decision-making process guides the ultrasonographer in explaining to the parents the type of cleft which affects the child.

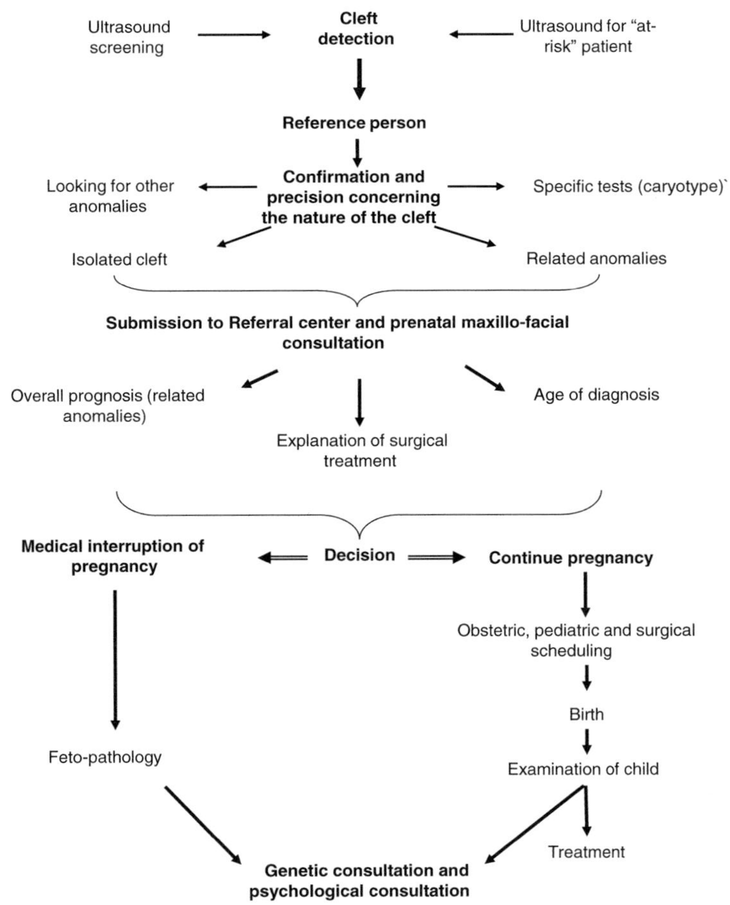

Fig. 3.17 Diagnostic and treatment decision making process

3.2 Pierre-Robin Syndrome

Retrognathia is the first warning sign related to Pierre Robin Syndrome seen on ultrasounds as early as the first trimester, and in each screening ultrasound done at 22 GW and 32 GW, which again emphasizes the importance of a systematic sagittal view.

3.2.1 How to Screen for Retrognathia

A first-trimester sagittal view makes it possible to visualize retrognathia and palate and mandibular abnormalities.

3.2.1.1 Retrognathia and the Inferior Facial Angle

The inferior facial angle evolves throughout pregnancy, due to suckling by the fetus, which modifies the angle by a few degrees.

In Pierre Robin syndrome, the inferior facial angle is less than 49°.

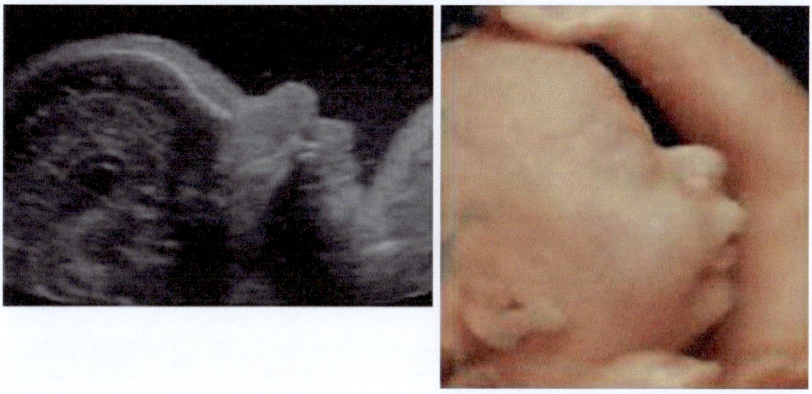

Fig. 3.18 Retrognathia at 32 GW. Note the absence of nasal bones in this case

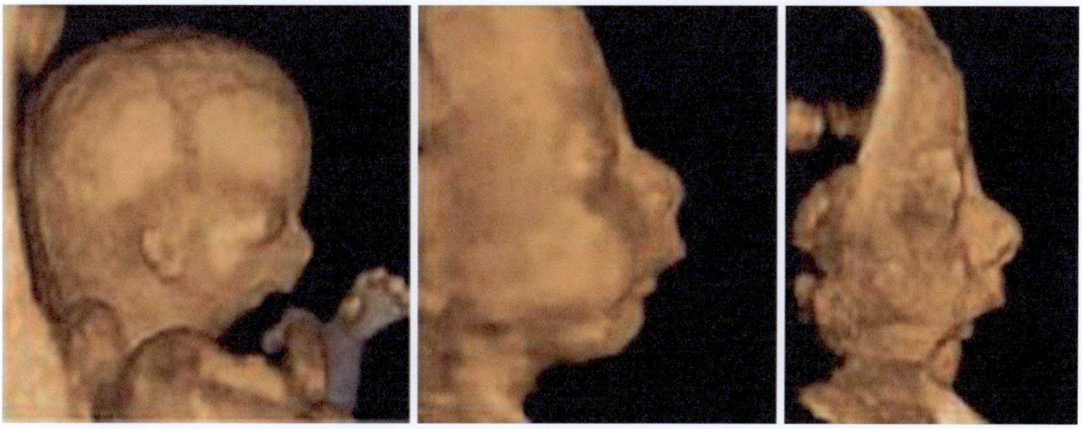

Fig. 3.19 Evolution of retrognathia (Pierre Robin syndrome) during a pregnancy. 12 GW, 24 GW and 32 GW. Retrognathia is most visible at the beginning of the pregnancy

3.2.1.2 Axial View of the Jaws

From the 12th week, the shape of the mandible in Pierre Robin syndrome is very different.

A normal mandible forms a near triangle, in a V shape, while a PRS mandible forms the rounded arc of a circle.

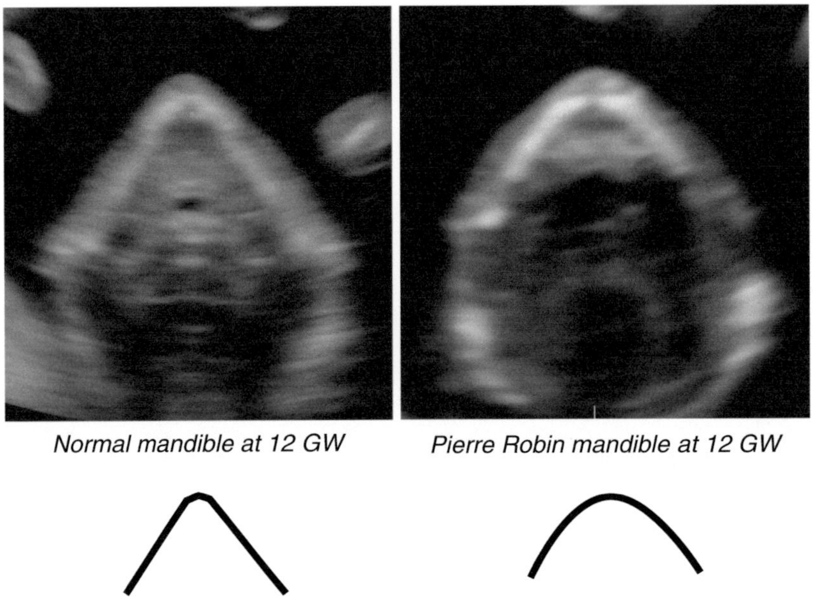

Fig. 3.20 Shape of a normal mandible and of a mandible in case of Pierre-Robin Syndrome

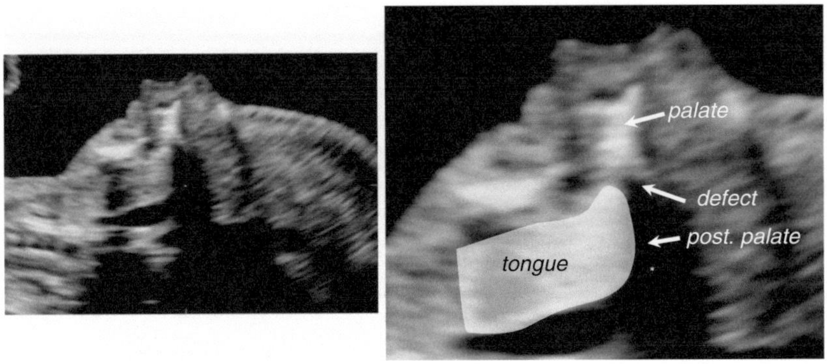

Fig. 3.21 Retrognathia with glossoptosis and posterior cleft palate at 15 GW

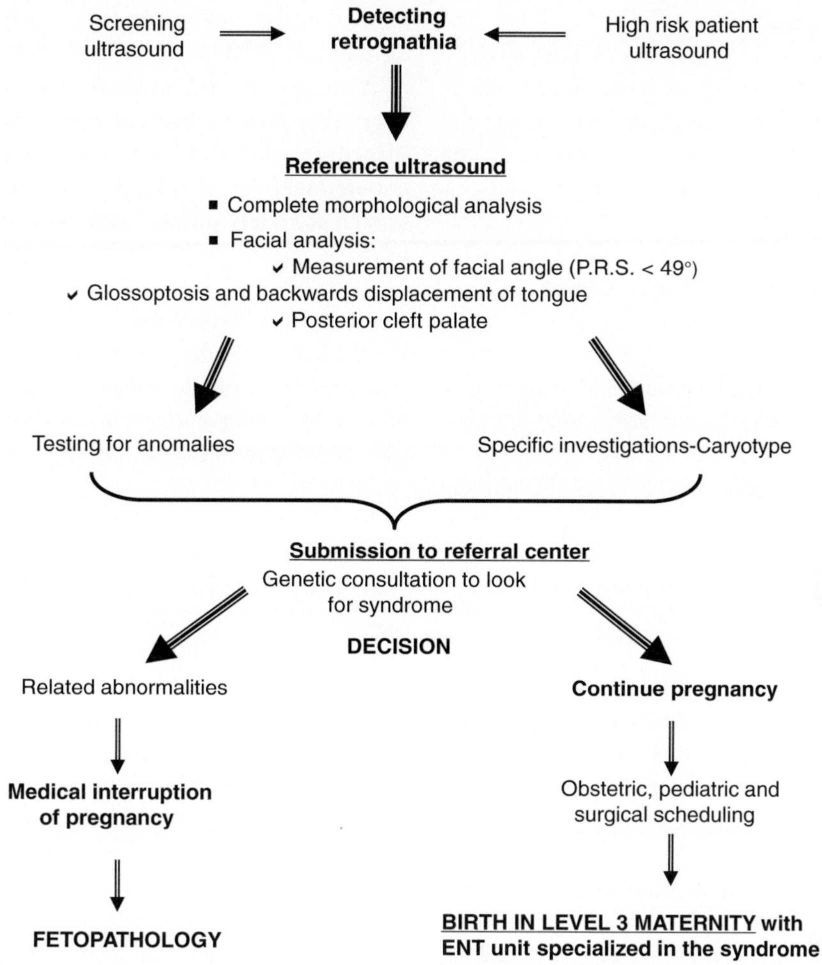

Fig. 3.22 Diagnostic and decision-making process in case of Pierre-Robin syndrome

3.2.1.3 Glossoptosis

Verticalization of the tongue can be seen as early as the first trimester. Its intra-buccal position is displaced backwards in comparison to the normal position, where the anterior extremity of the tongue is located right behind the jaws.

3.2.2 What to Look for in Cases of Retrognathia

Retrognatia is the warning sign, and leads immediately to a search for glossoptosis and cleft posterior palate.

The diagnostic trio will lead to a complete morphological examination due to the large number of congenital syndromes linked to the trio, the four most important of which, in order of frequency are:

3.2.2.1 Stickler Syndrome

Stickler syndrome is a hereditary vitreoretinopathy with ocular problems, a Pierre Robin sequence that is more or less complete, bone abnormalities and neuro-sensory deafness (10 % of cases).

Incidence at birth is estimated at around 1/7,500.

Ocular anomalies can include juvenile cataract, myopia, strabismus, vitreoretinal or chorioretinal degeneration, retinal detachment, and chronic uveitis. Bone abnormalities include discrete platyspondylia and often voluminous epiphyseal damage (Martine Le Merrer-Orphanet).

The diagnostic ultrasound cannot guarantee parents the absence of Stickler syndrome on the second and third trimester ultrasounds.

3.2.2.2 The 22q Deletion

Monosomy 22q11 or DiGeorge Syndrome is characterized by a combination of several malformations, including thyroid and parathyroid hypoplasia, conotruncal congenital heart disorders, and slight but distinctive facial dysmorphism.

The velocardiofacial syndrome, in its full form, combines congenital conotruncal heart disorders, cleft palate or velopharyngeal incompetence, facial dysmorphism, and learning difficulties. (N. Philip)

A diagnostic ultrasound that reveals retrognathia should lead to a search for heart disorders, and similarly, any discovery of a conotruncal heart disorder should lead to a search for cleft palate, and even a posterior cleft.

3.2.2.3 Fetal Alcoholism

Fetal alcohol syndrome (**FAS**), which is sometimes called alcoholic embryopathy, is the alcoholic intoxication of the embryo or fetus resulting from the mother drinking during pregnancy.

- Intrauterine growth deficiency. The growth deficiency is global, and touches weight, height and cranial perimeter. The children are born weak, fragile and often prematurely.
- Craniofacial dysmorphism. FAS leads to a specifically shaped baby head, with small palpebral fissures, a marked nasal hollow, a bump between the eyes, low and unstuck ears, microretrognathia (small mandible that is too far back), a thin and convex upper lip, anteverted nostrils (excessively turned toward the front), long and bulging philtrum.
- Malformations touch around 25 % of cases, and can impact the brain, heart, urinary tract, etc.
- Psychiatric troubles appear after birth, sometimes during childhood.
- Immediate alcohol withdrawal syndrome with tremors and trouble sleeping.
- Hyperactivity during childhood with irritability, attention deficit, difficulties concentrating.
- Mental retardation. Studies have demonstrated that a mother consuming three glasses of alcohol a day leads to a child's loss of an average of 7 IQ points.
- Poor muscle tone, memory and fine motor skills.

There are several approaches to the diagnosis on ultrasound, including the facial dysmorphism and microretrognathia, growth deficiencies—an excellent biometric marker—and of course the heart and the brain.

3.2.2.4 Treacher-Collins Syndrome

Treacher-Collins syndrome is a congenital disorder affecting craniofacial development, characterized by bilateral and symmetric oto-mandibular dysplasia without limb anomalies, associated with a variety of head and neck abnormalities. The annual incidence is estimated at 1/50,000 live births. Children show characteristic facial dysmorphism, with bilateral and symmetric hypoplasia of the malar bones and the infra-orbital opening (80 % cases) or the mandible (78 %) (retrognathia, retrogenia), which leads to dental malocclusion, characterized by anterior open bite. Predominant hypoplasia of the soft tissues can be observed at the malar level, around the lower orbital edge and on the cheeks. One can also observe complex anomalies in the temporomandibular joint responsible for limitation in buccal opening that can be of variable severity, down-slanting palpebral fissures (89 %) and colobomata of the lower eyelids at the 1/3 external-1/3 middle union (69 %) with absence of eyelids on the external third of the lower eyelid. The palate is ogival, and cleft palate could be observed (28 %). External ear anomalies such as anotia or microtia, atresia of the external auditory canals and anomalies in the ossicular chain are often present (60 %) and lead to conduction deafness. Intelligence is generally normal. Respiratory and nutrition difficulties can be encountered during early years, resulting for the narrowness of the upper respiratory passages and the limitations in the buccal opening. (M-P Vasquez)

The diagnostic ultrasound revealing any retrognathia should look for abnormalities in the external earls and the semi-circular canals, the most frequent related sign, which often only touches one ear and supports a presumed Treacher-Collins syndrome.

3.2.2.5 Pierre Robin Syndrome: Non-isolated Syndrome

GROUP 1 – The 4 most frequent syndromes

Stickler syndrome	34 %
22q.11 deletion	11–15 %
Fetal alcohol syndrome	10 %
Treacher-Collins syndrome	5 %

GROUP 2 – the 26 rarest syndromes

Orofacial digital syndrome
Nager syndrome
Silver Russel syndrome
Trisomy 22
Trisomy 21
Vaproate syndrome
Hemifacial microsomia
Cerebral ocular facial syndrome
Marshal syndrome
Cerebral costo mandibular syndrome
Kabuki syndrome
Atelosteogenesis type 3
Beckwith Weidemann syndrome
Distal anthrogryposis
Bilateral femoral dysgenesis
Catel-Manzke syndrome
Klinefelter syndrome
Non-ocular Stickler syndrome
Miller Dieker syndrome
Spondyloepichyseal syndrome
Dysplasia
Moebius syndrome
Pterygium popliteal syndrome
Soethre Chotzen syndrome
Adam sequence
Diastrophic dysplasia syndrome
Larsen syndrome

Table 3.22 shows the importance of having a pluridisciplinary team in helping with the choice of continuing pregnancy as well as in the continued research for syndromes that could lead to a medical interruption of the pregnancy.

Dysmorphism

4

4.1 How to Define Objective Facial Dysmorphism

The study of fetal face anomalies using precise, reproducible, anthropometric criteria helps avoid the too-frequent subjectivity found in an approach that is not truly quantified. The discovery of dysmorphism and its characterization is one way that is frequently used to uncover various pathological fetal syndromes.

4.2 Tools for This Study

They were described in Chap. 1 and consist of:

Volume ultrasound images that enable:

- Capturing rigorously strict views using multiple plane views and their orthogonality principle.
- Demonstrative imagery with surface rendering modes
- An evaluation of the bone structures with the "maximum" mode.

Thus, the study of the **sagittal view**, especially with a study of the facial angles, is a key moment in the study of the fetal profile, which is very often altered in cases of facial dysmorphism.

The **axial view** will contribute key information regarding the shape and size of the maxillary bones and particularly the mandible, and will, among other things, enable a precise measurement of the inter-orbital diameter.

The **coronal view** will, as shown in other chapters of this book, be an important tool to evaluate labioalveolar-palate defects.

A panel of measurements covers the various facial parameters.

For a clear presentation, we decided to subdivide the most common dysmorphic pathologies that impact the face into five categories:

- Major chromosomal anomalies
- Pathologies impacting primarily the upper face
- Pathologies impacting primarily the mid-face
- Pathologies impacting primarily the lower face
- Unclassifiable pathologies

4.3 Dysmorphism of Major Chromosomal Anomalies

Screening for the major chromosomal anomalies currently depends on quality screening during the first trimester. In this chapter, we will identify the principal characteristics of the dysmorphism related to the major chromosomal anomalies.

4.3.1 Trisomy 21

There is frequently a delay in the ossification of the nose bones (Figs. 4.1 and 4.2), the forehead could

J.-M. Levaillant et al., *Normal and Abnormal Fetal Face Atlas*, DOI 10.1007/978-3-319-43769-9_4

be domed, sometimes with a prefrontal edema (Fig. 4.3), the features could appear flat without the upper facial angle truly being modified (Figs. 4.4 and 4.5a, b), the mouth is small, with corners that to not exceed the wings of the nose (Fig. 4.6a, b), the palpebral fissures are directed up and behind (Fig. 4.7), the ears are small and low (Fig. 4.8), and the tongue could be protrusive (Fig. 4.9).

Nevertheless, in a certain number of cases, the profile could appear normal (Fig. 4.10).

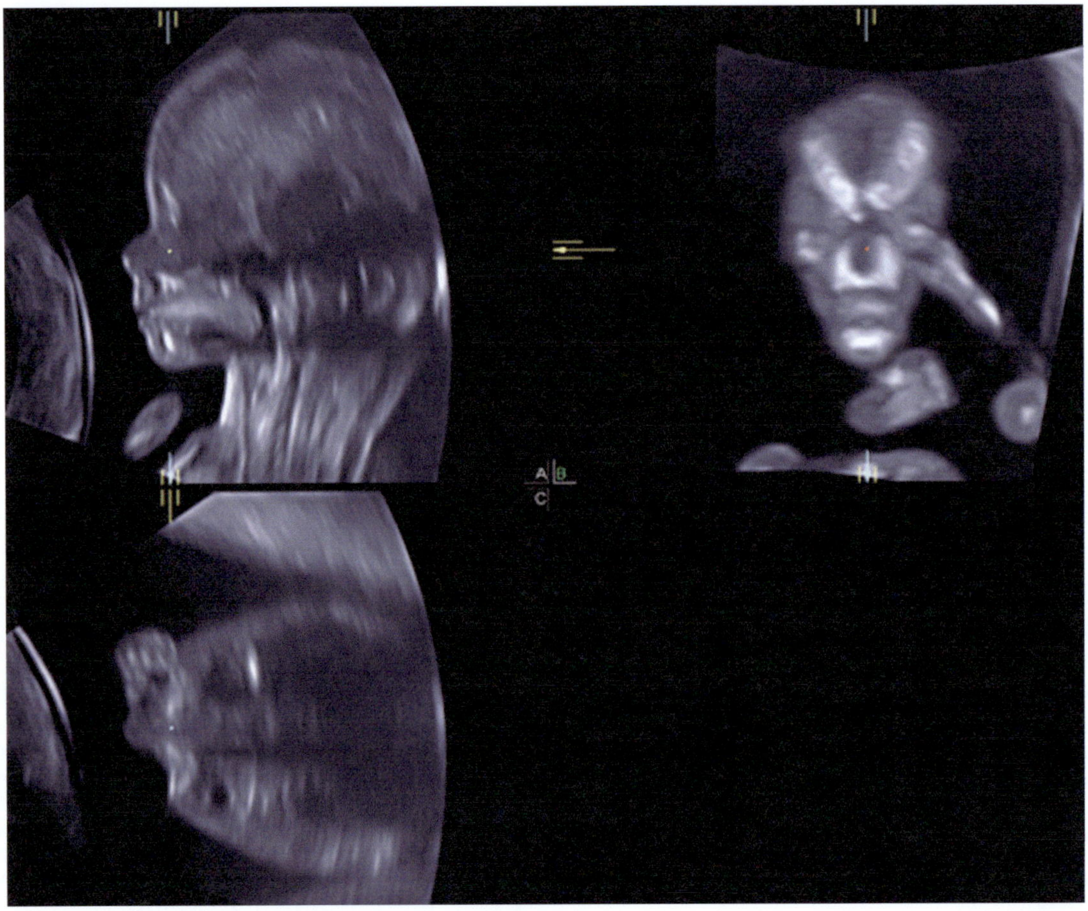

Fig. 4.1 Nasal bone not visible at 13 GW (multiplanar view)

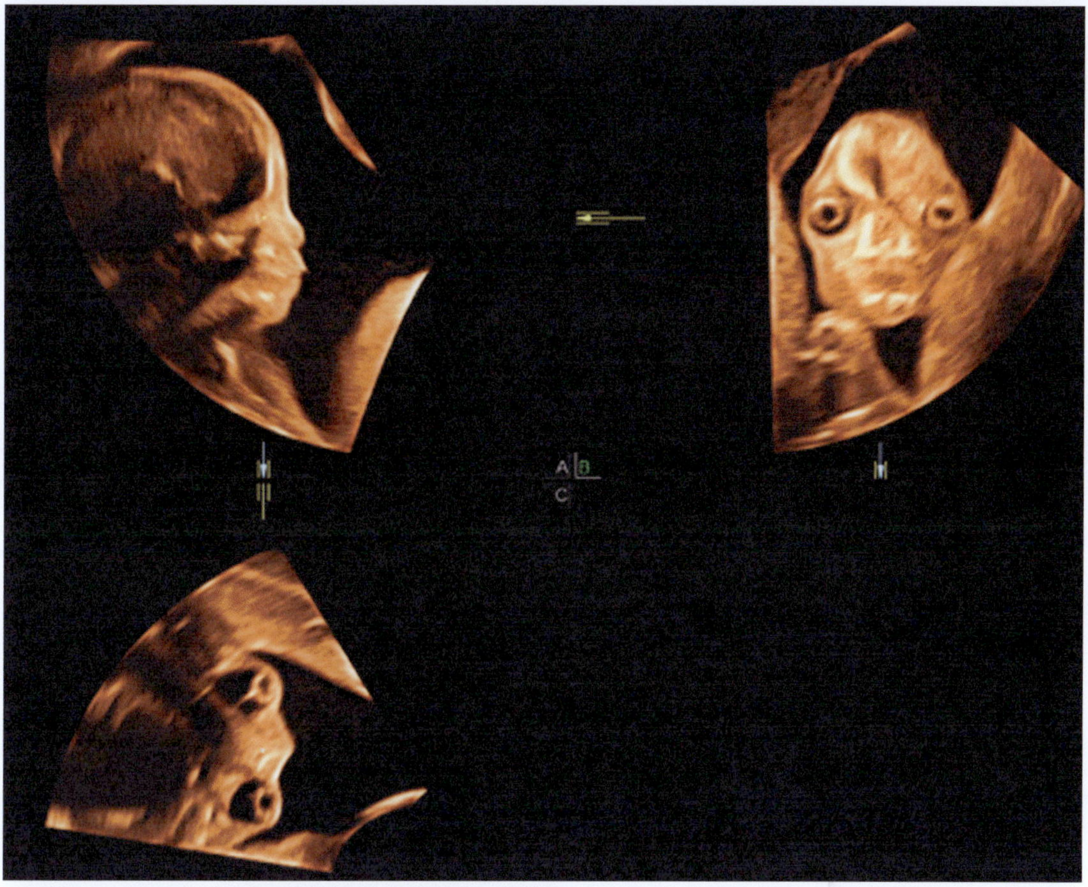

Fig. 4.2 Nasal bone not visible at 15 GW (multiplanar view)

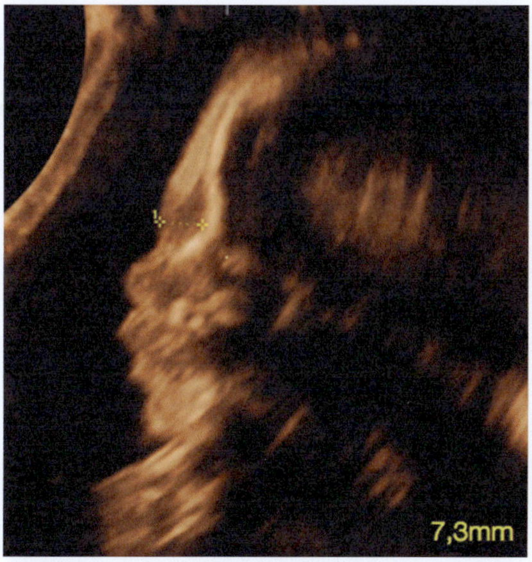

Fig. 4.3 Prefrontal edema 22 GW

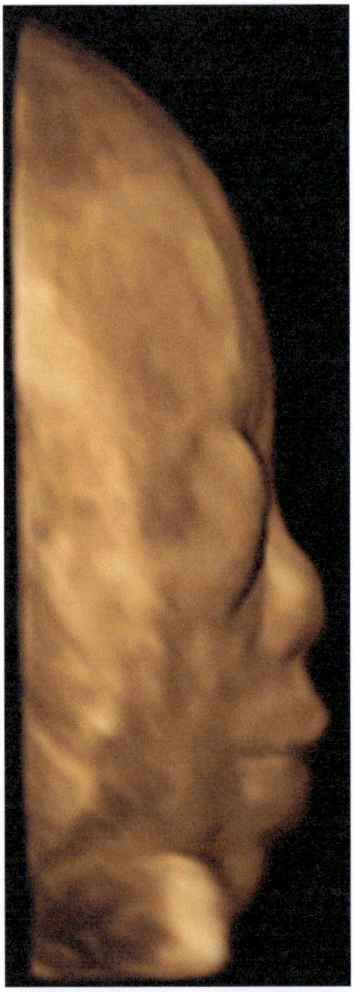

Fig. 4.4 Flat features

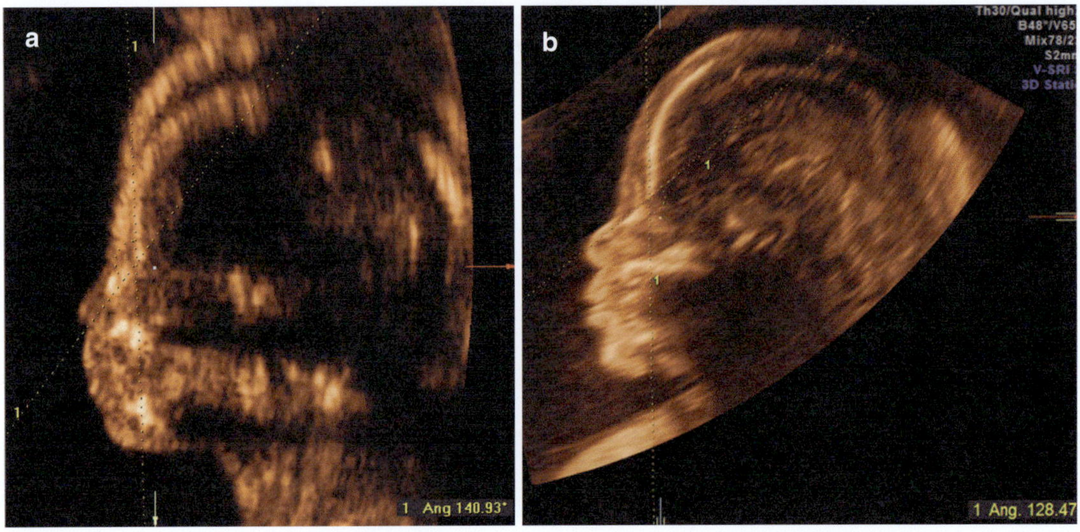

Fig. 4.5 (**a**) Increased upper facial angle 18 GW; (**b**) Normal facial angle at 22 GW

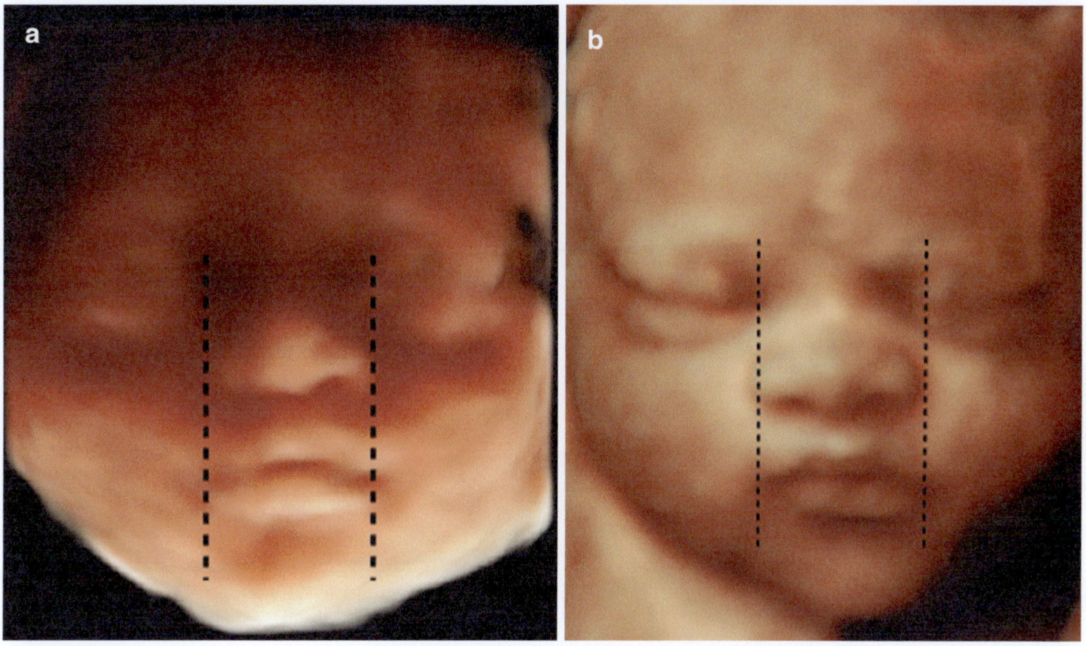

Fig. 4.6 Small mouth: (**a**) 22 GW; (**b**). 32 GW

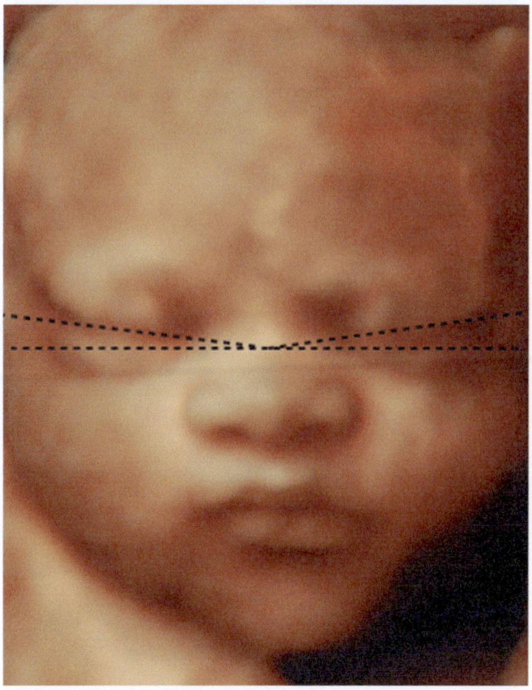

Fig. 4.7 Palpebral fissures oriented up and back

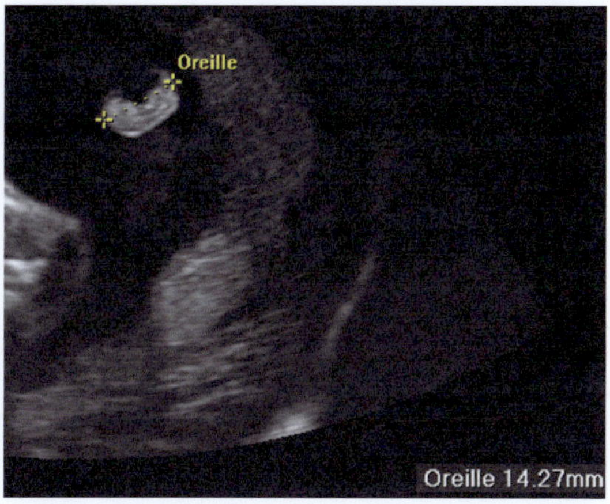

Fig. 4.8 Small ear at 26 GW

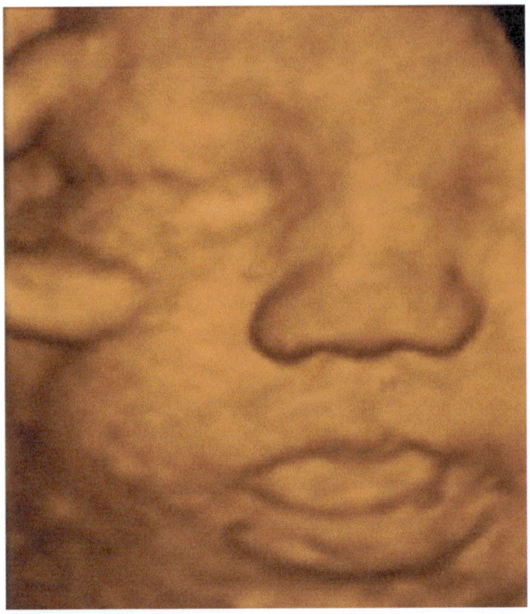

Fig. 4.9 Protrusive tongue

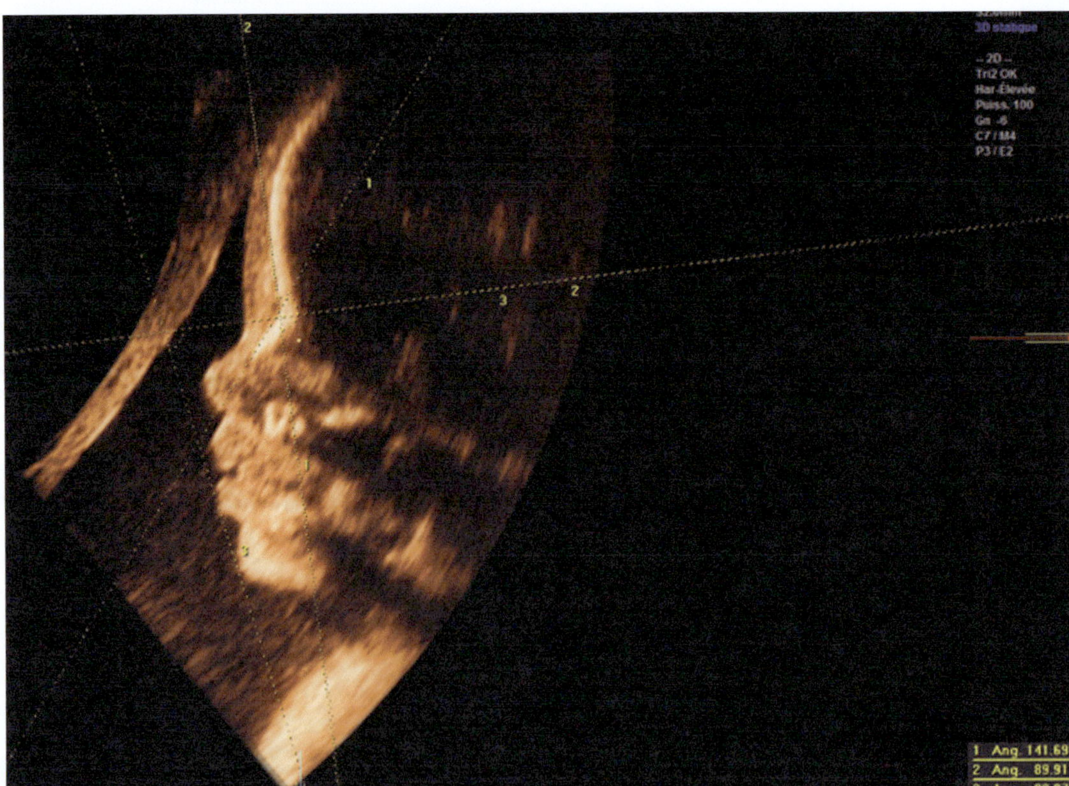

Fig. 4.10 Trisomy 21, normal profile

4.3.2 Trisomy 18

The diagnosis of this chromosomal anomaly is usually based on a collection of morphological anomalies.

In these case, the dysmorphism associates microretrognathia, sloping forehead, and small low ears. Cleft palate is frequent (cf. Clefts)

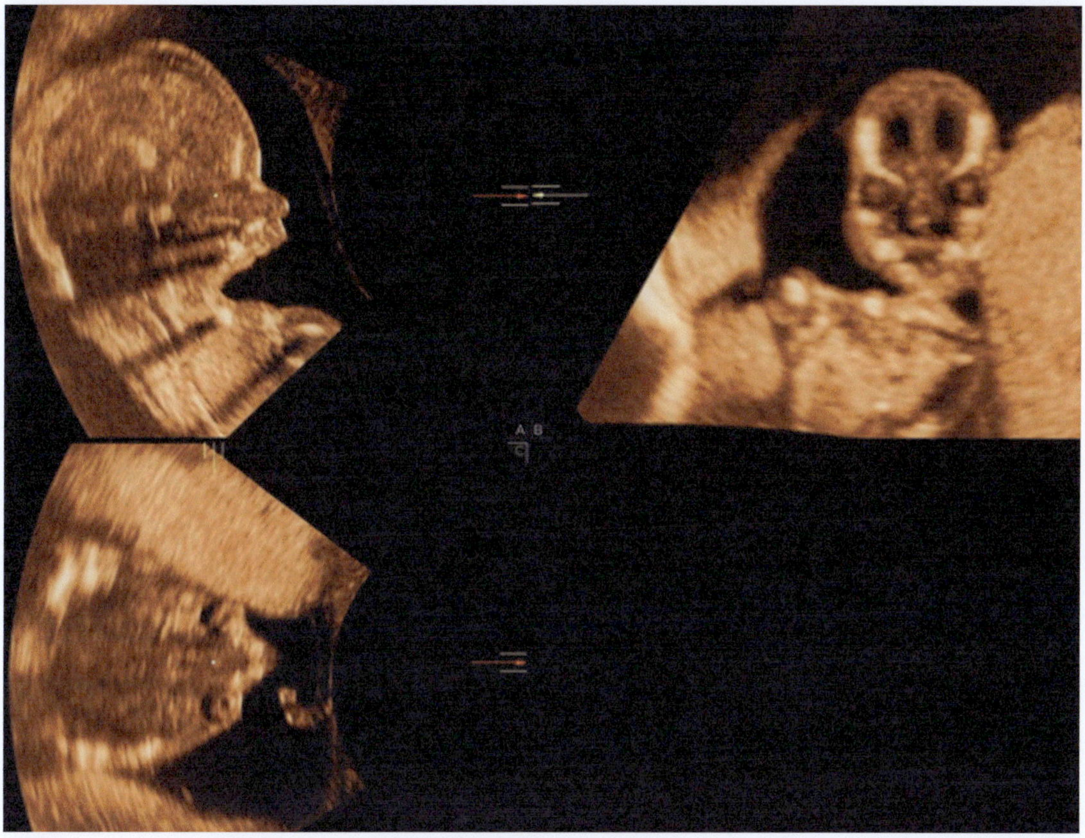

Fig. 4.11 Microretrognathia at 13GW (multiplanar image)

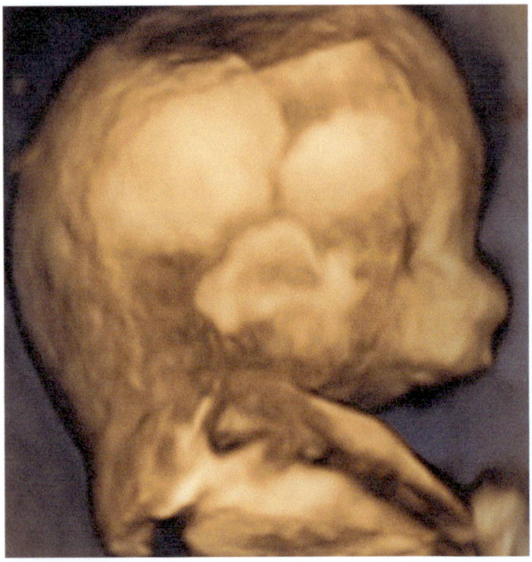

Fig. 4.12 Microretrognathia at 13GA, rendered image

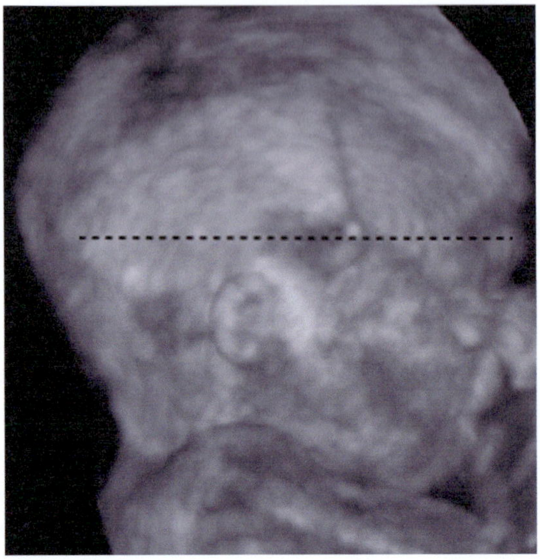

Fig. 4.14 Small ear

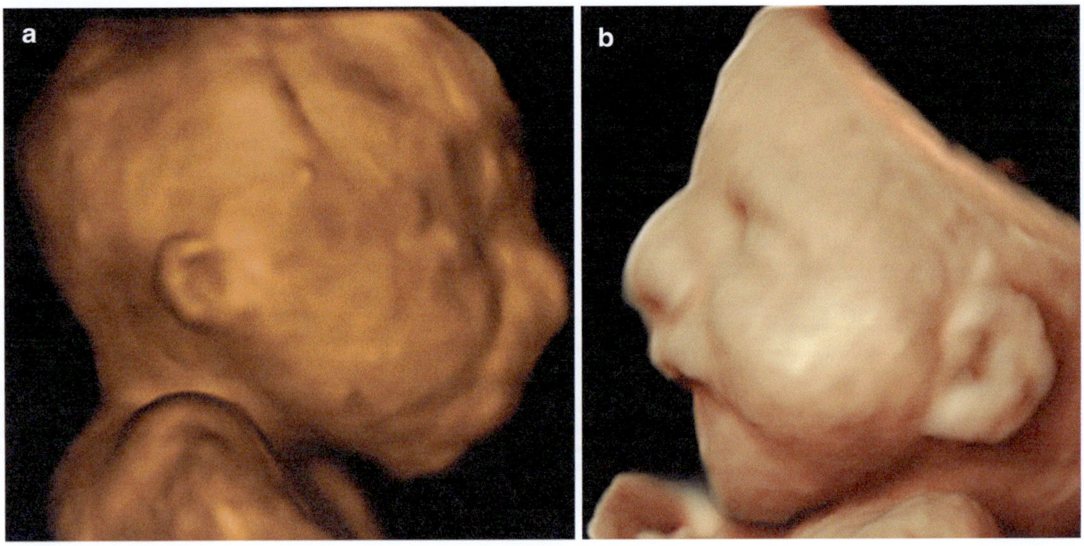

Fig. 4.13 (**a**) Sloping forehead and (**b**) microretrognathia

4.3.3 Trisomy 13

Dysmorphism in the case of trisomy 13 is often very spectacular, covering a wide range of possibilities from a simple median incisor (Fig. 4.15) to cyclopia (Figs. 4.16 and 4.17), with a full range of intermediary aspects possible including more or less hypotelorism (Fig. 4.18), single nostrils (Figs. 4.18, 4.19, and 4.20). Clefting is found in 60–80 % of cases, and is both unilateral and bilateral. Median clefts are characteristic (Figs. 4.21 and 4.22).

These anomalies are related to cerebral medial line anomalies (alobar, semi-lobar or lobar holoprosencephaly) that are present in 50 % of cases.

Facial dysmorphism related to other chromosomal anomalies are often rougher, as seen in the below examples corresponding to:

- A partial trisomy 8 with an everted lower lip, a round nose and retrognathia (Figs. 4.23 and 24).
- Distal monosomy 6P, with related hypertelorism, very slight retrognathia, palpebral fissures oriented down and out, low-set ears (Fig. 4.25).

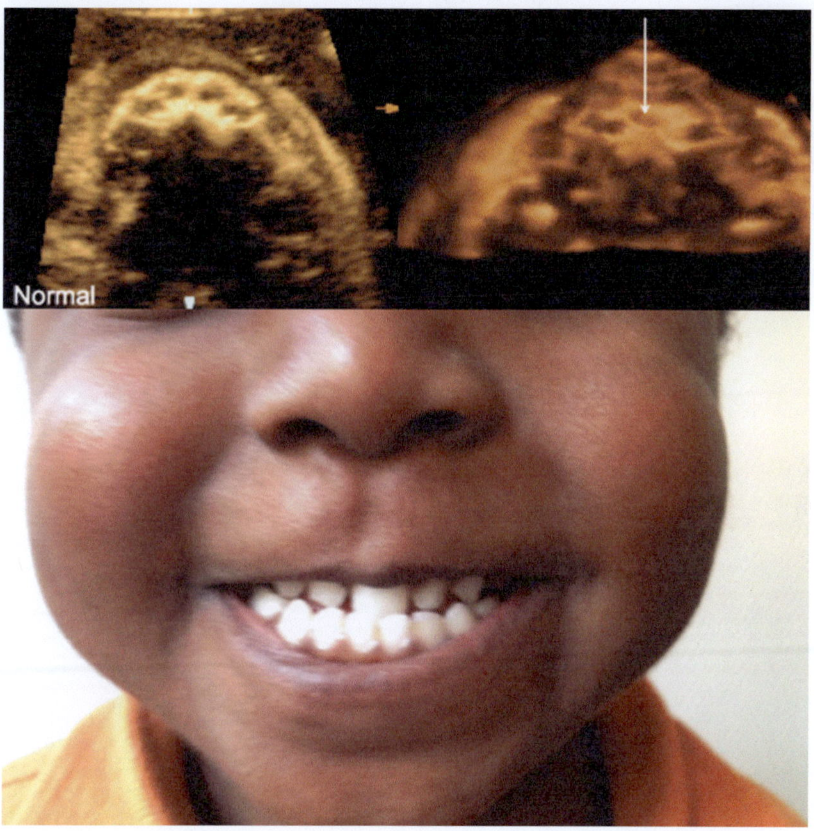

Fig. 4.15 Median incisor

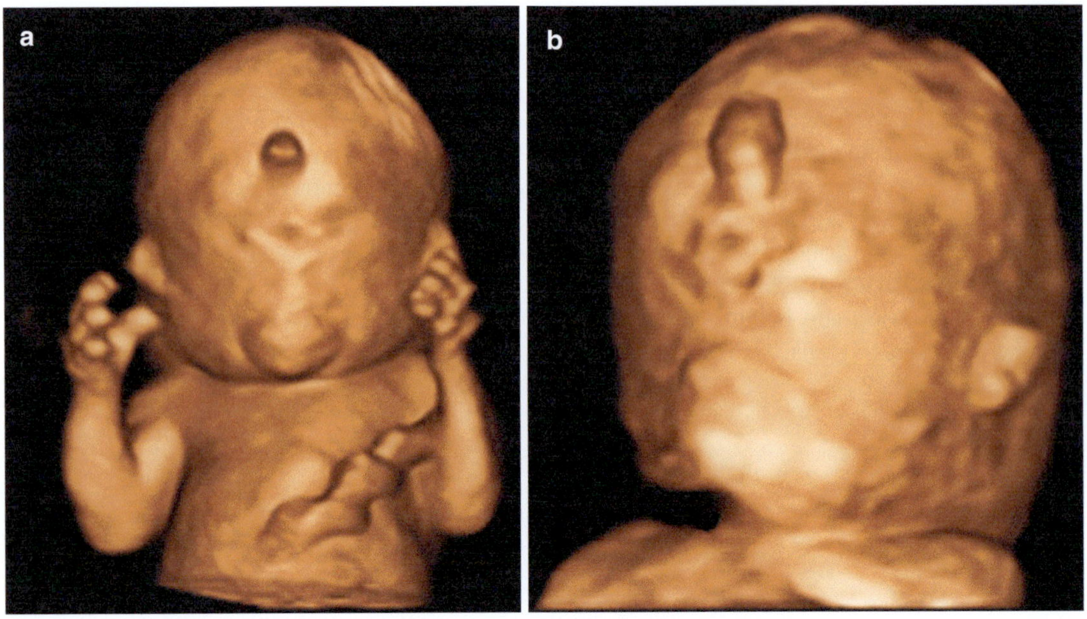

Fig. 4.16 (**a**, **b**) Cyclopia, proboscis 13GW

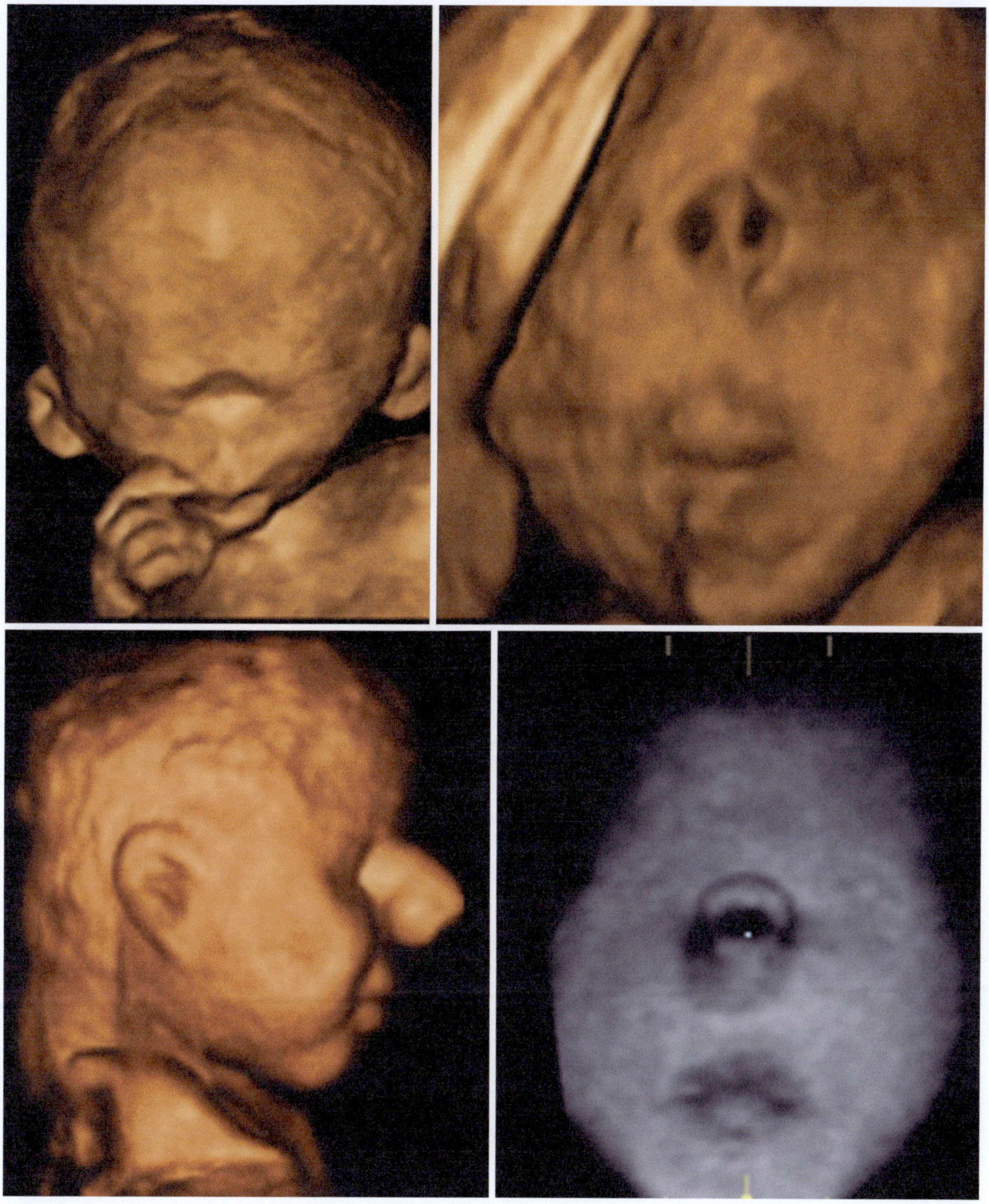

Fig. 4.17 Cyclopia, single nostril. 22 GW

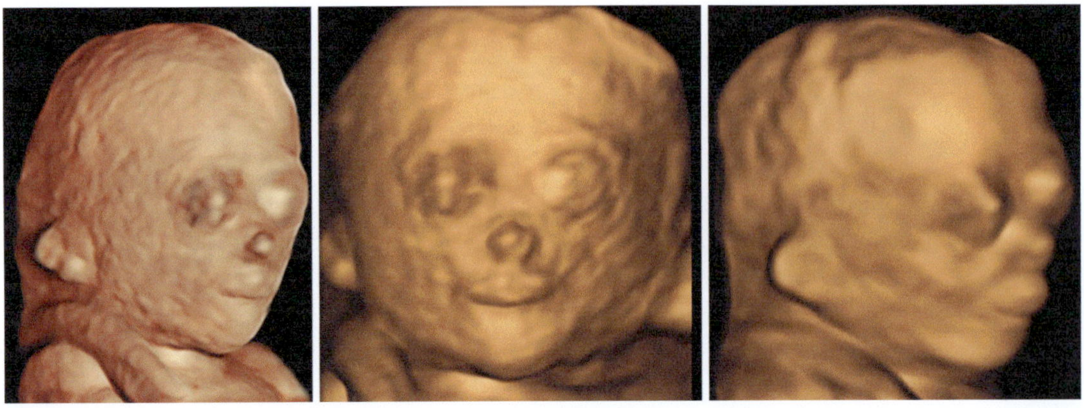

Fig. 4.18 Hypotelorism, single nostril 13GW

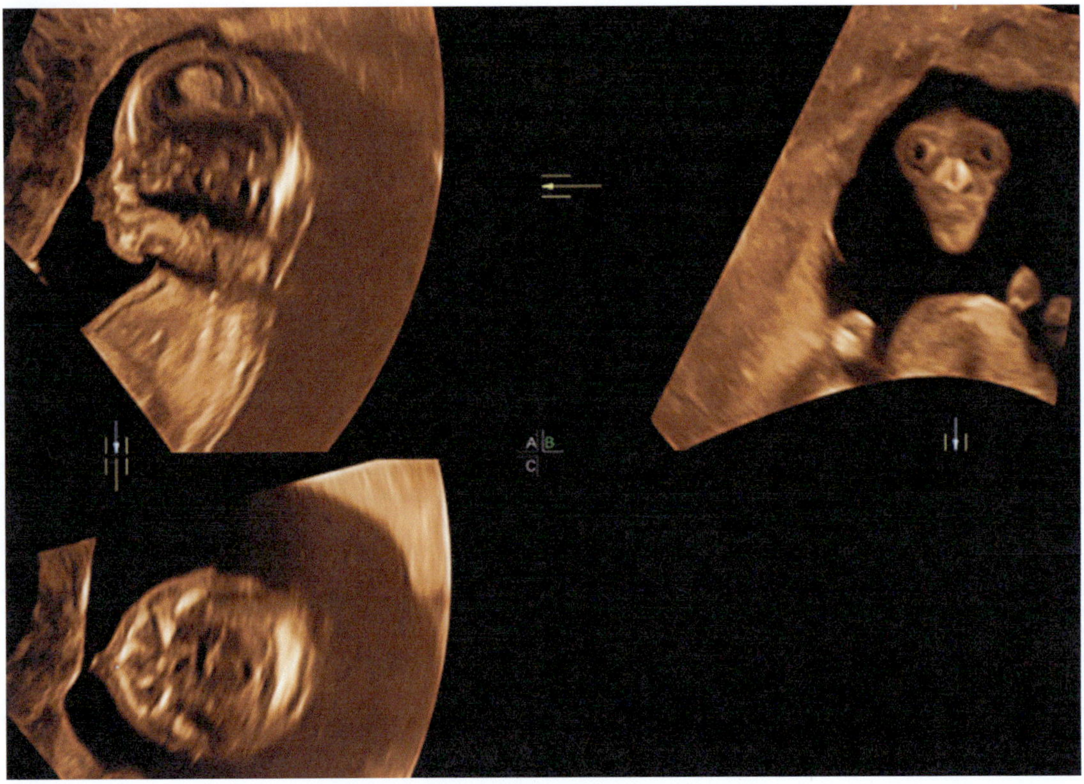

Fig. 4.19 Single nostril 13GW (multiplanar image)

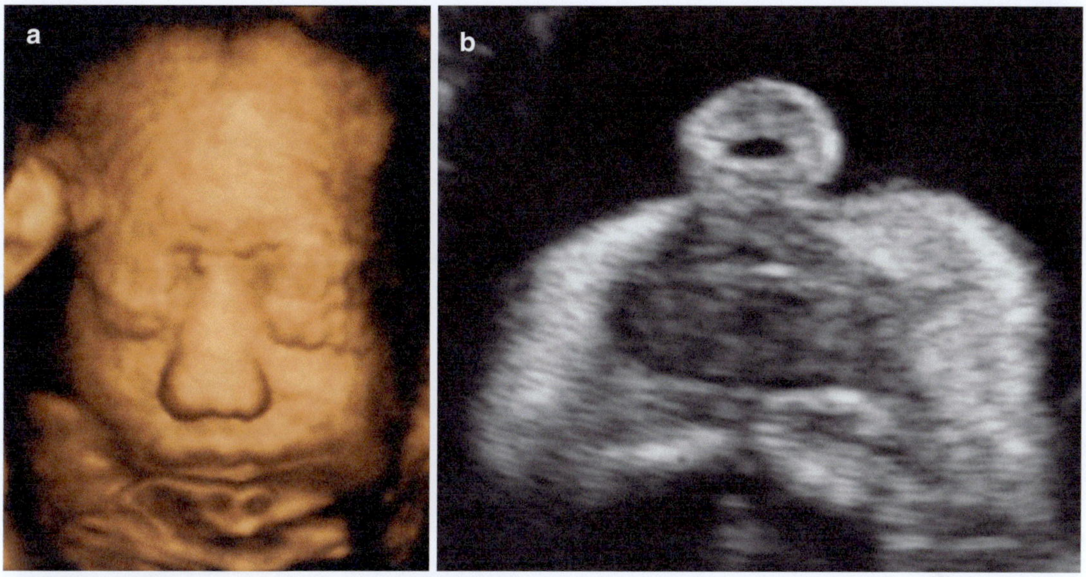

Fig. 4.20 Single nostril 22 GW:(**a**): Render 3D, (**b**): 2D

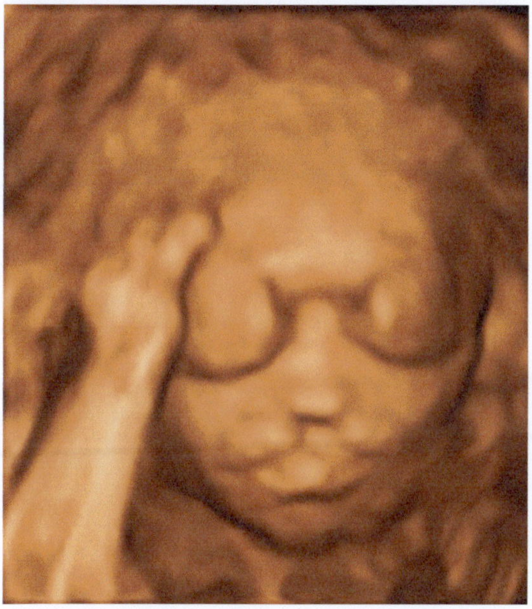

Fig. 4.21 Median cleft 14 GW

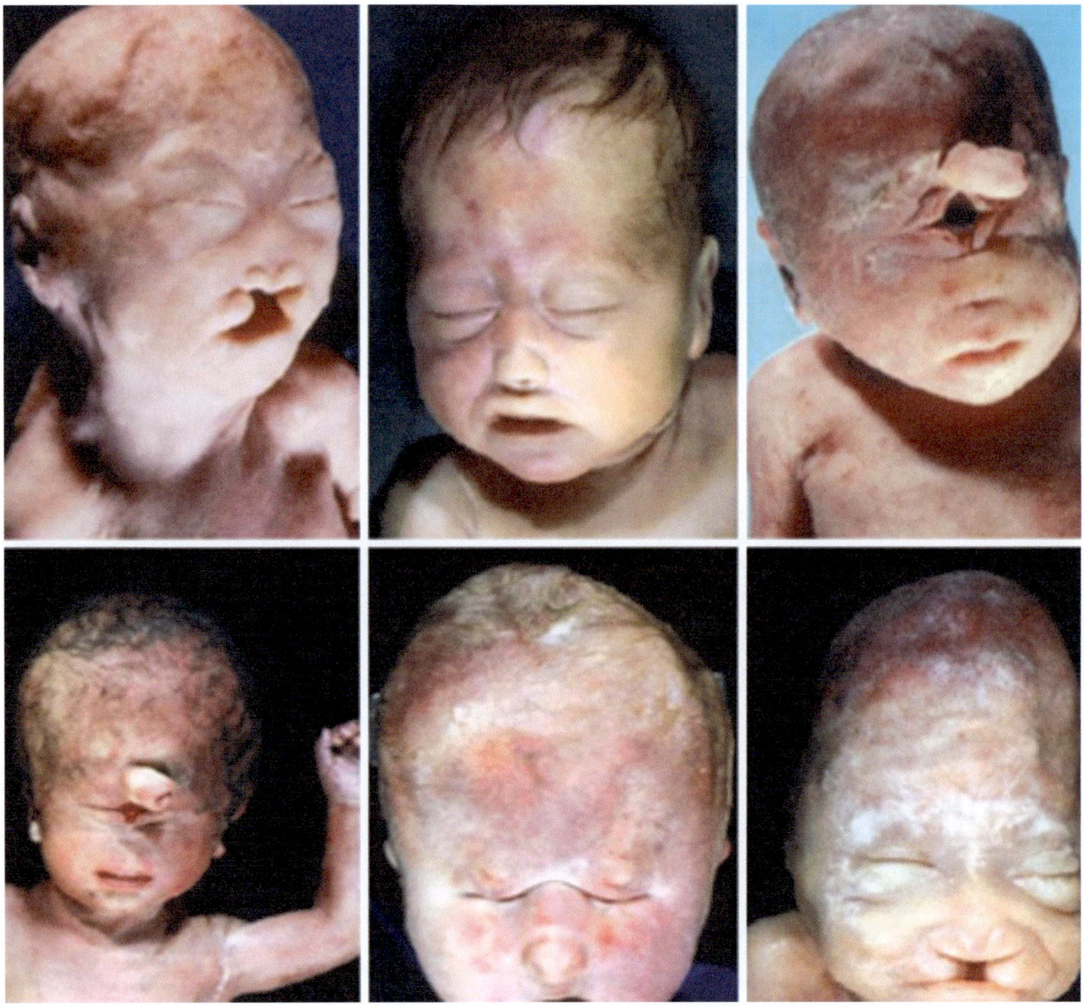

Fig. 4.22 Various fetal facial dysmorphic features related to trisomy 13

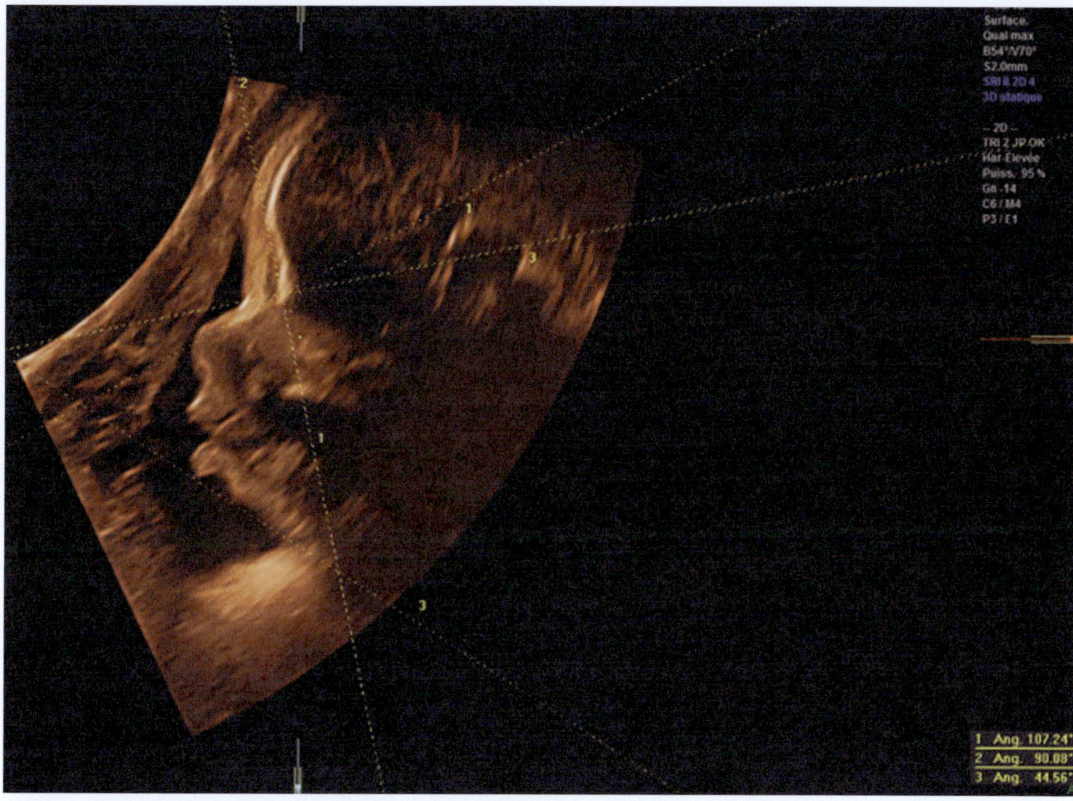

Fig. 4.23 Partial trisomy eight profile

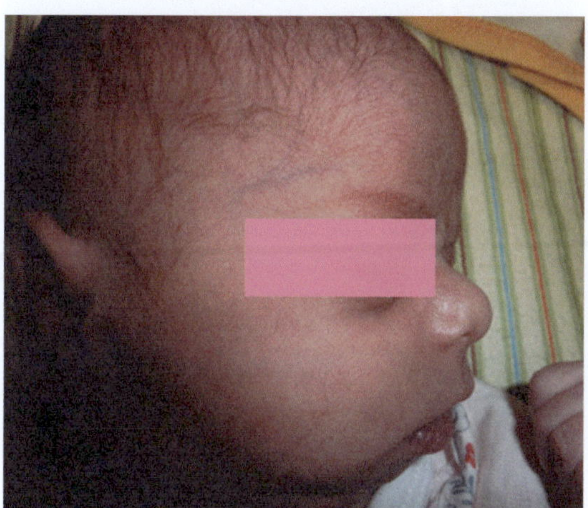

Fig. 4.24 Partial trisomy eight (post-natal aspect)

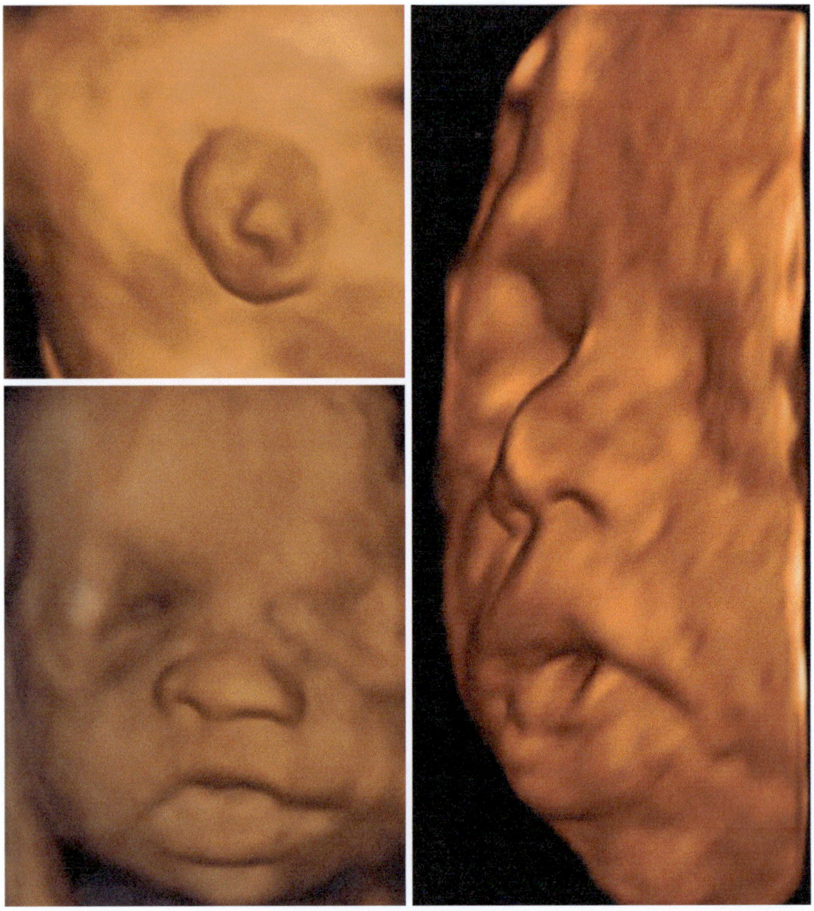

Fig. 4.25 Distal monosomy 6p

4.4 Upper Face Pathologies

4.4.1 Wolf-Hirschhorn Syndrome

Wolf-Hirschhorn syndrome results from a microdeletion, with an incidence of 1/50,000. The deletion occurs on the short arm of chromosome 4 at the 4p.16.3 region. This deletion is sporadic in the majority of cases. Research for this deletion should be signaled to geneticists when they test the amniotic fluid.

The dysmorphism is characteristic:

- On the profile, the upper facial angle measurement is barely inferior to normal measurements. However, the glabella appears prominent with a frequently not very echogenic aspect (the prefrontal tissues), which results in a characteristic "Greek Helmet" look (Fig. 4.26).
- The corners of the lips appear to be falling.
- There can also be hypertelorism, dysplastic ears with periauricular tags, cleft lip or labio-alveolar-palatal clefting.

This dysmorphism occurs in a context of early IUGR with, in particular, severe microcephaly, mental retardation could be severe with walking and language acquisition occurring more or less late.

4.4.2 Craniostenosis

These anomalies, resulting from the premature fusing of one or several sutures, often lead to severe dysmorphism. The overall incidence is estimated at 1/2000 and around 100 different forms have been described. The role of an ultrasonographer consists of labeling as best as possible the type of anomaly and then looking for other markers to determine whether or not the craniostenosis is related to a syndrome. It is useful to remember that the first sign signaling craniostenosis could be an enlargement of a healthy contralateral suture.

Craniostenoses are classified based on the type of suture involved (Figs. 4.27, 4.28, 4.29, and 4.30).

Below are a few examples of an approach to these complex pathologies.

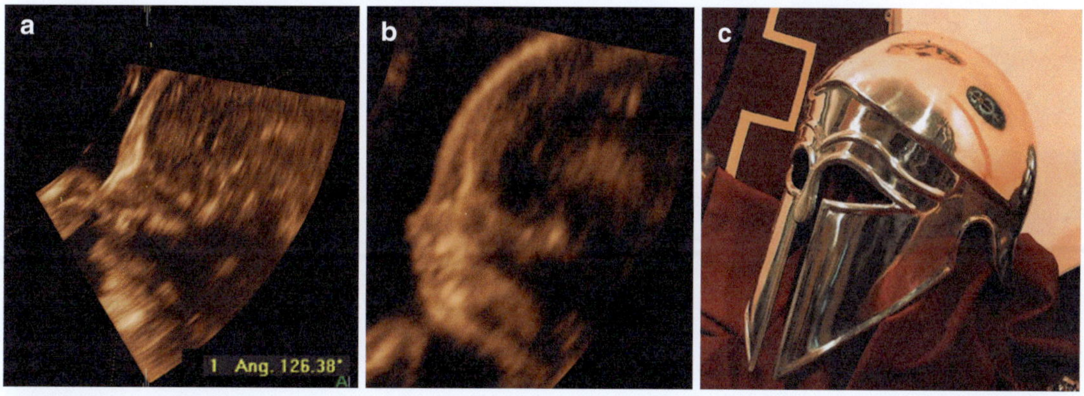

Fig. 4.26 "Greek Helmet" look

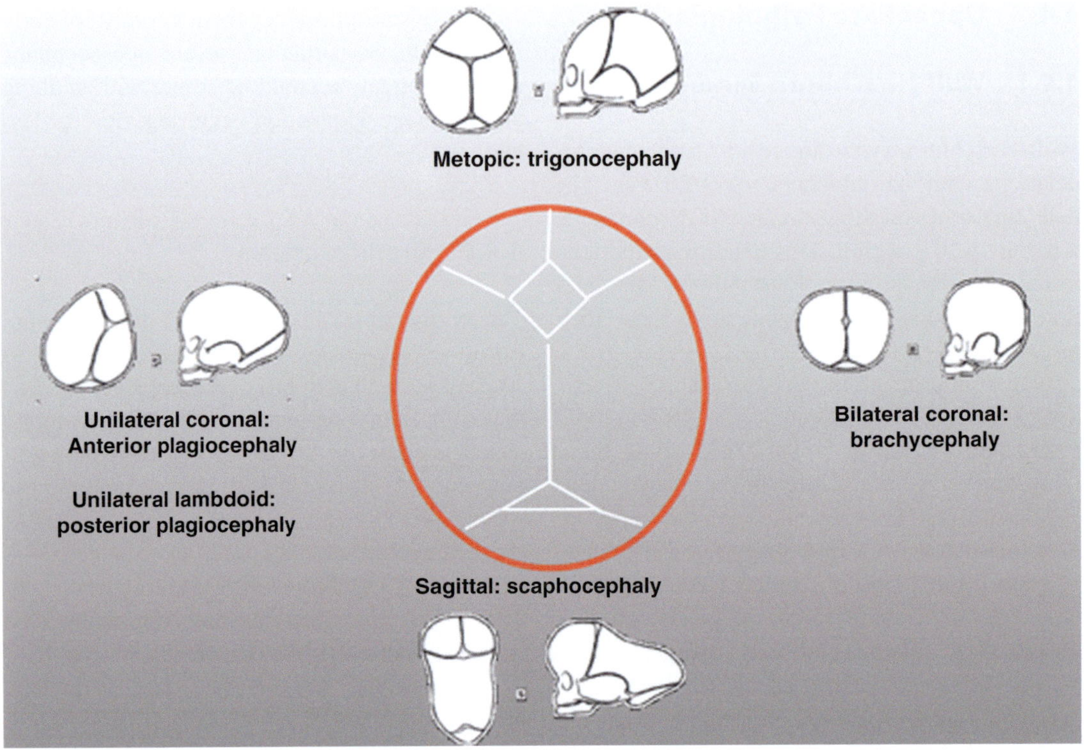

Fig. 4.27 Classification of craniostenoses

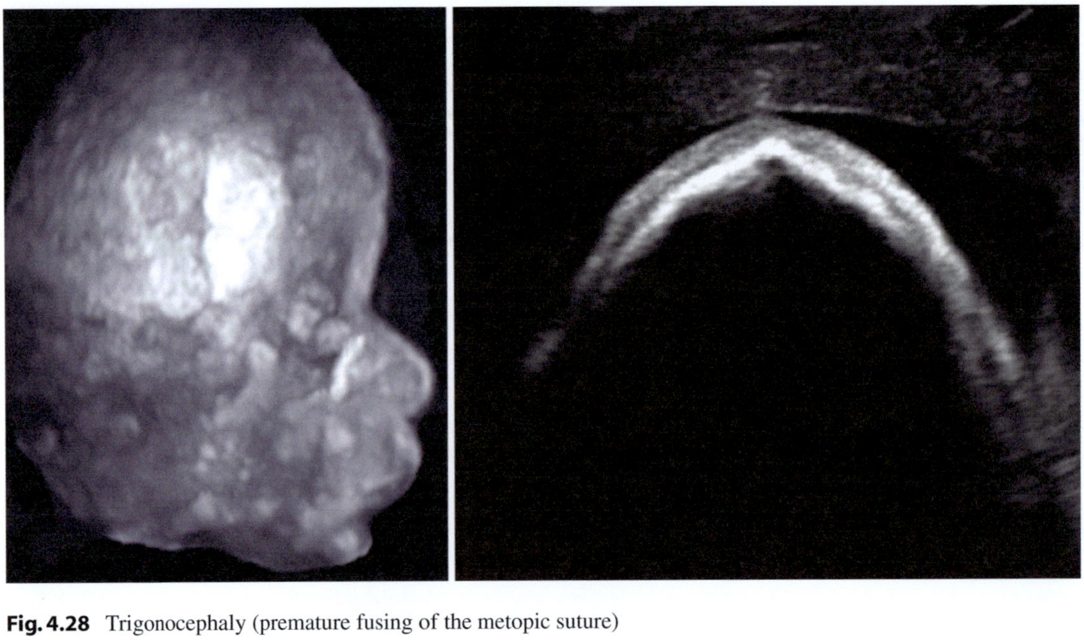

Fig. 4.28 Trigonocephaly (premature fusing of the metopic suture)

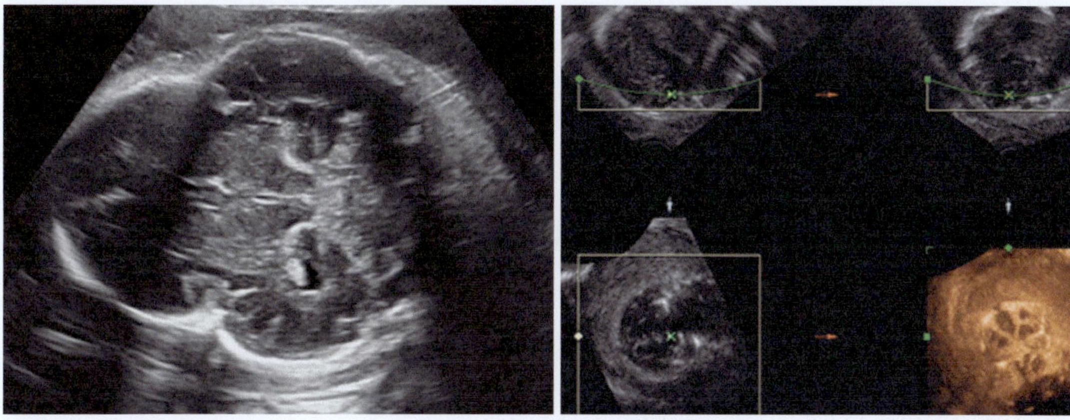

Fig. 4.29 "Cloverleaf" skull with intra cranial marks (premature fusing of coronal sutures)

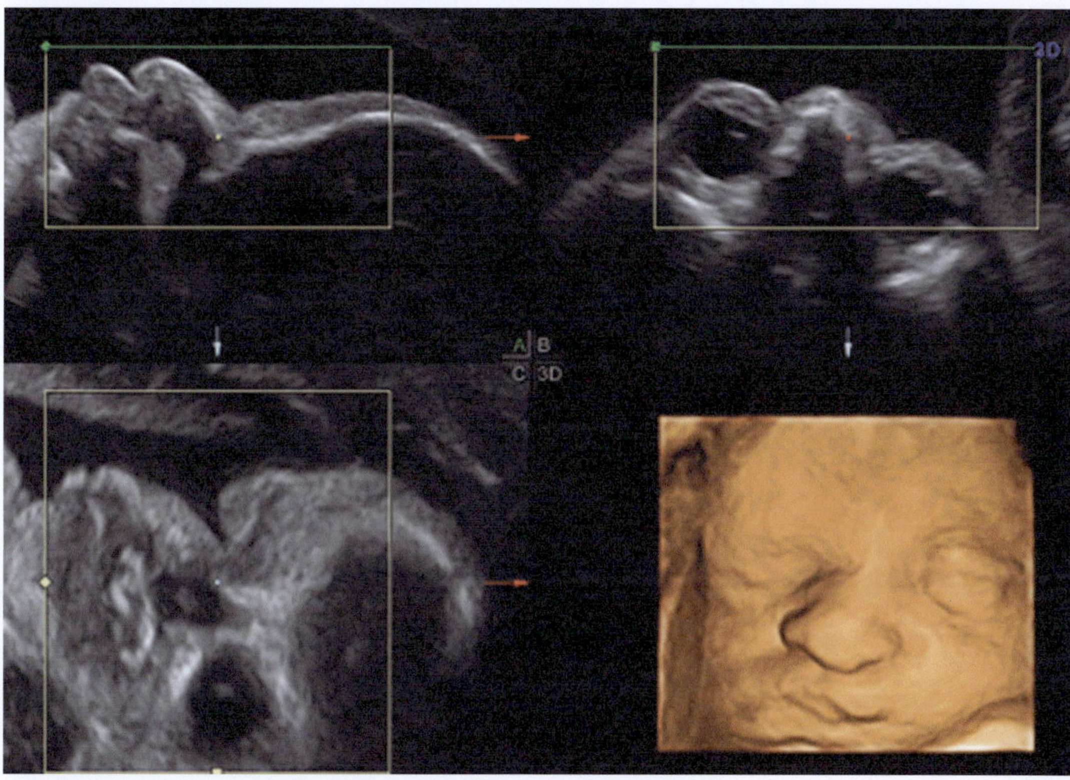

Fig. 4.30 Asymmetric craniostenosis (premature fusing of a coronal suture)

4.4.2.1 Examples of Syndrome Craniostenosis

Crouzon syndrome This syndrome results in the premature closing of the coronal sutures. Facial dysmorphism is characterized by branchycephaly, with a very domed appearance, frontal bumps, moderate hypoplasia of the mid-level face (Figs. 4.31, 4.32, 4.33, and 4.34), proptosis related to shallow orbits (Fig. 4.33), slight hypertelorism (Fig. 4.33), and a straight mandible (Figs. 4.33 and 4.34). Intelligence is generally normal. Complications such as atrophy of the optic nerves and deafness can occur if not caught and treated early. It is an autosomal dominant genetic disorder.

Apert syndrome This results in the premature closing of the coronal sutures, with broad bone defects that can be seen on the median line between the two fontanels. Dysmorphism is characterized by branchycephaly, a very marked nasal bridge (Fig. 4.35), moderate mid-level face hypoplasia, hypertelorism and a prominent forehead. The hands show syndactyly that leaves the thumb free, producing "mitten" hands (Fig. 4.36), however syndactyly of the feet impacts all the toes.

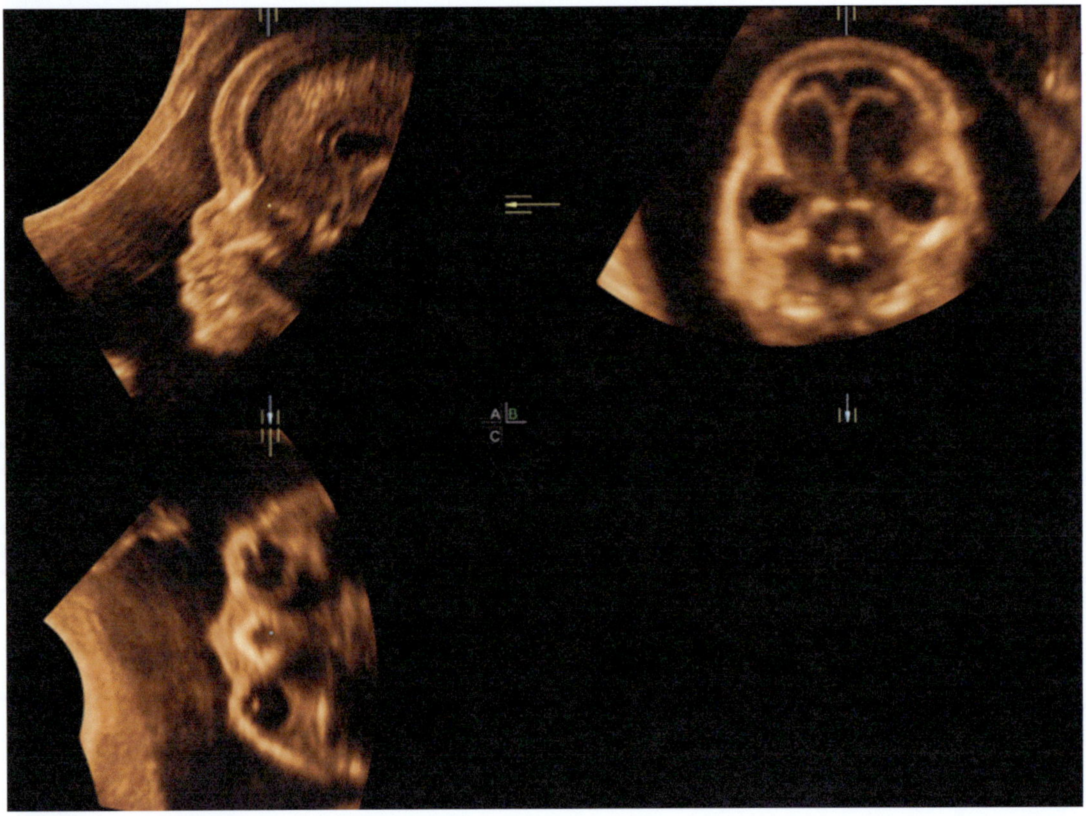

Fig. 4.31 Crouzon syndrome (multiplanar image)

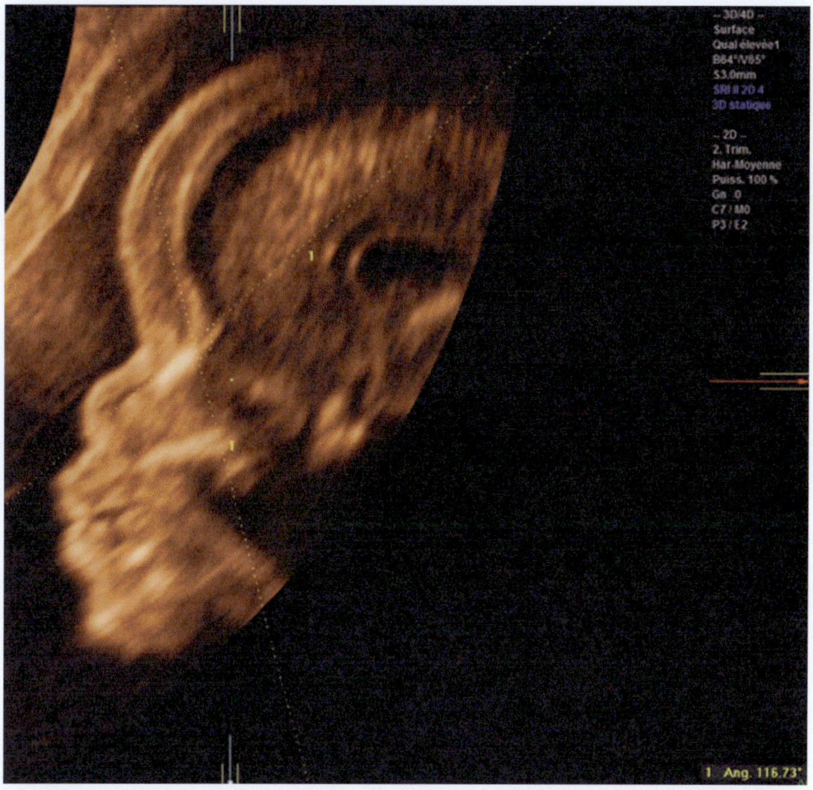

Fig. 4.32 Crouzon syndrome: profile

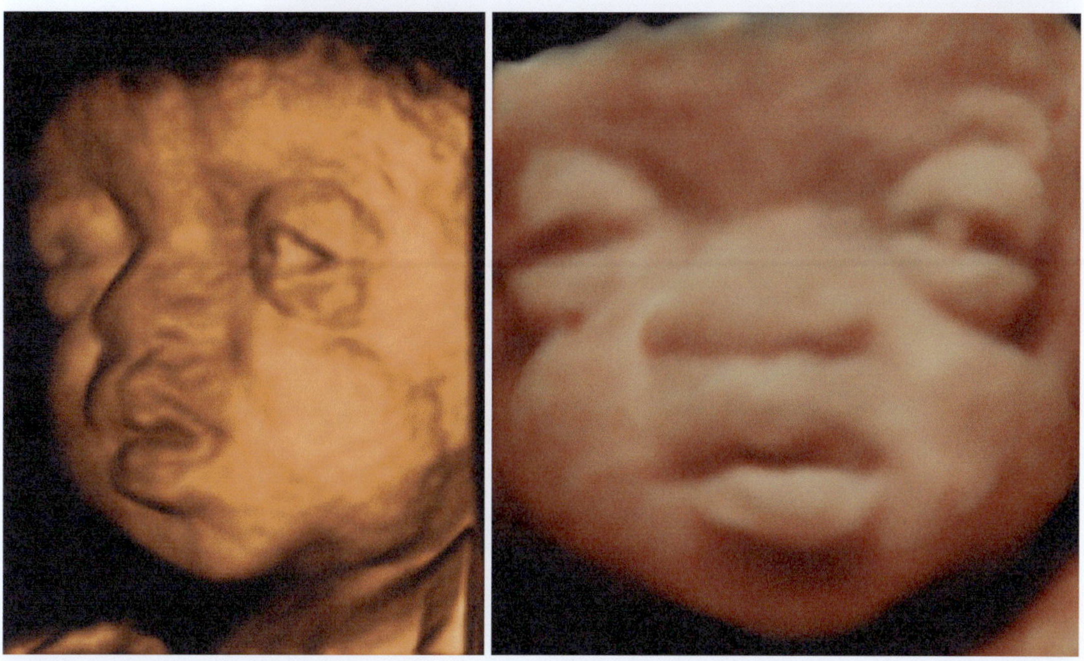

Fig. 4.33 Crouzon syndrome: "render"

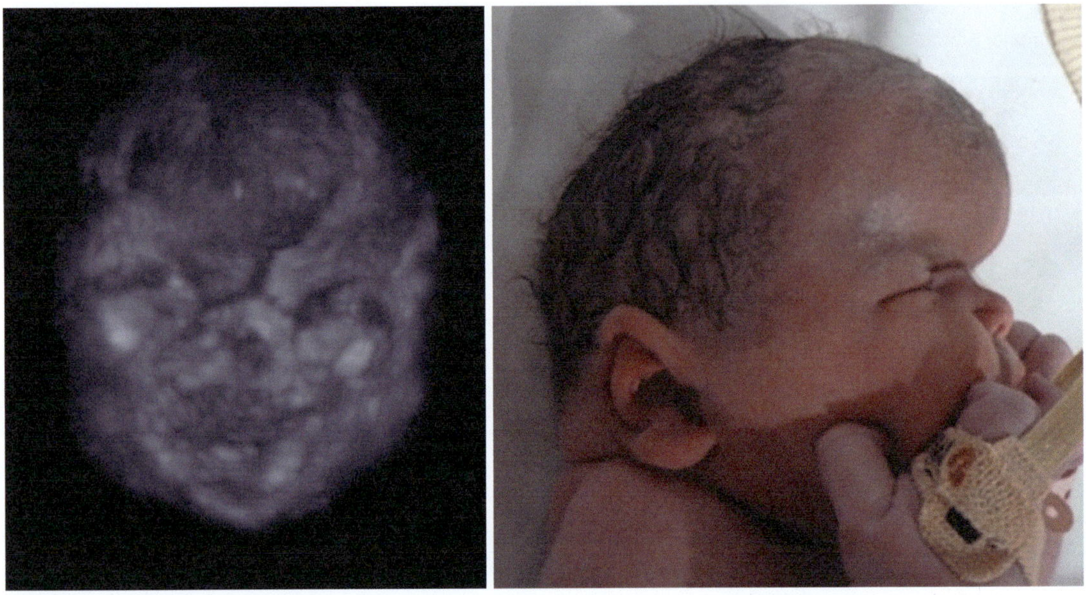

Fig. 4.34 Crouzon syndrome, with very wide metopic suture. (**a**) Maximum mode; (**b**) Post-natal

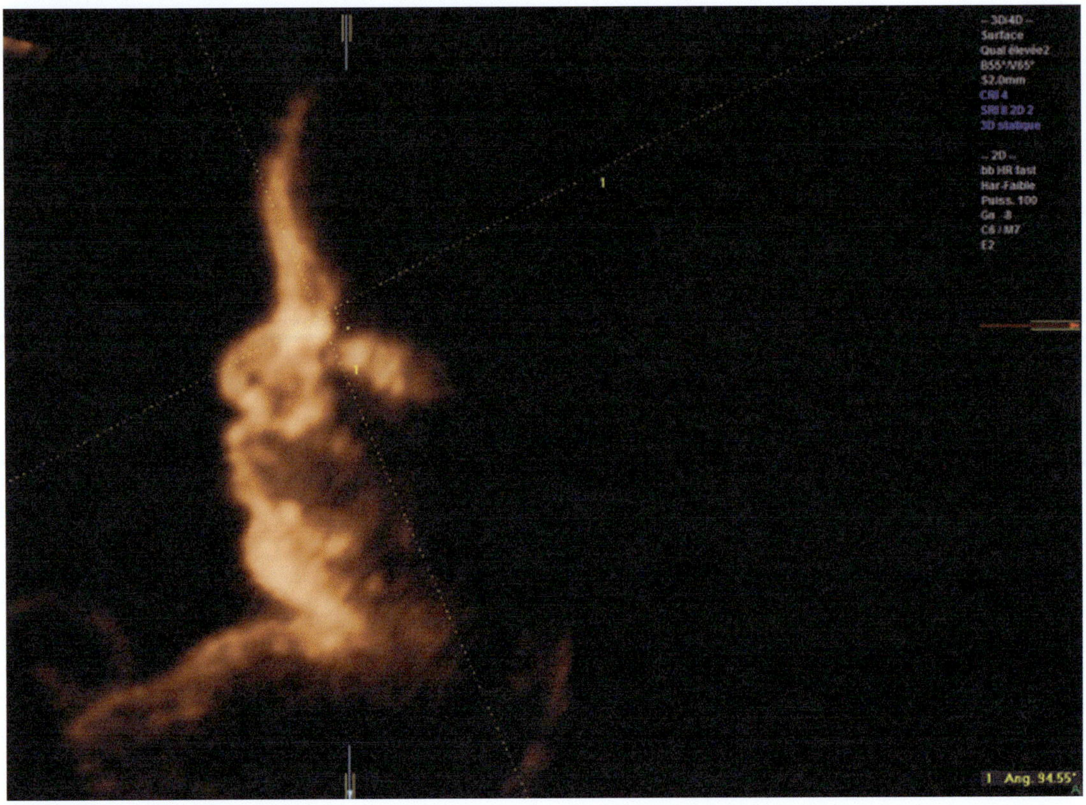

Fig. 4.35 Apert syndrome: profile

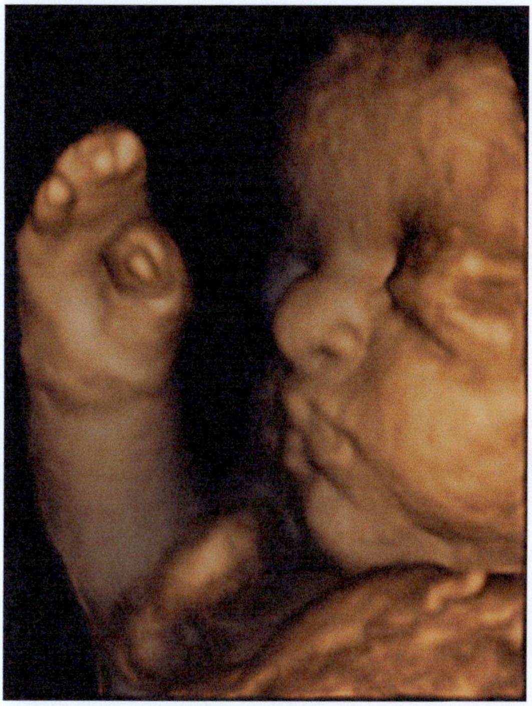

Fig. 4.36 Apert syndrome: dysmorphism and syndactyly producing "mitten" hands

4.4.2.2 DiGeorge Syndrome

This syndrome is linked to a microdeletion at 22.q.1.1 and occurs at a frequency of 1/4000. Transmission is dominant; one parent is carrier of the deletion in 10–20 % of cases, most frequently this deletion is "de novo".

The main markers are:

- Conotruncal heart disorders (75 %) of cases
- Hypoplasia or absence of the thymus (in 83 % of cases if there is a heart malformation, and in 31 % of cases if the deletion is isolated)
- Palate defects (76 % of cases: submucosal cleft palate and cleft palate, velopharyngeal incompetence, etc.)
- Renal anomalies (35 % of cases)
- Limb anomalies (hyperlax hands or fingers, polydactyly)
- A broad range of psycho-motor development anomalies (slow learning with IQ variations from normal to seriously retarded, psychiatric disorders in 10 % of cases).

The dysmorphism is very moderate, association a "tubular"-looking nose (Figs. 4.37, 4.38, and 4.39), a relatively narrow mouth (Fig. 4.39), and minimal external ear anomalies (Fig. 4.40).

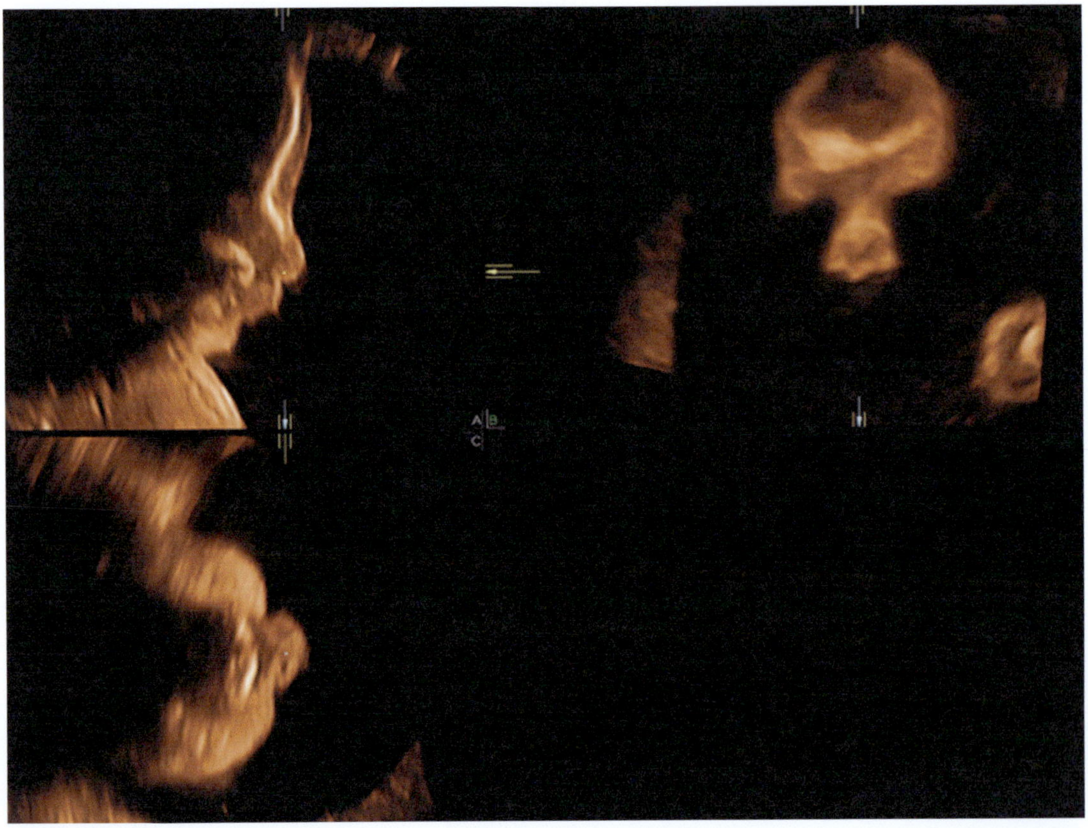

Fig. 4.37 DiGeorge syndrome: multiplanar image, tubular nose (image B)

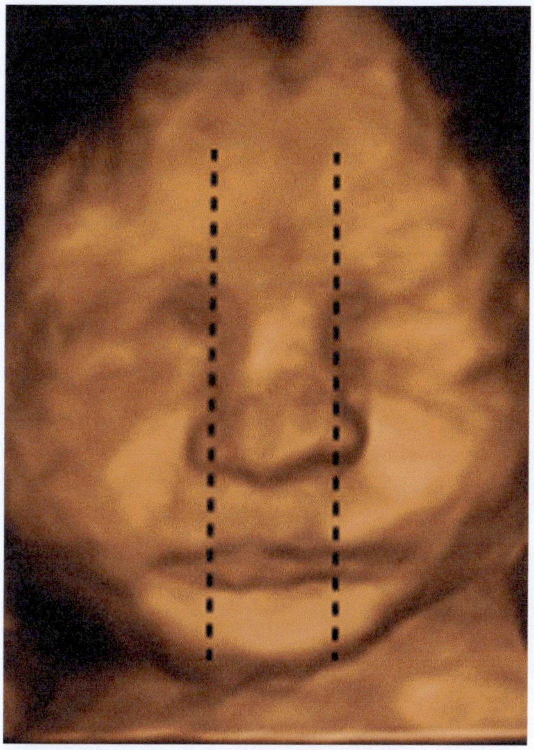

Fig. 4.38 DiGeorge syndrome: tubular nose, "render"

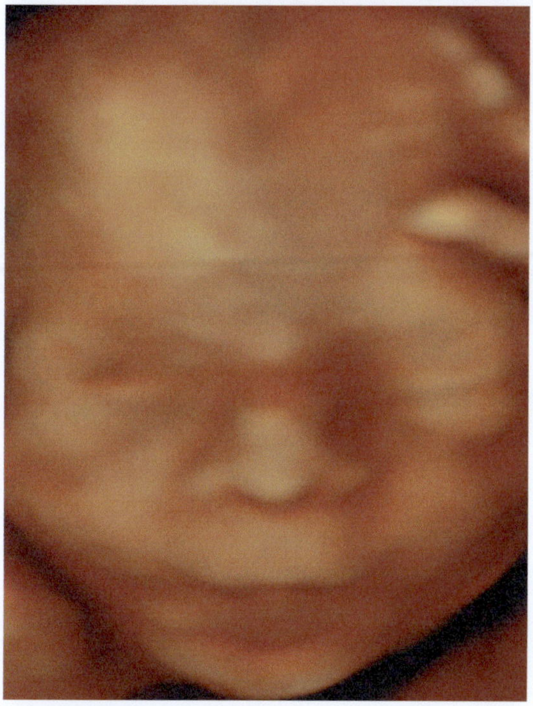

Fig. 4.39 DiGeorge syndrome: tubular nose, small mouth (Render HD-Live)

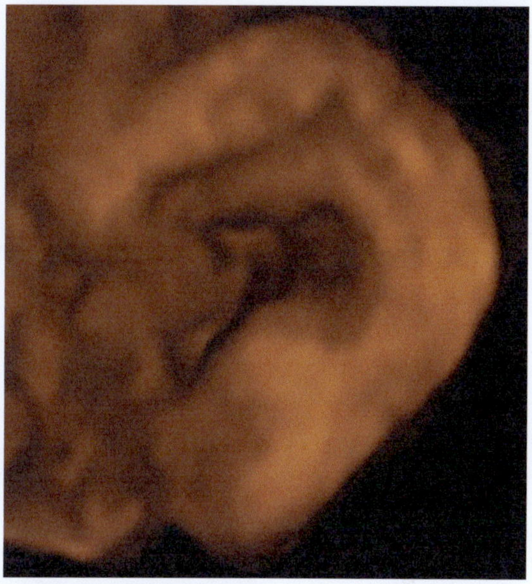

Fig. 4.40 DiGeorge syndrome: ear: excessively folded helix

4.4.2.3 Microcephaly

Microcephaly is a clearly defined biometric anomaly: the cranial circumference is smaller than −3 SD or Z-Score : −3. The frequency of 1/6200 to 1/8500 is certainly underestimated. It can be isolated, related to holoproscencephaly, or associated with chromosal anomalies, a number of genetic syndromes, and other morphological anomalies.

Three conditions are required to consider microcephaly:

- The **dating of the pregnancy** and the **biometric definition** must be precise, and the **measuring technique** must be irreproachable.

When microcephaly is discovered, on must:

- Correlate the cranial circumference with other biometric parameters.
- Look for a vascular context.
- Test for infections.
- Do karyotype testing for a 4p- deletion
- Do a family survey
- Do a detailed ultrasound, particularly of the cerebral structures.

Dysmorphism primarily takes the form of a sloping forehead (Figs. 4.41, 4.42, and 4.43) and a disproportion between the size of the face and the skull (Figs. 4.42 and 4.43).

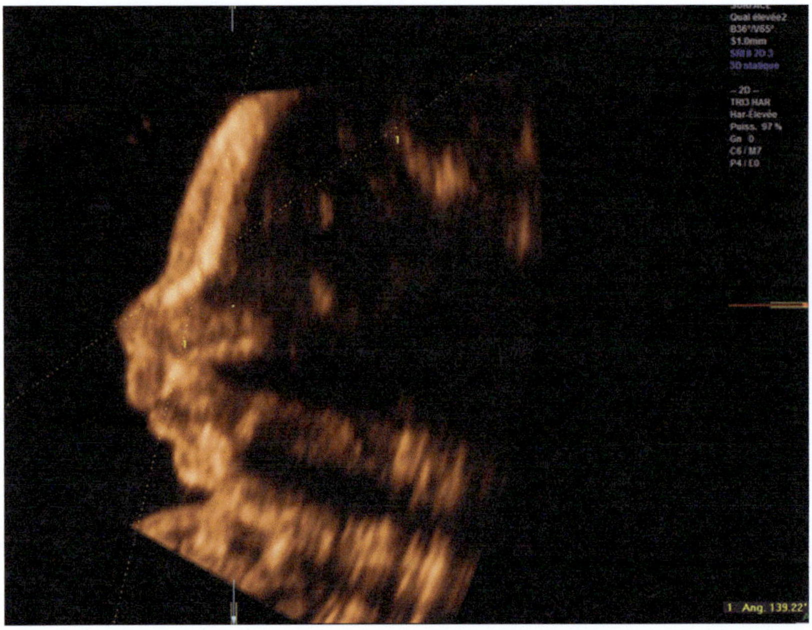

Fig. 4.41 Microcephaly: normal-sized face, sloping forehead

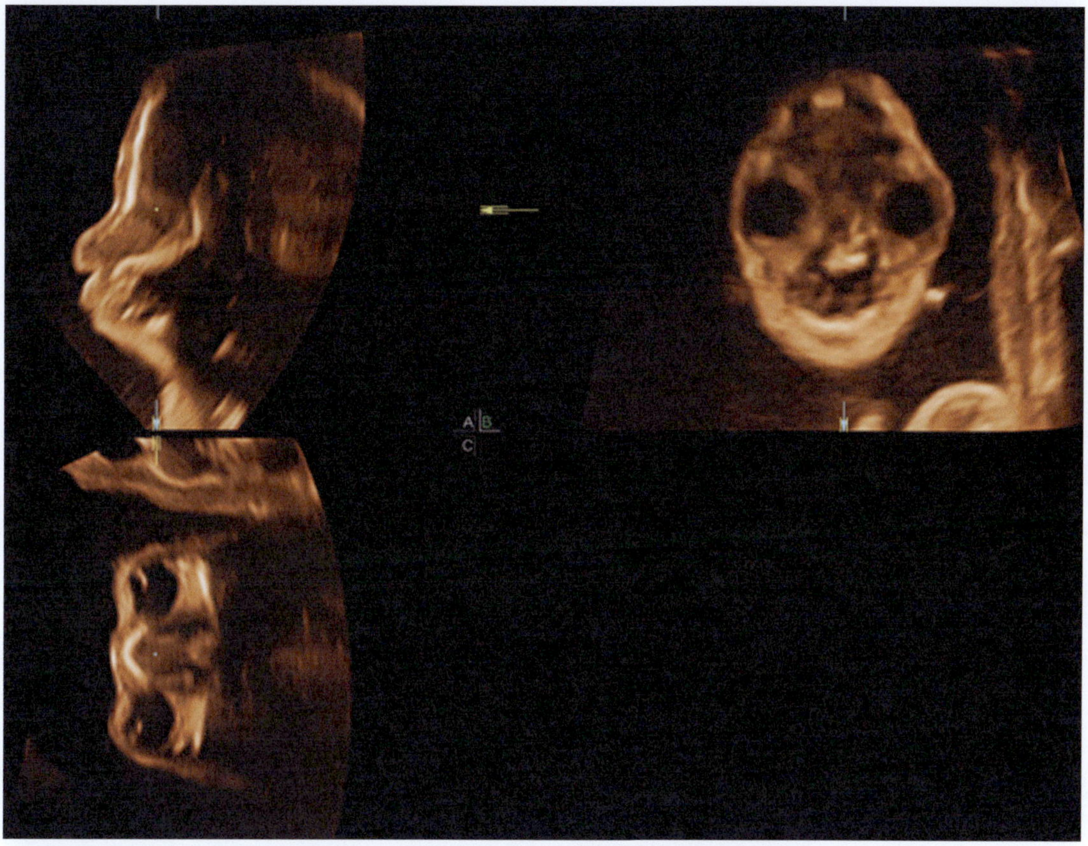

Fig. 4.42 Microcephaly: sloping forehead, narrow skull with retrognathia

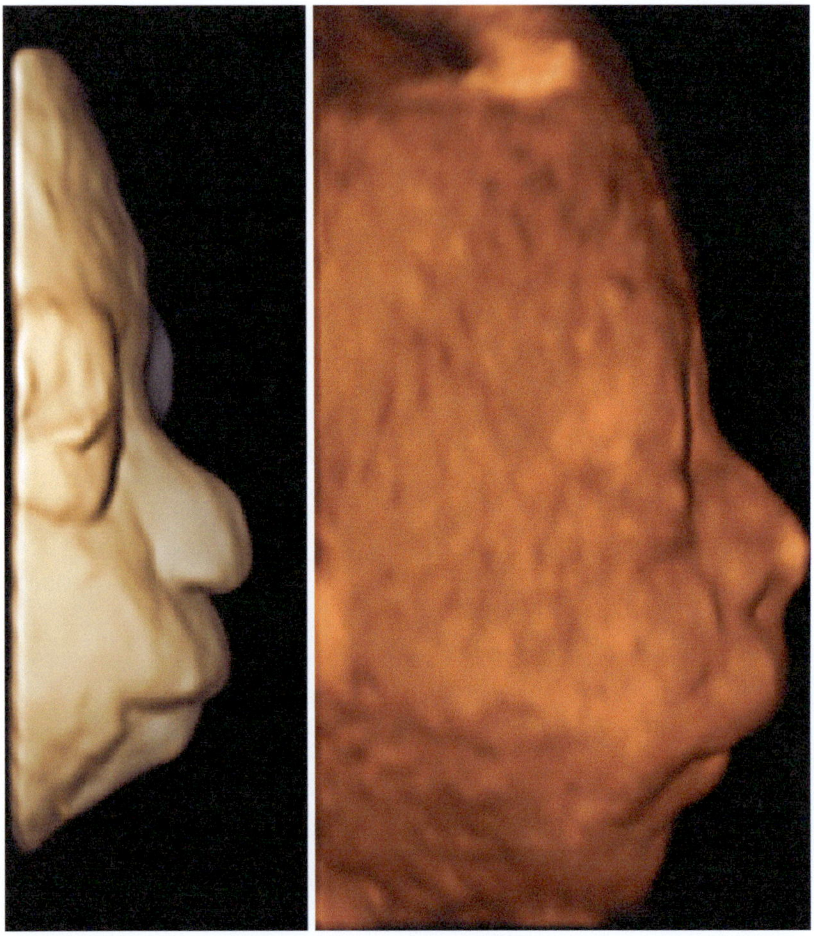

Fig. 4.43 Microcephaly: "render"

4.5 Mid-Face Pathologies

4.5.1 Achondroplasia

Achondroplasia is the most frequent osteochondrodysplasia, with an estimated frequency between 1/15,000 and 1/26,000. It is related to a mutation of the FGFR3 gene. This bone pathology becomes apparent starting at 26 gestational weeks, the diagnosis most frequently made during the third trimester ultrasound showing very short limbs, in particular rhizomelia contrasting with macrocrania (very marked frontal bossing), and a relatively narrow thorax (Fig. 4.45). In addition to these markers, the ultrasound will show a characteristic appearance of the upper femoral metaphyses, which appear to be slender like "ice cream cones," leading to an "opening" of the upper femoral angle (Fig. 4.44), short trident hands, and platyspondyly (Fig. 4.45).

Dysmorphism combines marked frontal bossing and a very "hollowed" nasal bridge with an abnormally closed upper facial angle (Figs. 4.46, 4.47, 4.48, and 4.49).

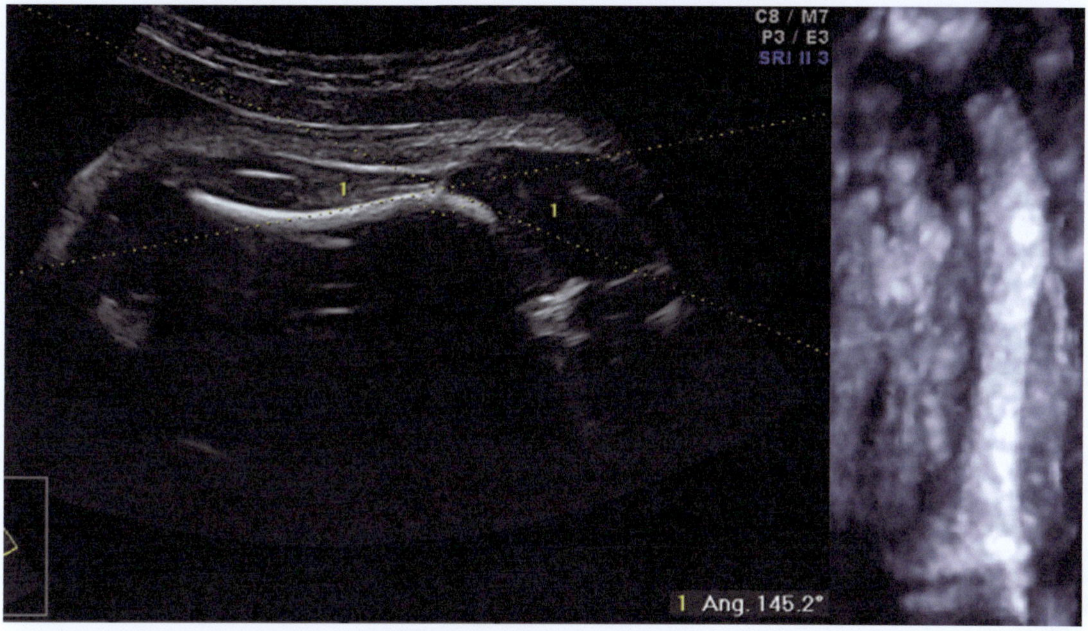

Fig. 4.44 Achondroplasia: slender upper femoral metaphysis (2D and 3D)

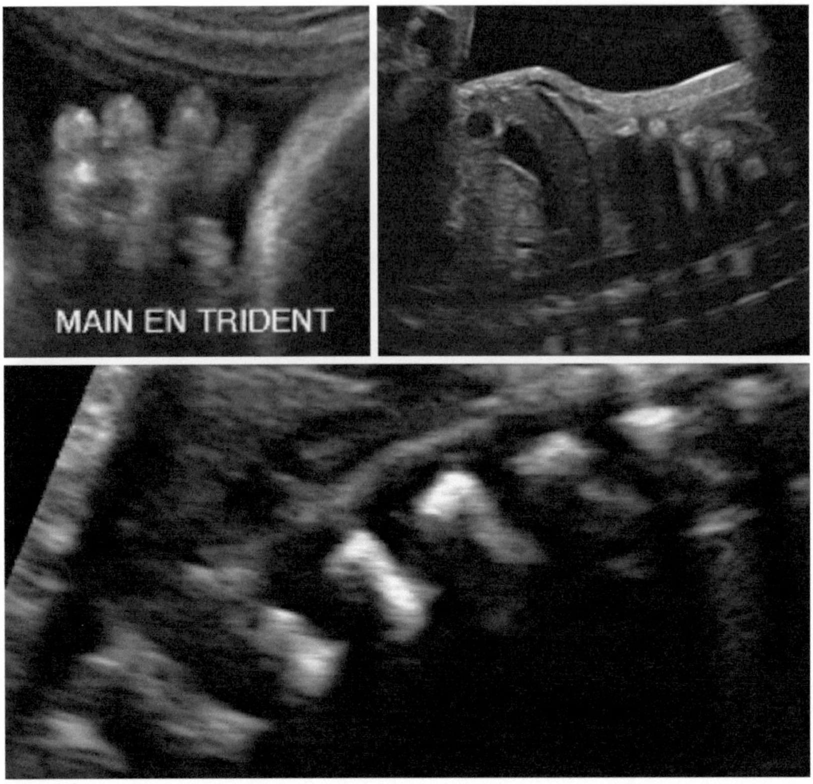

Fig. 4.45 Achondroplasia: trident hand, narrow thorax, platyspondyly

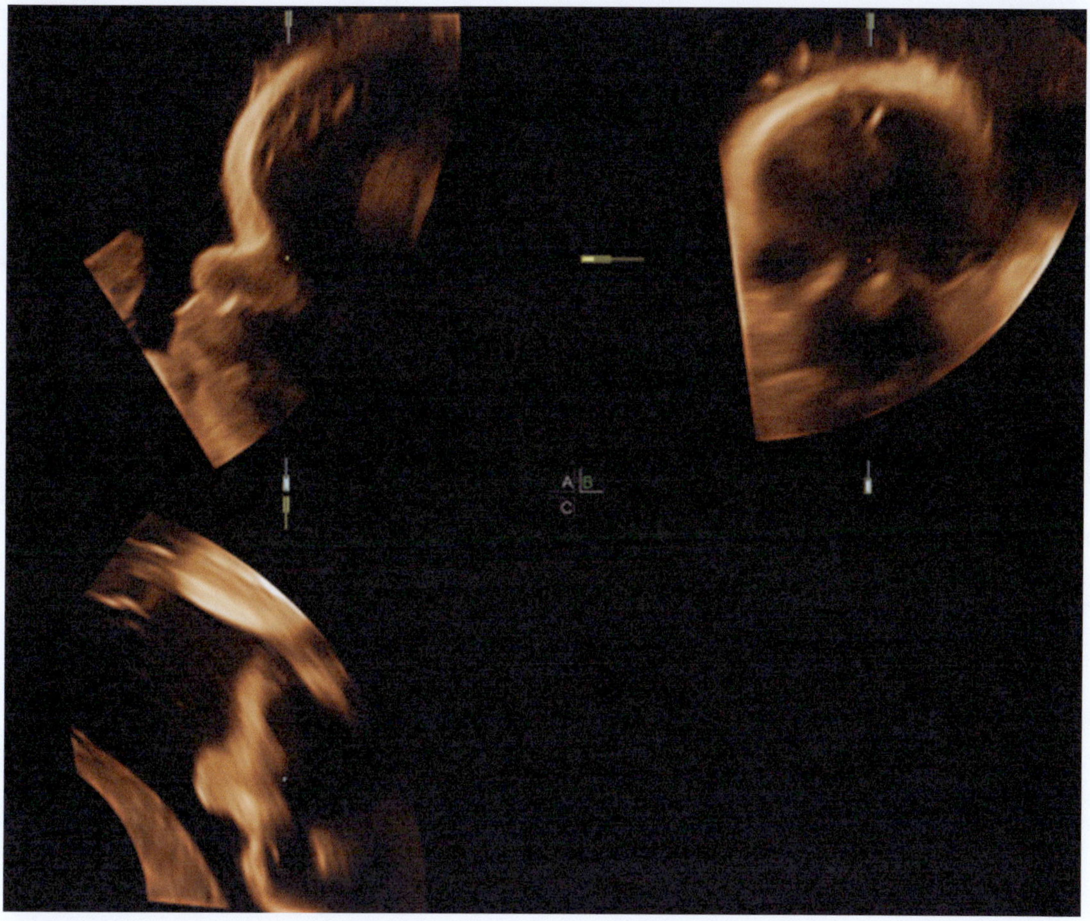

Fig. 4.46 Achondroplasia: multiplanar view

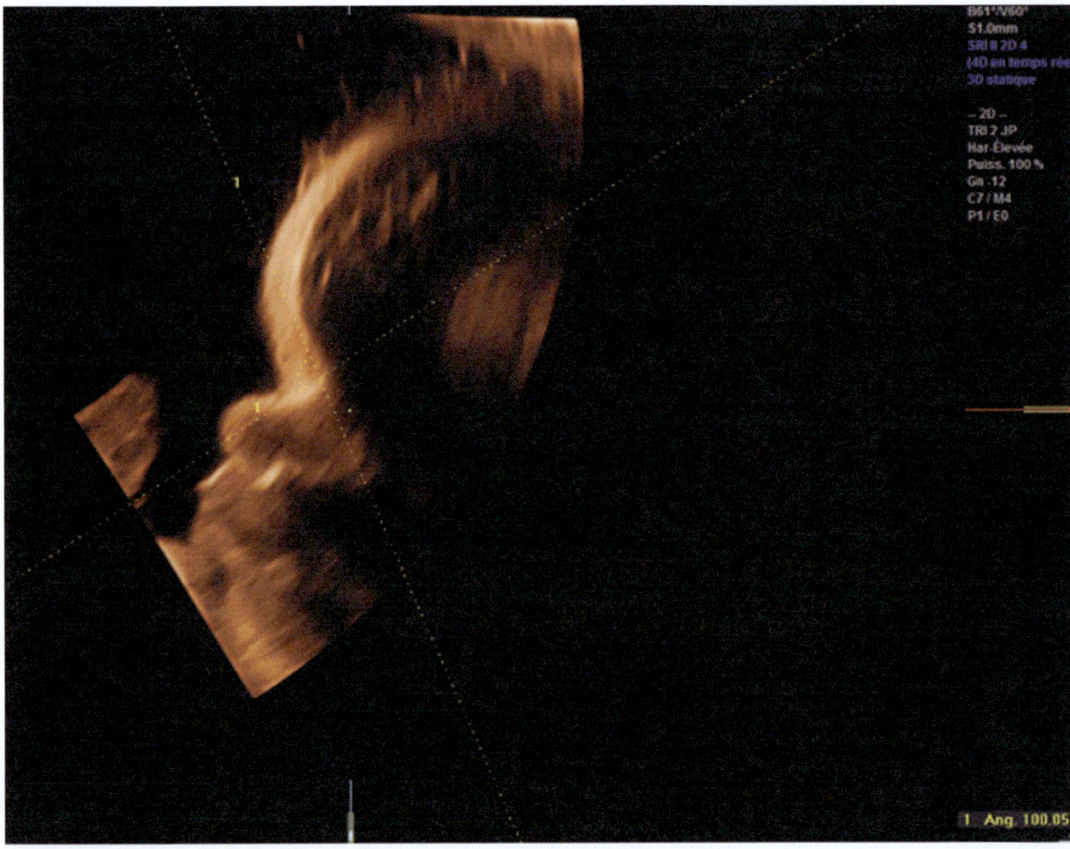

Fig. 4.47 Achondroplasia: closed upper facial angle

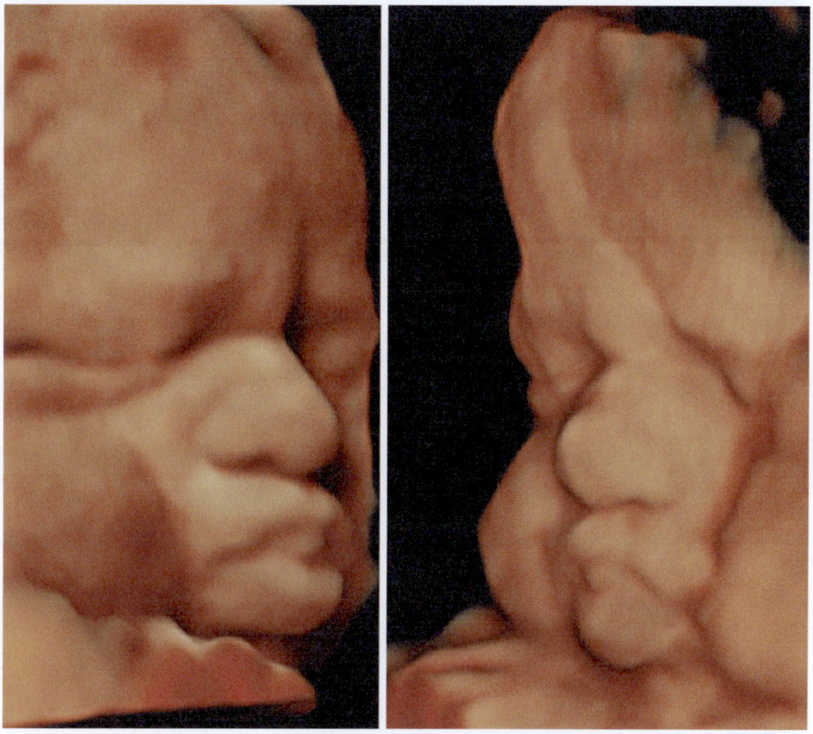

Fig. 4.48 Achondroplasia: dysmorphism "Render HD Live"

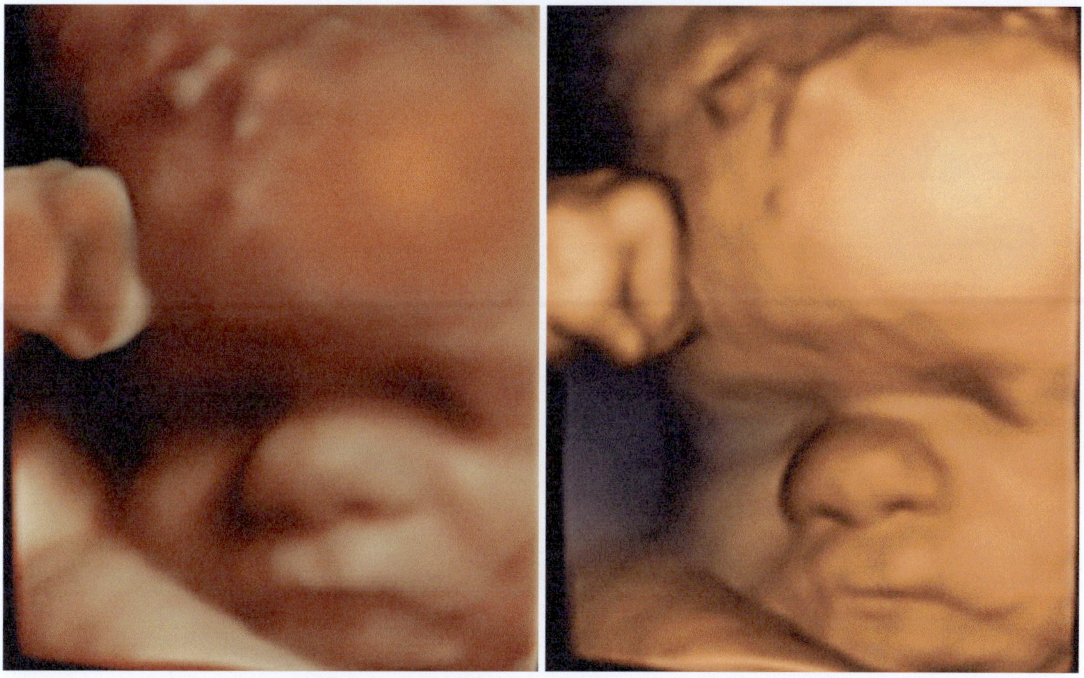

Fig. 4.49 Achondroplasia: dysmorphism "Render"

4.5.2 Maxillo-Facial Dysplasia: Binder Syndrome

This anomaly has an estimated incidence of 1/10,000, but is probably underestimated. A few family recurrences suggest autosomal recessive transmission or low-penetrating dominance. Binder-like dysmorphism can be found in cases of mothers taking phenytoin and anti-vitamin K type anticoagulants, or with lupus.

This sequence is considered to be an allelic variation of punctate chondrodysplasias.

Dysmorphism takes the form of a flattened midface that translates into an abnormally open upper facial angle, a short nose with a flattened root, a slightly protruding mandible, and a little developed upper maxilla (Figs. 4.50, 4.51, 4.52, 4.53, and 4.54). Tests should include karyotype, maternal blood test for lupus, a detailed ultrasound looking for punctate epiphyses (Figs. 4.55 and 4.56), a scan could be done to confirm the ultrasound results. Discovery of a brachytelephalangy would suggest a benign form of chondrodysplasia punctate (Fig. 4.57).

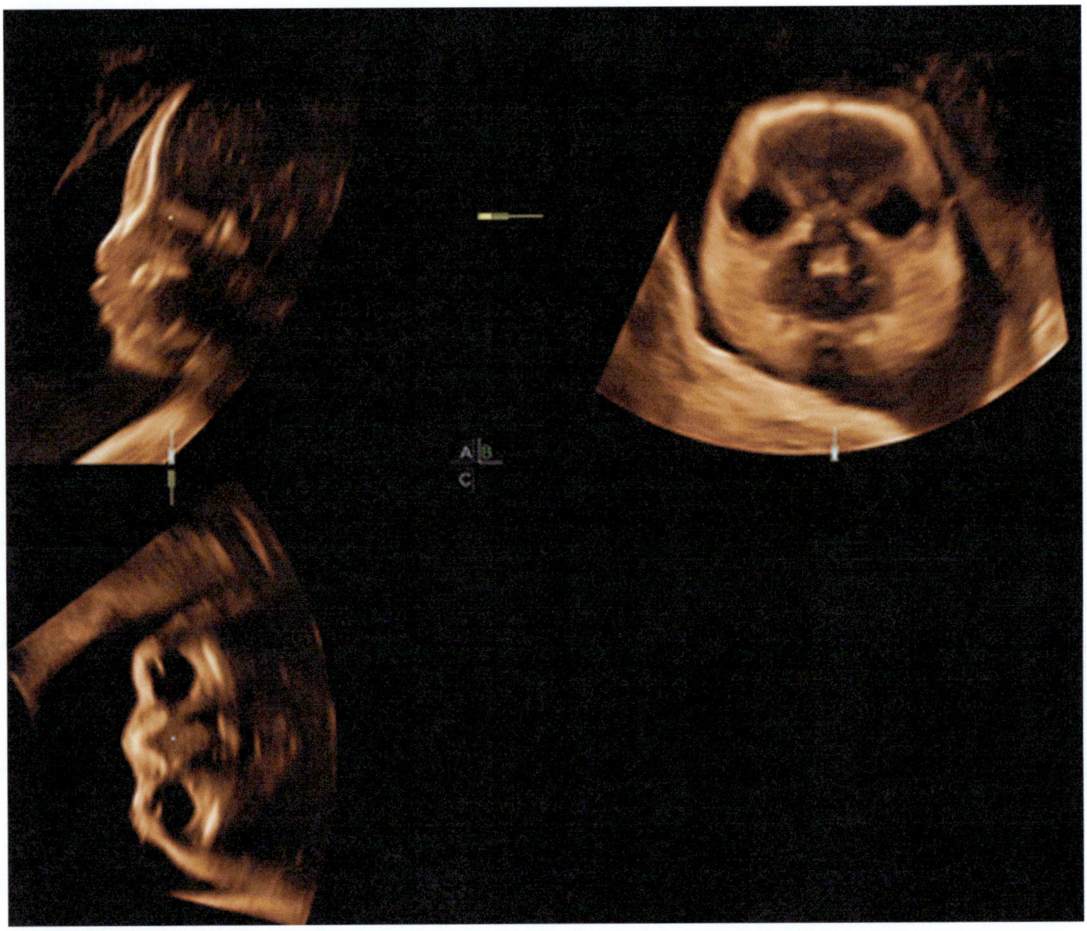

Fig. 4.50 Binder, multiplanar view

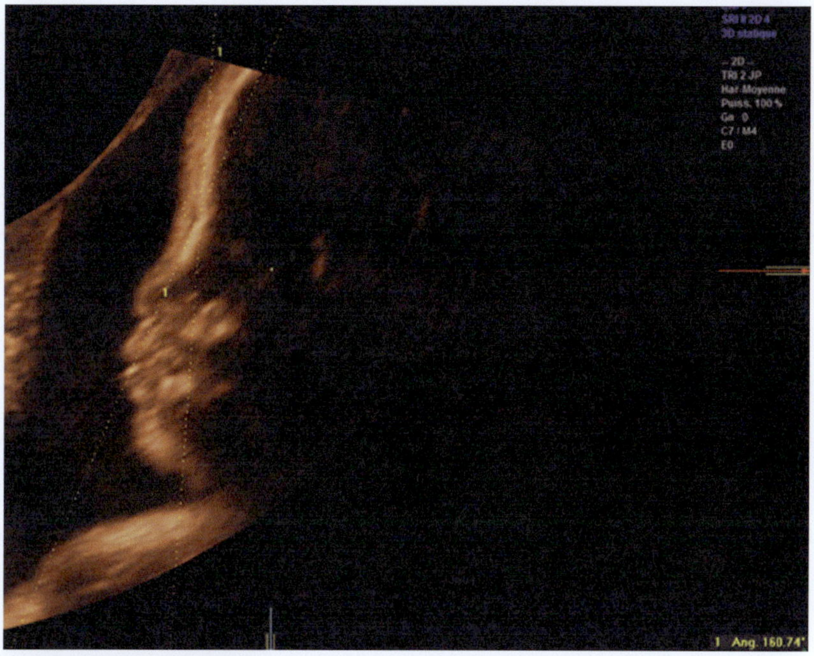

Fig. 4.51 Binder, open upper facial angle

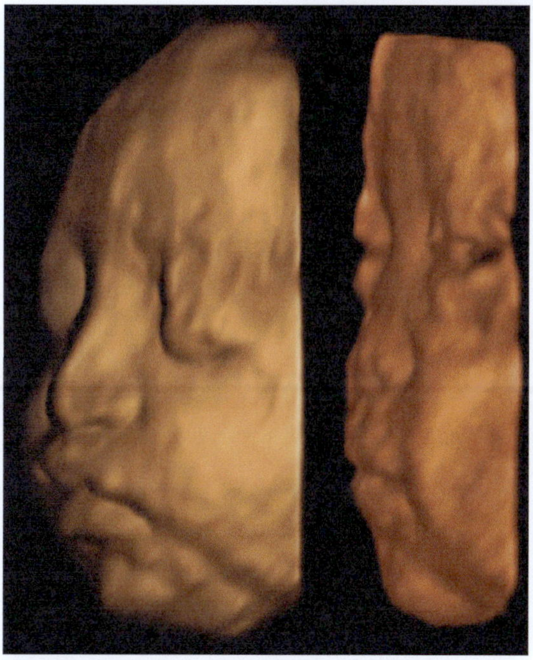

Fig. 4.52 Binder, rendered mode

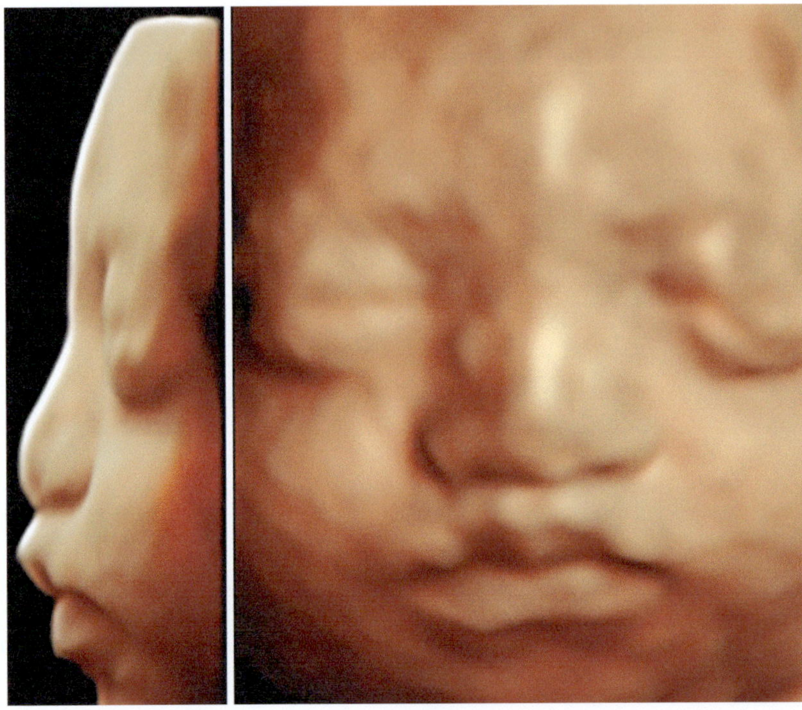

Fig. 4.53 Binder, Render HD-live mode

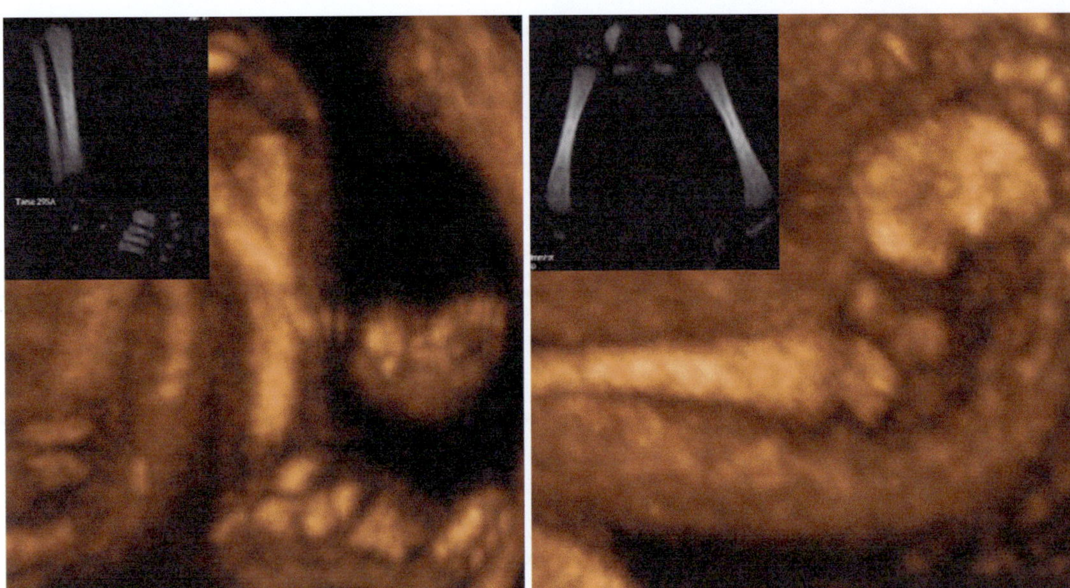

Fig. 4.54 Punctate calcifications: prenatal ultrasound and corresponding post-natal x-ray

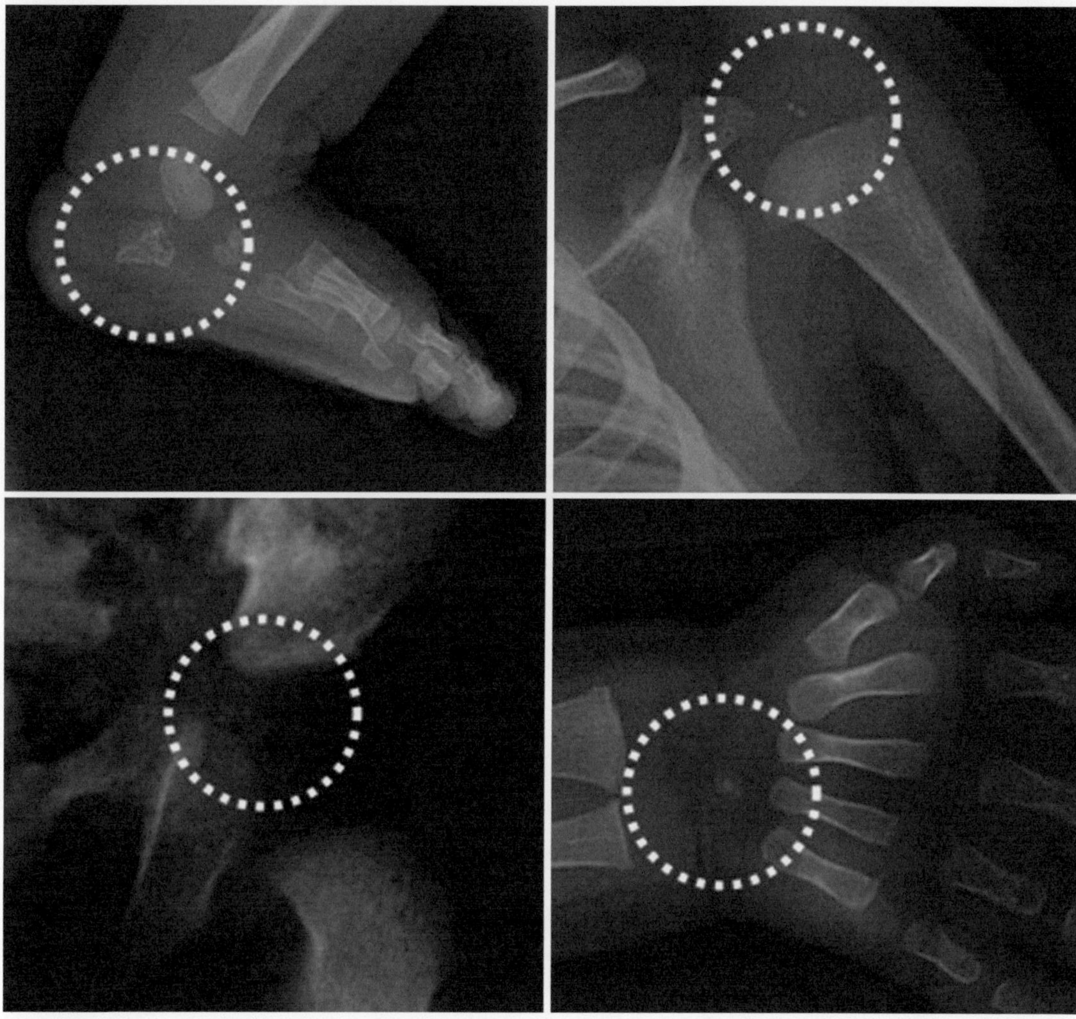

Fig. 4.55 Punctate calcifications – post-natal appearance

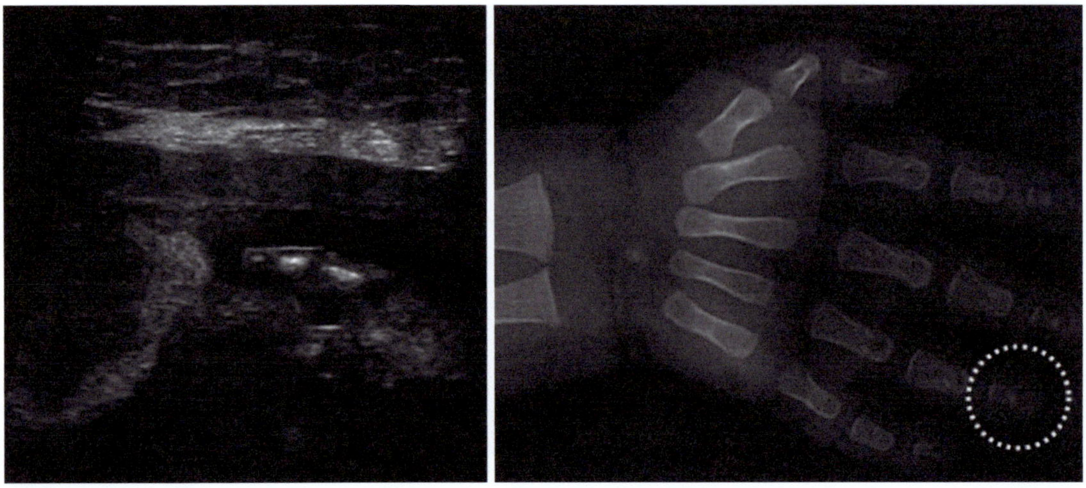

Fig. 4.56 Brachytelephalangy: prenatal ultrasound appearance and corresponding post-natal x-ray

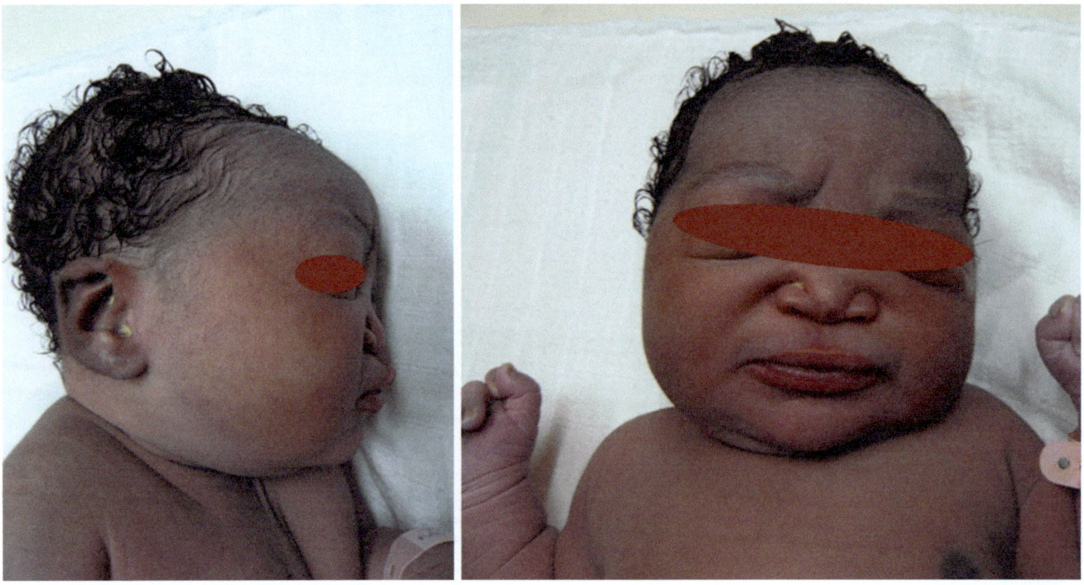

Fig. 4.57 Binder, post-natal appearance

4.5.3 Thanatophoric Dysplasia

This is the most frequent of fatal osteochondrodysplasias, with an estimated 1/20,000 to 1/60,000.

Currently, a diagnosis is made most often during the first trimester, although it could be made during the second trimester. It combines very short limbs with femurs shaped like old-fashioned telephone receivers, a narrow thorax that looks like a champagne cork on a sagittal view, platyspondyly with H-shaped vertebrae, and macrocrania. Hydramnios is common (Fig. 4.58).

Dysmorphism is marked by macrocrania, mid-face hypoplasia translated by a marked nasal bridge with a reduced upper facial angle (Figs. 4.59, 4.60, 4.61, and 4.62).

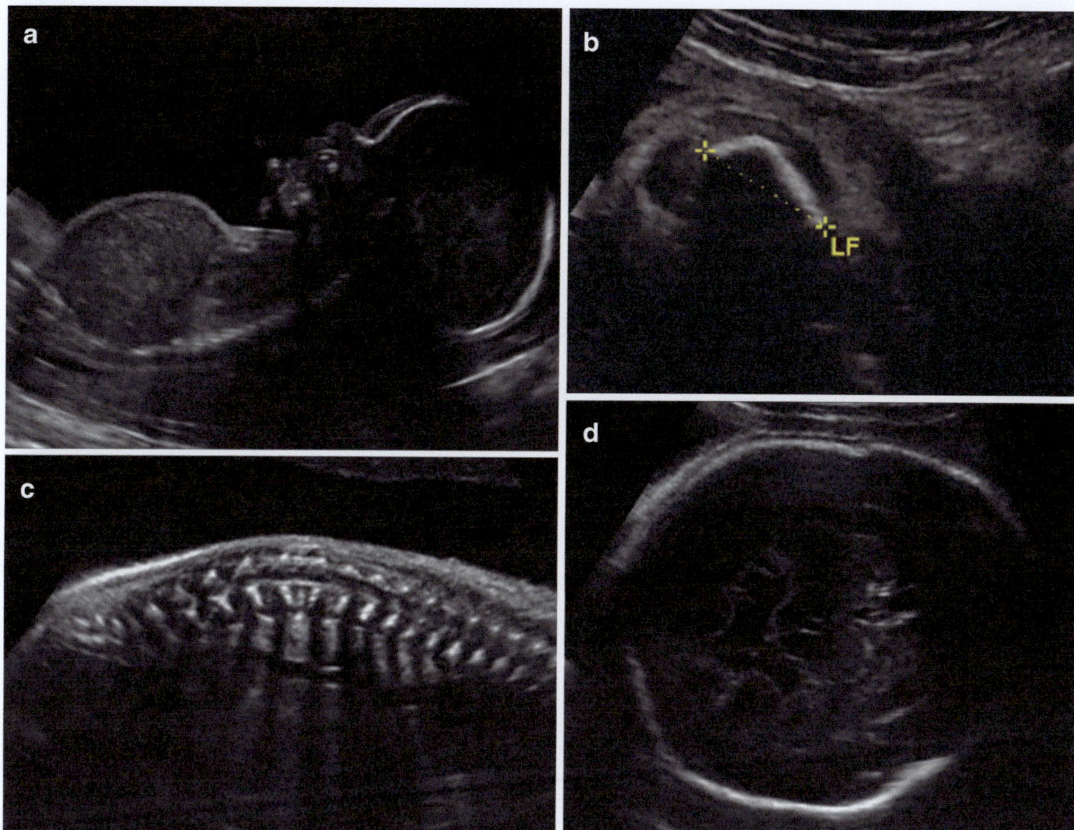

Fig. 4.58 Thanatophoric dysplasia: (**a**) narrow thorax. (**b**) short curved femur. (**c**) platyspondyly. (**d**) macrocrania

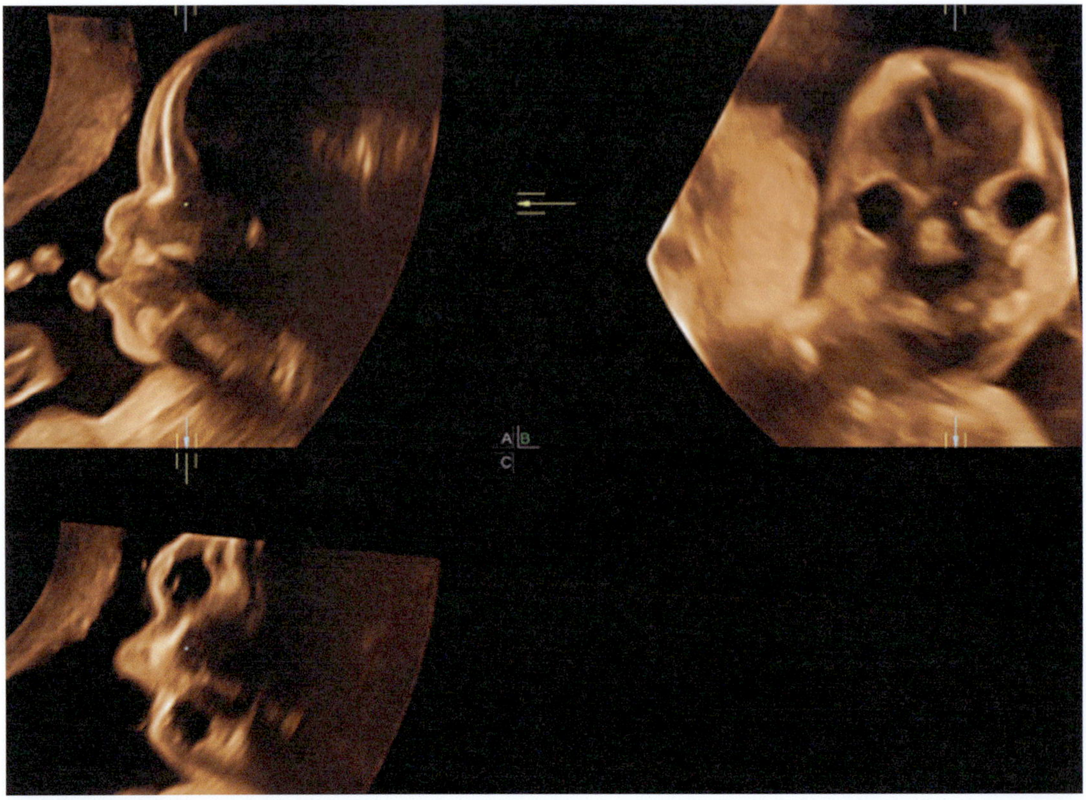

Fig. 4.59 Thanatophoric dysplasia: multiplanar view

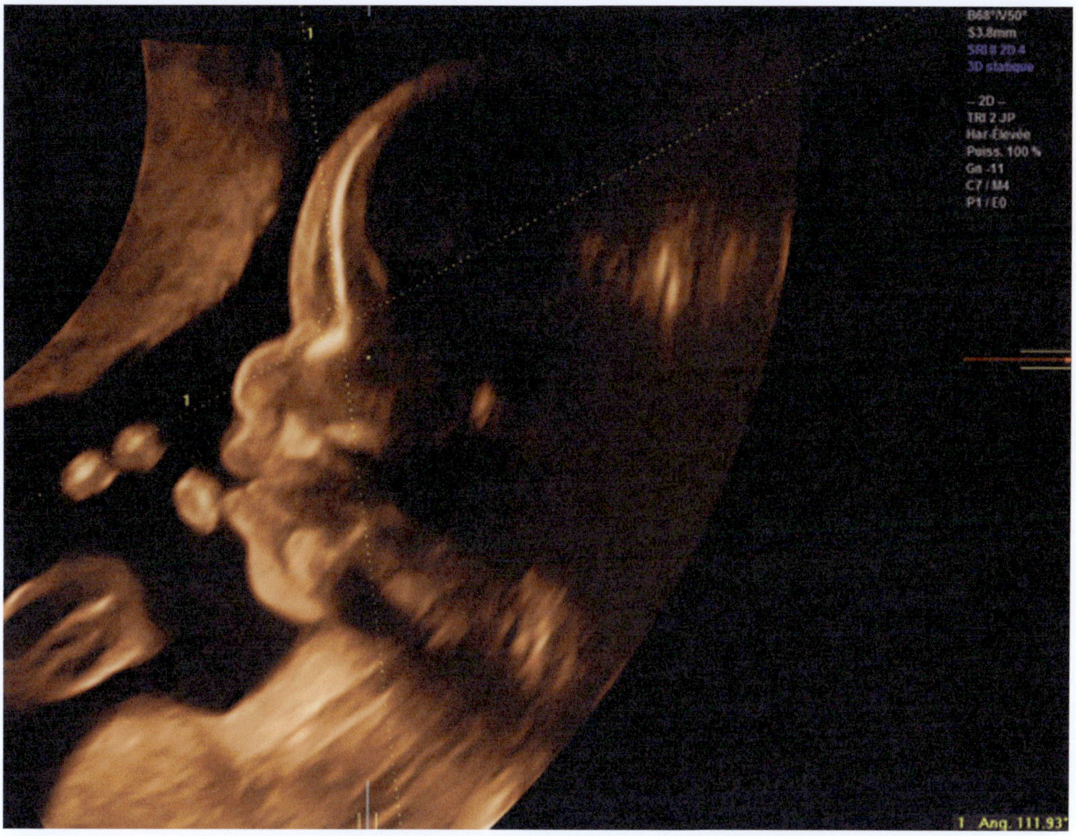

Fig. 4.60 Thanatophoric dysplasia: reduced upper facial angle

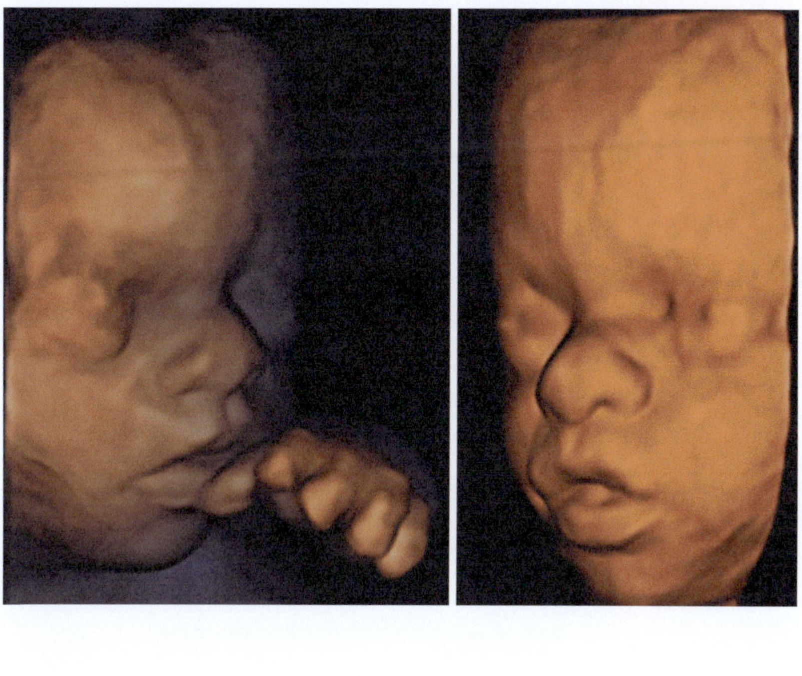

Fig. 4.61 Rendered image of thanatophoric dysplasia

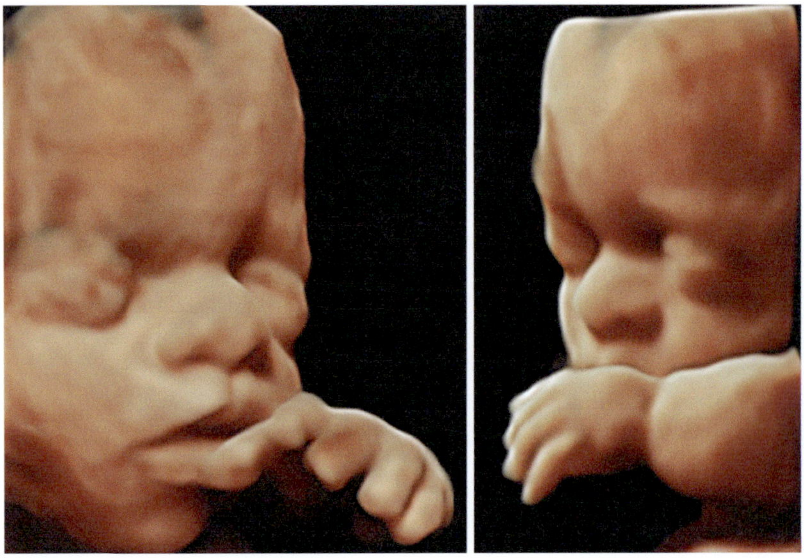

Fig. 4.62 HD live rendered image of thanatophoric dysplasia

4.5.4 Prader-Willi Syndrome

This rare syndrome affects between 1/10,000 and 1/25,000 births, its origin being a genetic deletion on the paternal chromosome 15 (at 15q11-q13). It is:

- Either de novo
- or part of a maternal uniparental disomy.

Prader-Willi syndrome is considered with the discovery of hydramnios and slow growth related to diminished active movements, fetal cardiac disorders, and genital anomalies, and a detailed study of the face could then support the hypothesis.

Dysmorphism in this syndrome is characterized by:

- A thin, slightly turned up upper lip
- Almond-shaped eyes, narrow forehead
- A closed upper facial angle

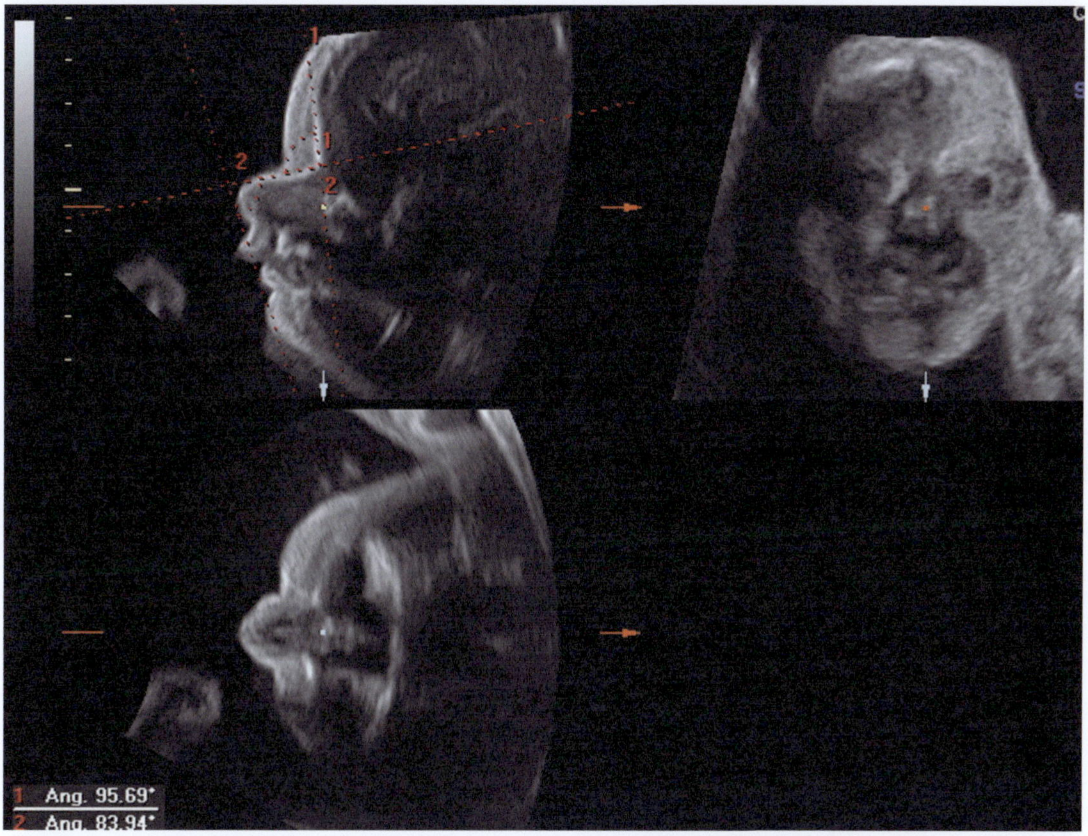

Fig. 4.63 Prader-Willi Syndrome: closed upper facial angle, thin turned up upper lip

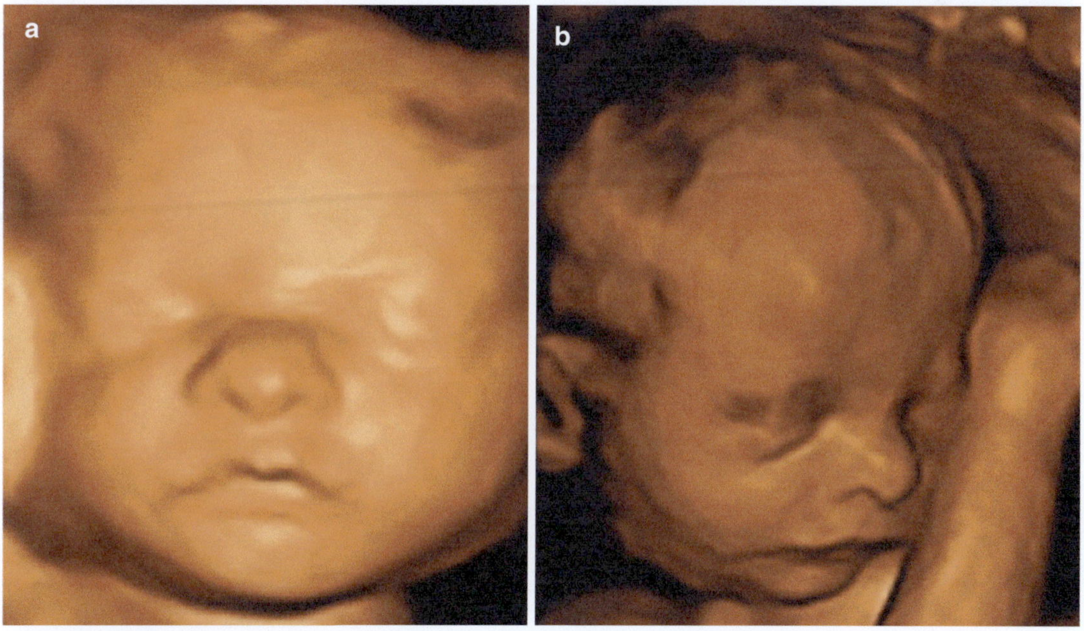

Fig. 4.64 Prader-Willi Syndrome: rendered image

4.5.5 Otopalatodigital Syndrome

This is a very rare syndrome (only around 30 cases recorded), two types of which are described: type 1 and type 2. This genetic disease results from mutations in the *FLNA* gene associated with skeletal dysplasia (curved long bones, limb anomalies, including fan-shaped toes, bent fingers, syndactyly, polydactyly, etc.), auditory deficit, and characteristic facial dysmorphism.

In this section, we are presenting a type 1 otopalatodigital syndrome with marked dysmorphism associated with hypertelorism, a protruding supra orbital ridge, a small flat nose, and palpebral fissures oriented downward and outward (Figs. 4.65 and 4.66).

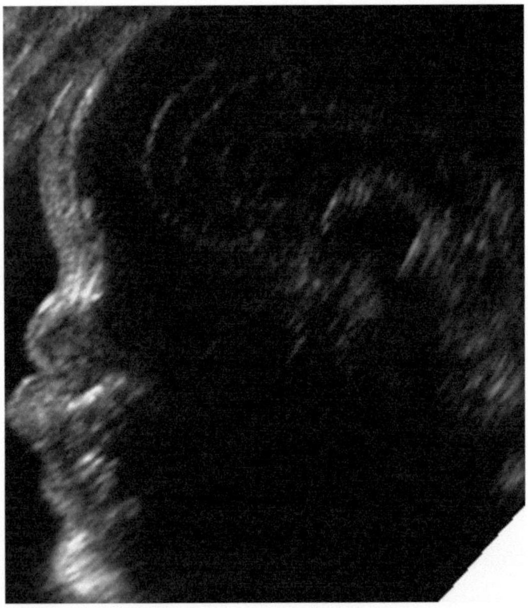

Fig. 4.65 Otopalatodigital syndrome: profile

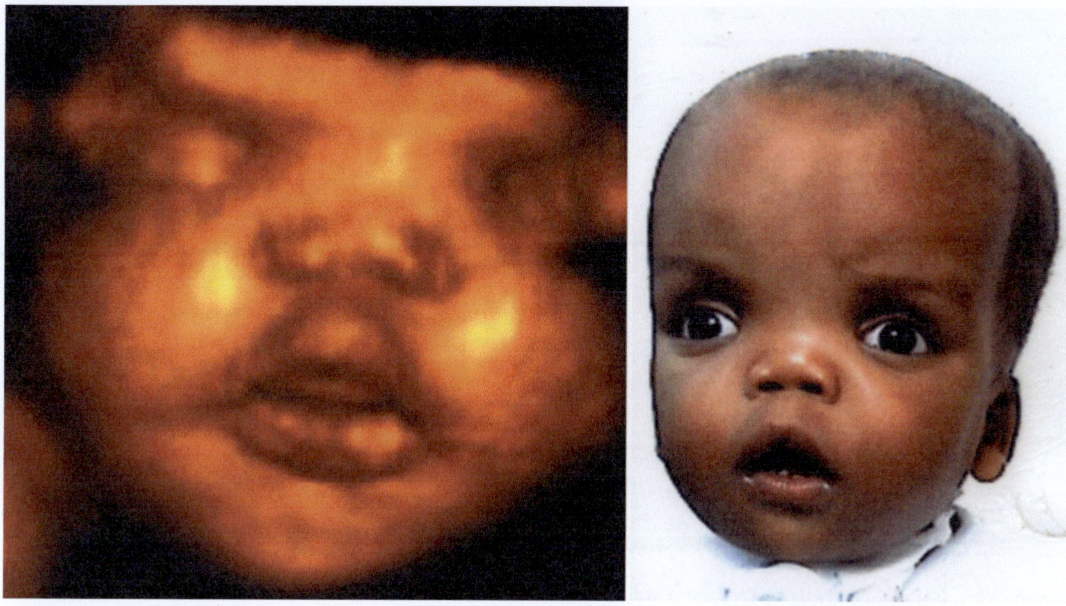

Fig. 4.66 Otopalatodigital syndrome: rendered image and post-natal appearance

4.6 Lower Face Pathologies

4.6.1 Otomandibular Dysplasias

These pathologies combine external ear anomalies that are often warning signs (Figs. 4.67 and 4.68), lower face anomalies at the mandibular level and mid-face anomalies at the malar level.

We can divide otomandibular dysplasias into two groups:

- Symmetric bilateral forms
  - Isolated:
    Mandibulofacial dysostosis
    Francheschetti syndrome or Treacher Collins syndrome
  - Associated: Acrofacial dysostosis
    Nager or Miller Syndroms with limb anomalies.
- Unilateral or asymmetric forms
  - Isolated
    First arch syndrome or first and second arches syndrome
    Hemifacial microsomy
  - Associated: Goldenhar syndrome

The facial dysmorphism that these syndromes share is in external ear anomalies, with preauricular tags (or fibrochondromas (Fig. 4.67), auricular anomalies (Fig. 4.68).

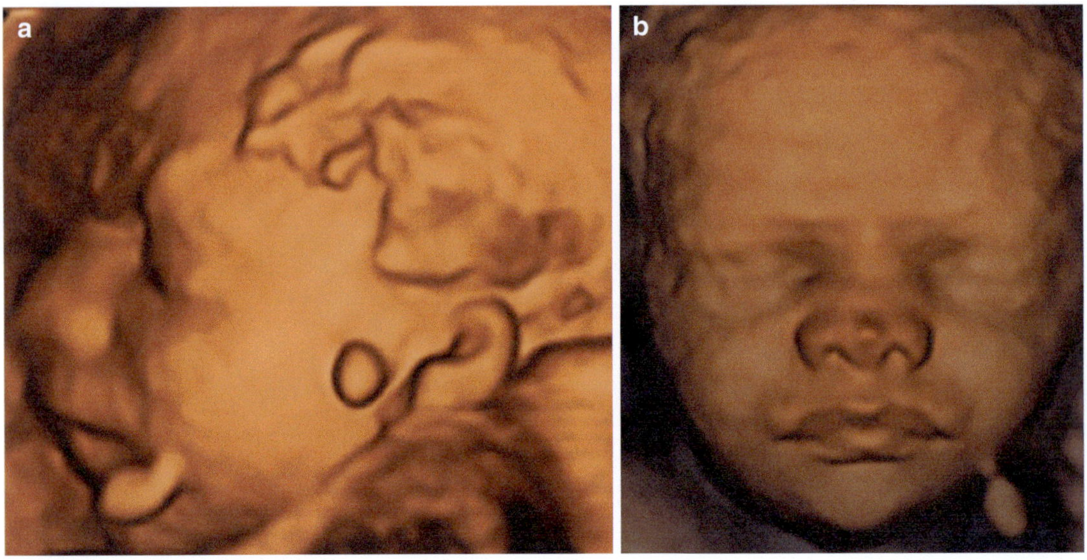

Fig. 4.67 (**a**, **b**) Francheschetti or Treacher-Collins syndrome

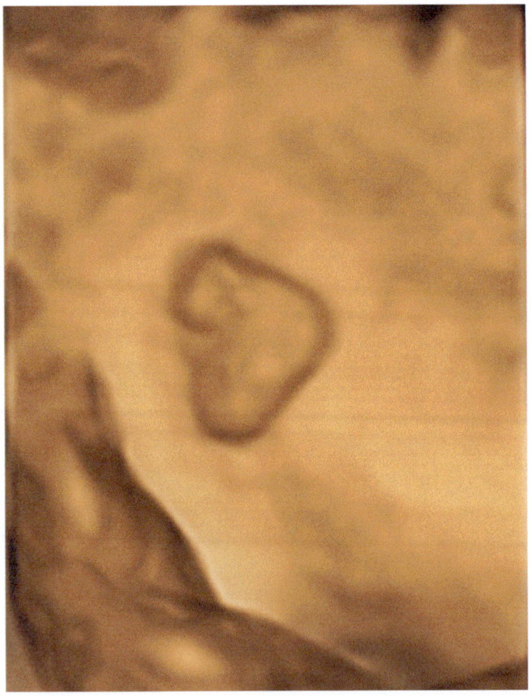

Fig. 4.68 Francheschetti syndrome: auricular anomalies

4.6.1.1 Francheschetti or Treacher-Collins Syndrome

Incidence is estimated at 1/50,000. The major markers of this syndrome other than ear anomalies (auricular deformities, preauricular tags, internal ear anomalies) include facial bone anomalies: hypoplastic malar bones, absent zygomatic arches, mandibles that lose their usual angulation between the ramus and the horizontal branch, impacting the temporomandibular joint that can lead to serious limitations in buccal opening (Figs. 4.70, 4.71, and 4.72), micrognathia leading to a decrease in the lower facial angle measurement (Figs. 4.73 and 4.74).

Malar hypoplasia results in palpebral fissures oriented downward and outward (Fig. 4.69), and coloboma of the lower eyelids is frequent between the external and middle thirds.

There can be cleft palate (Fig. 4.75).

This syndrome can express differently within a same family. Family history can be found in 40 % of cases, 60 % of cases being a de novo mutation. The genetic origin corresponds to a mutation at 5q32 encoding a protein called "Treacle," which is essential for the survival of neural crest cells.

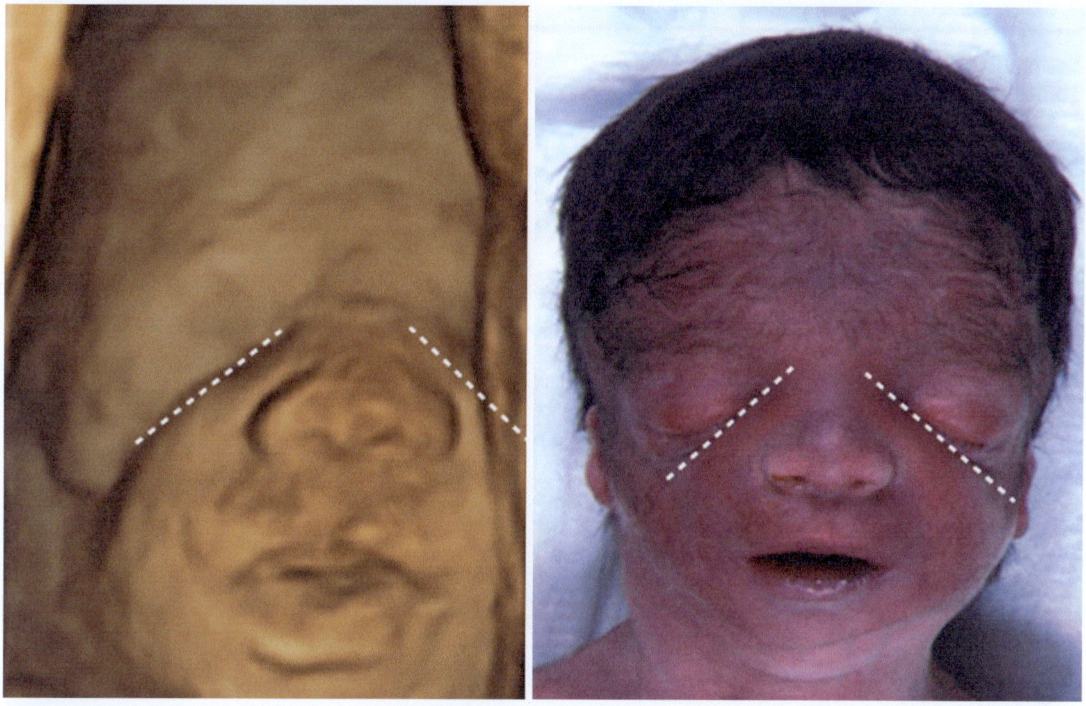

Fig. 4.69 Treacher-Collins syndrome: malar hypoplasia and palpebral fissures oriented downward and outward

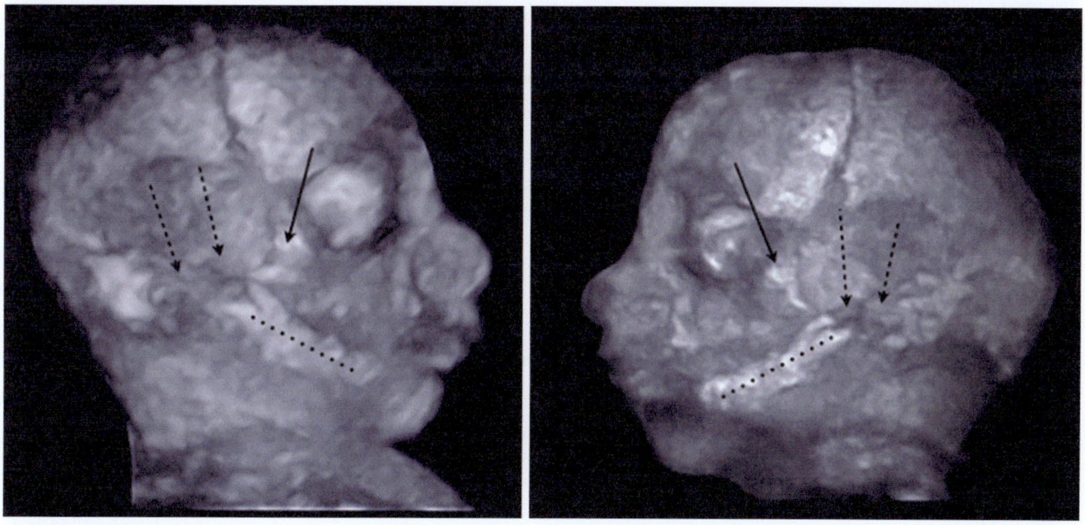

Fig. 4.70 Treacher-Collins syndrome: dotted line: unangulated mandible, dotted arrows: absence of zygomatic arch, *full-line arrow*: malar hypoplasia

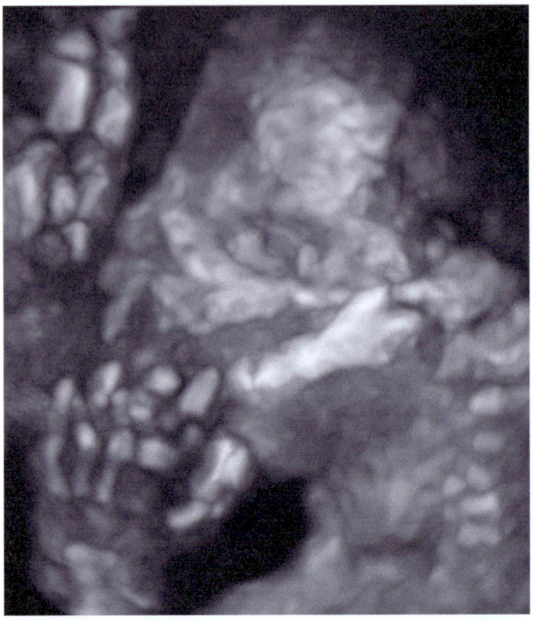

Fig. 4.71 Francheschetti syndrome: unangulated mandible, interruptions of zygomatic arch

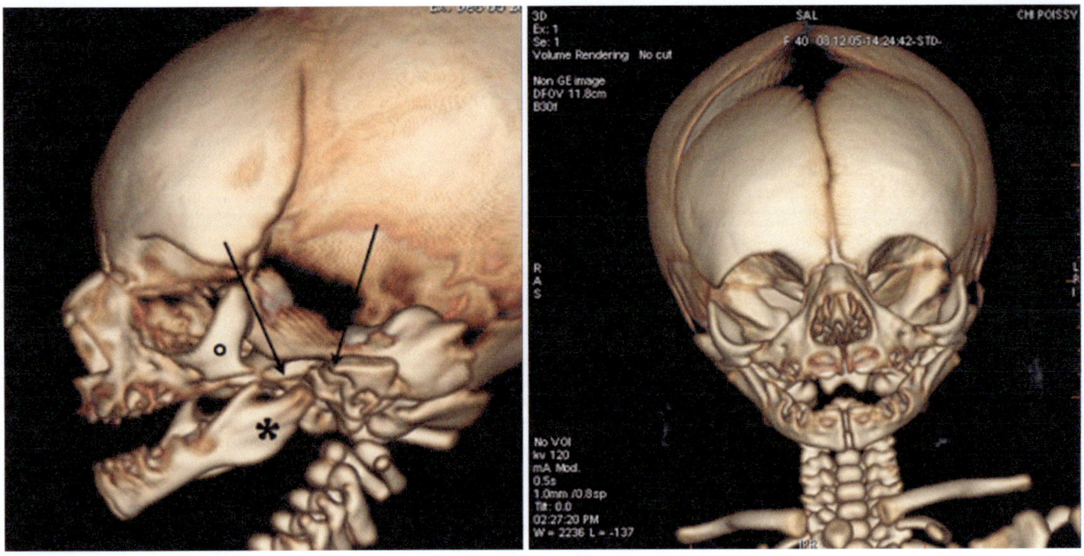

Fig. 4.72 Francheschetti syndrome: bone scan: malar hypoplasia, interrupted zygomatic arch, unangulated mandible

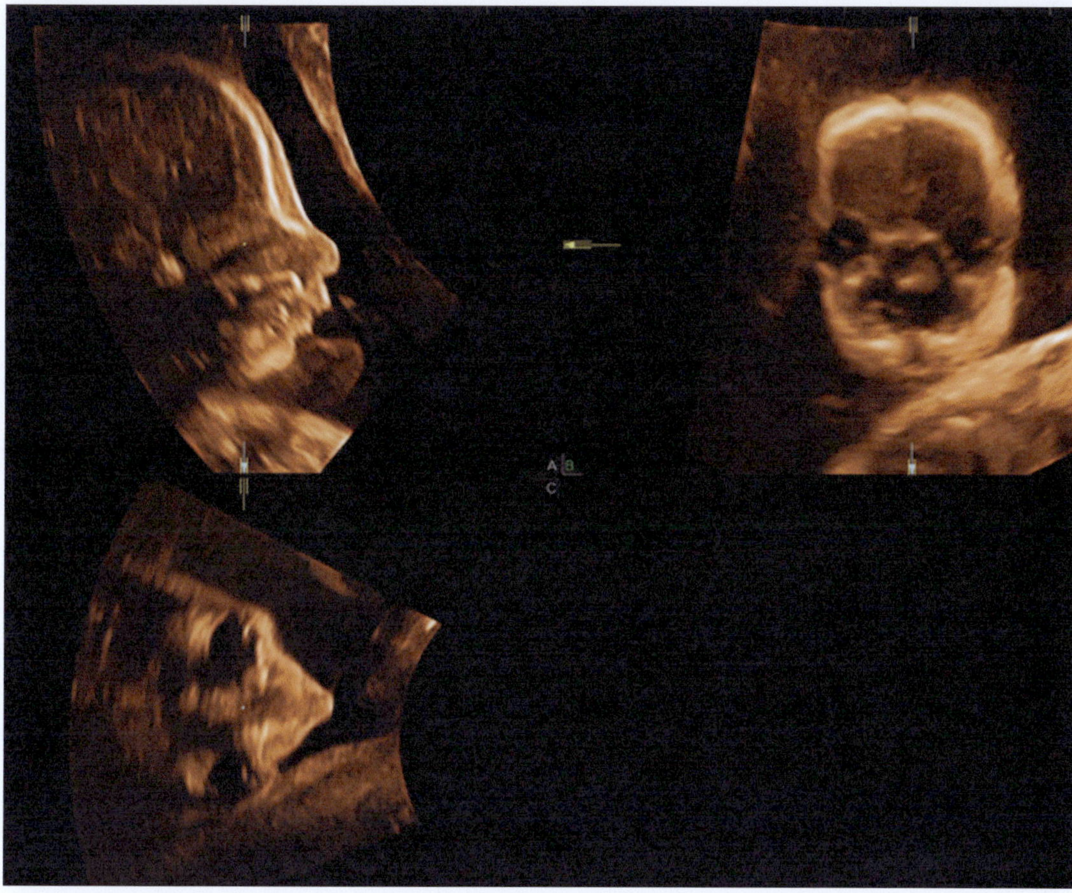

Fig. 4.73 Treacher-Collins syndrome: multiplanar view

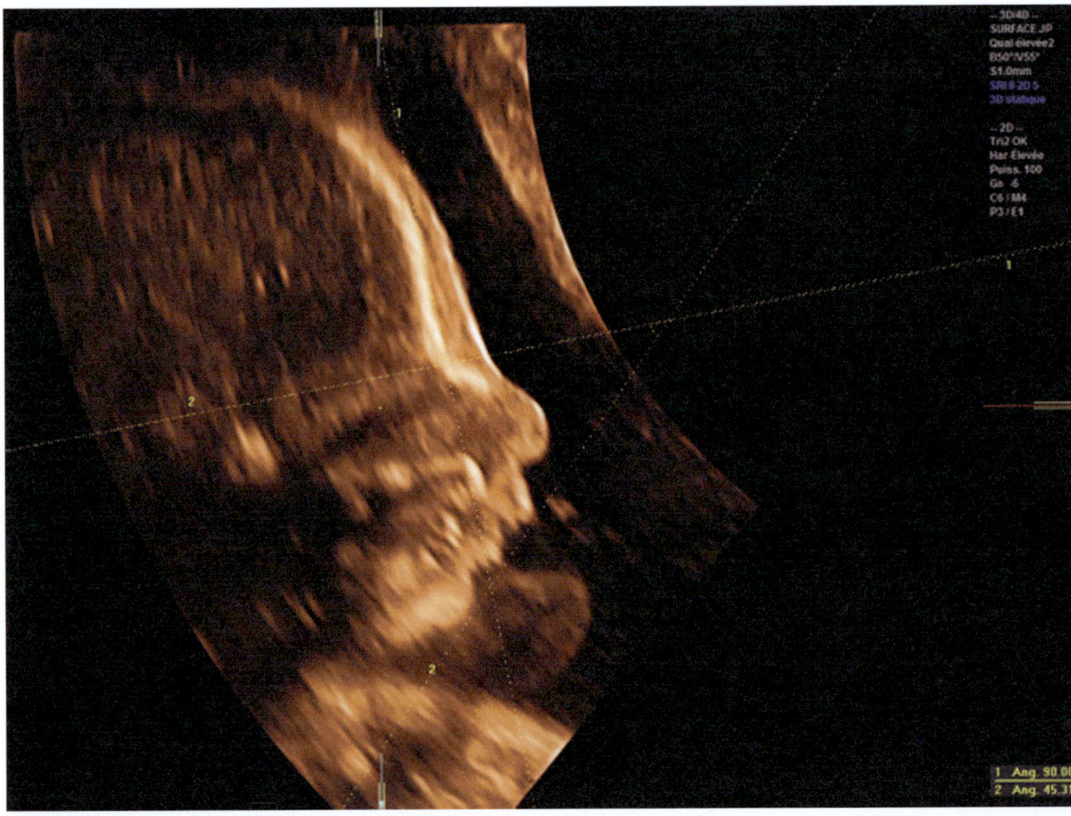

Fig. 4.74 Treacher-Collins syndrome: micrognathia, lower facial angle: 45°

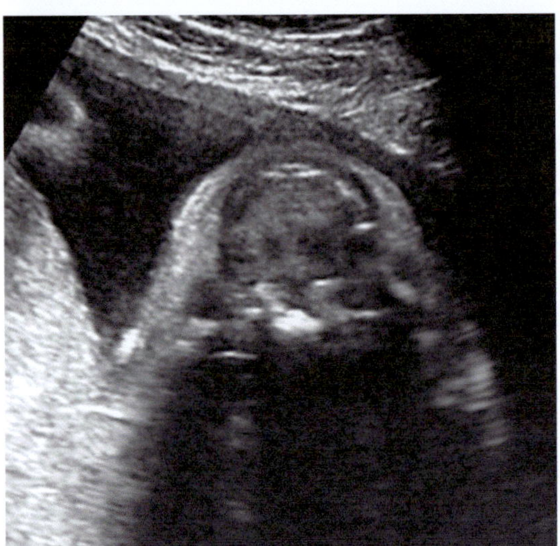

Fig. 4.75 Francheschetti syndrome: cleft palate (secondary palate)

4.6.1.2 Goldenhar Syndrome

This syndrome is part of the oculo-auriculo-vertebral spectrum of diseases, with an estimated incidence of 1/6000.

Facial dysmorphism in this syndrome includes external ear anomalies (fibrochondromas, auricular anomalies) (Fig. 4.76).

- Facial asymmetry impacts the mandible, the soft parts of the cheeks, and edges of the mouth, creating a lateral cleft that leads to an appearance of macrostomia (Figs. 4.77 and 4.78).
- There are related vertebra malformations (cervical vertebra fusion, hemi-vertebra, etc.)
- Limbal dermoids

The etiology of this syndrome remains unclear.

The differential diagnosis for the oculo-auriculo-vertebral spectrum is the branchio-oto-renal syndrome (BOR syndrome, Fig. 4.79) which associates facial markers with renal anomalies and family history. It is an autosomal dominant syndrome linked to a mutation of the EYA1 gene located on chromosome 8.

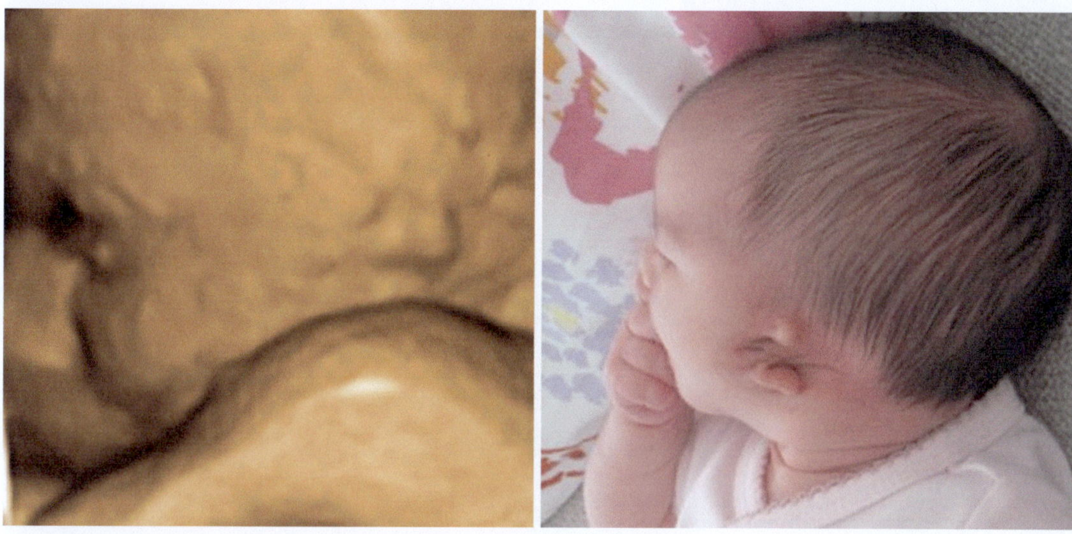

Fig. 4.76 Goldenhar syndrome: unilateral anomaly of the left ear, prenatal ultrasound and post-natal image

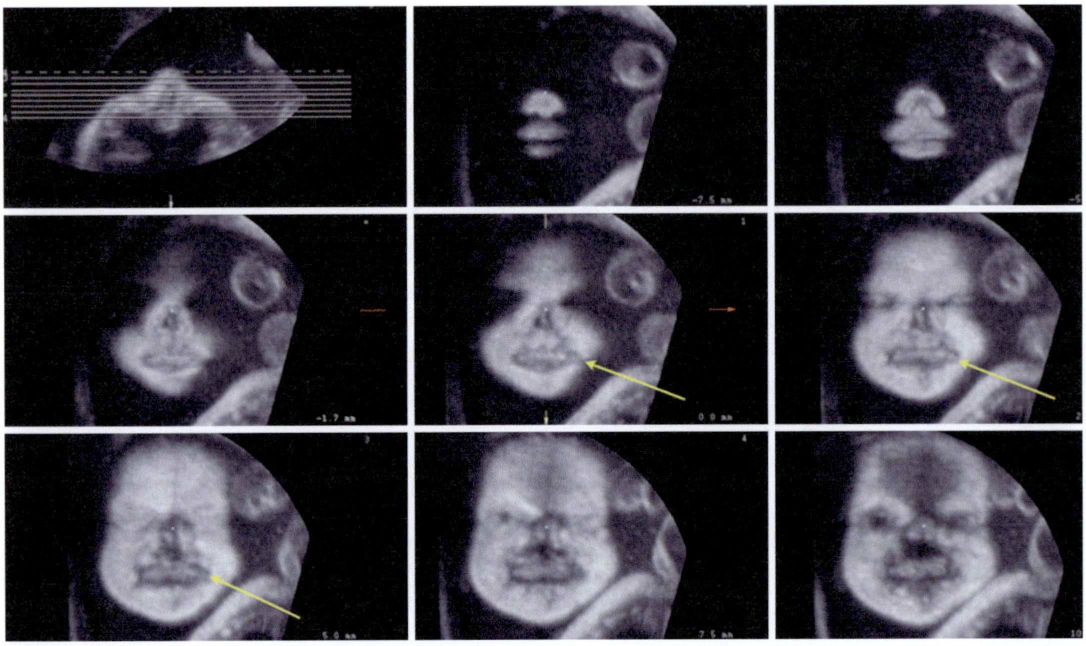

Fig. 4.77 Goldenhar syndrome, TUI, mouth asymmetry (*yellow arrows*)

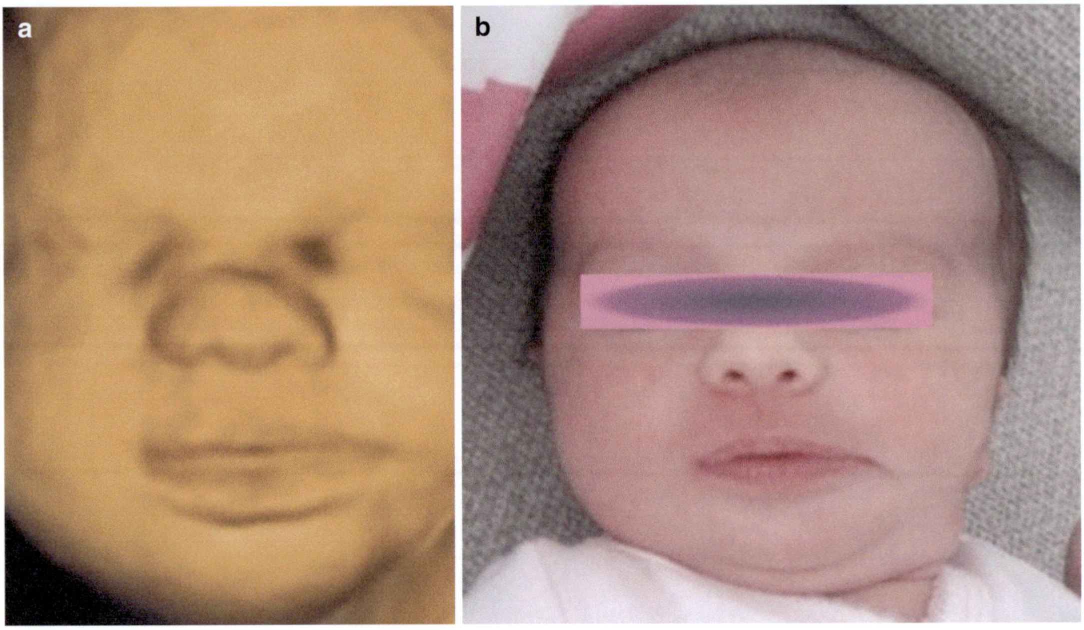

Fig. 4.78 Goldenhar syndrome mouth asymmetry. (**a**) prenatal, (**b**) post-natal

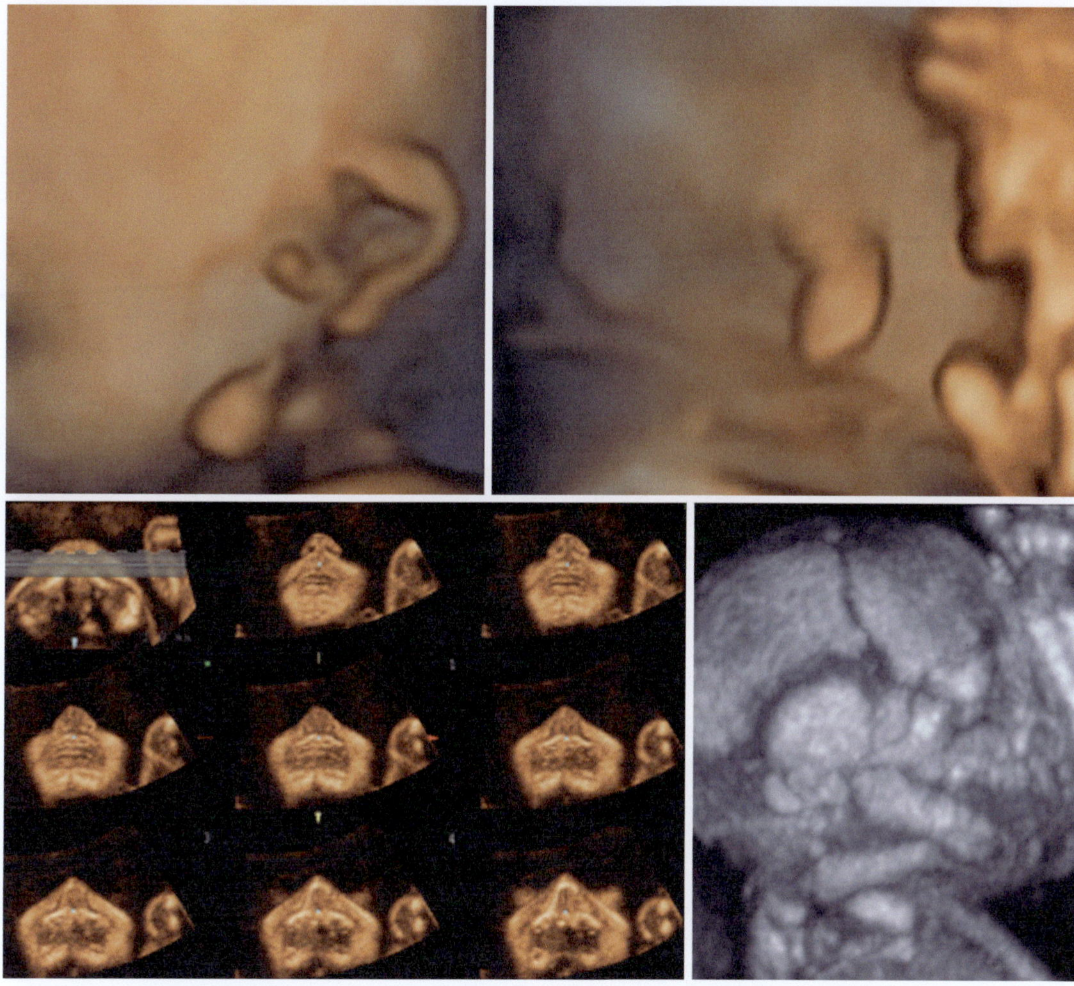

Fig. 4.79 BOR Syndrome: ear anomalies, facial symmetry, normal zygomatic arch and mandibles

4.6.2 Beckwith-Wiedemann Syndrome

The incidence of this syndrome is estimated at 1/13,000. It is a syndrome that results in excessive growth. The three major classic markers are macrosomia, macroglossia (Figs. 4.80, 4.81, and 4.82), and abnormal openings in the anterior abdominal wall, especially an omphalocele (Fig. 4.83). The minor markers are visceromegaly (particularly the kidneys, liver, spleen and pancreas, Fig. 4.84), ear anomalies (Fig. 4.85), neonatal hypoglycemia, mesenchymal dysplasia of the placenta in 20 % of cases (Fig. 4.86).

A diagnosis of Beckwith-Wiedemann is accepted when two major markers and one minor marker have been found.

The major risks linked to this syndrome are the development of tumors during the first 5 years of life (Wilms tumor or nephroblastoma, hepatoblastoma, corticoadrenaloma, neuroblastoma, rhabdomyosarcoma). This risk is estimated at between 7.5 % and 10 %. The genetics of this syndrome are complex, but recent progress has made it possible to better target children with tumor risk.

Facial dysmorphism is primarily marked by macroglossia, with a tongue that is constantly sticking out (Figs. 4.80 and 4.81), the lingual perimeter is greater than normal (normal: gestational age in weeks since last menstrual period X 2.75, according to D. Nyberg Fig. 4.82), subtle auricular anomalies that look like notches and anomalies in the shape of the anthelix (Fig. 4.85).

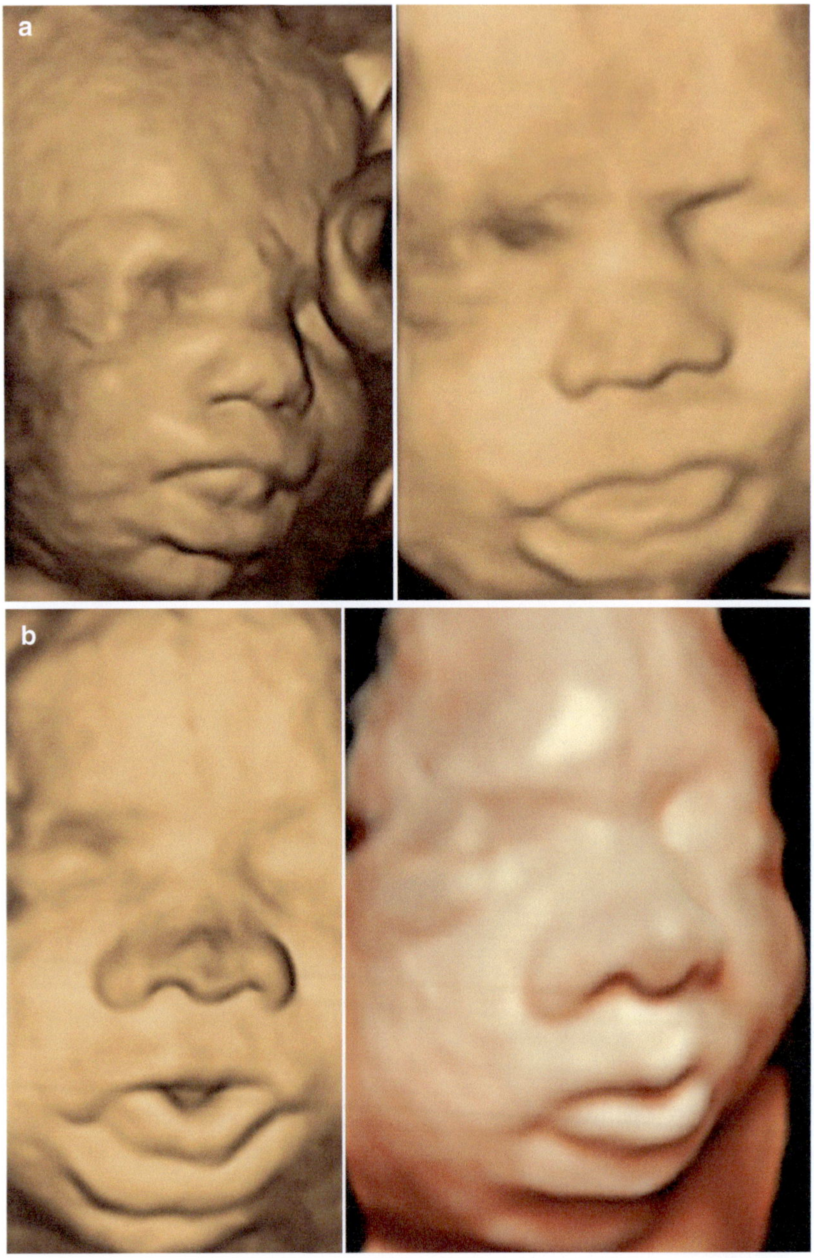

Fig. 4.80 (**a**) Beckwith-Widermann syndrome: macroglossia. (**b**) Beckwith-Wiedemann syndrome: macroglossia

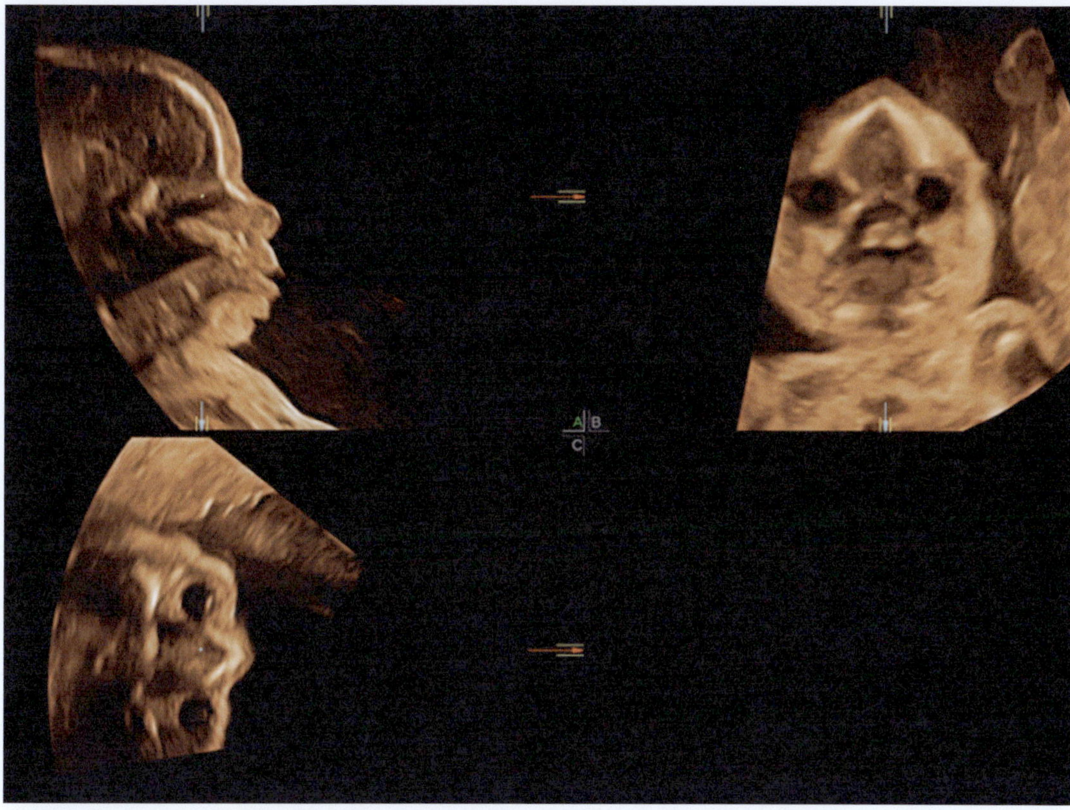

Fig. 4.81 Beckwith-Wiedemann syndrome: multiplanar image

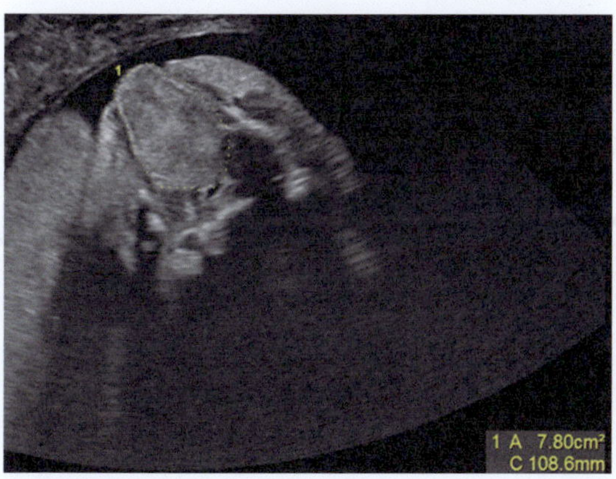

Fig. 4.82 Beckwith-Wiedemann syndrome: lingual perimeter

Fig. 4.83 Beckwith-Wiedemann syndrome: omphalocele

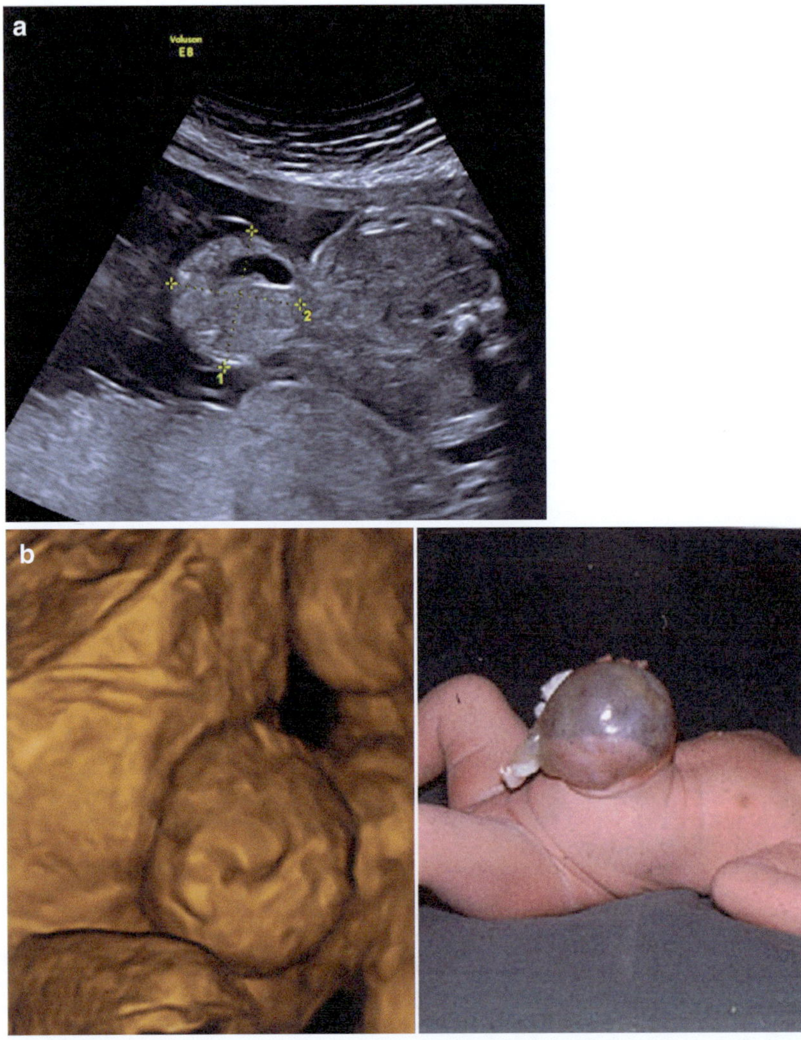

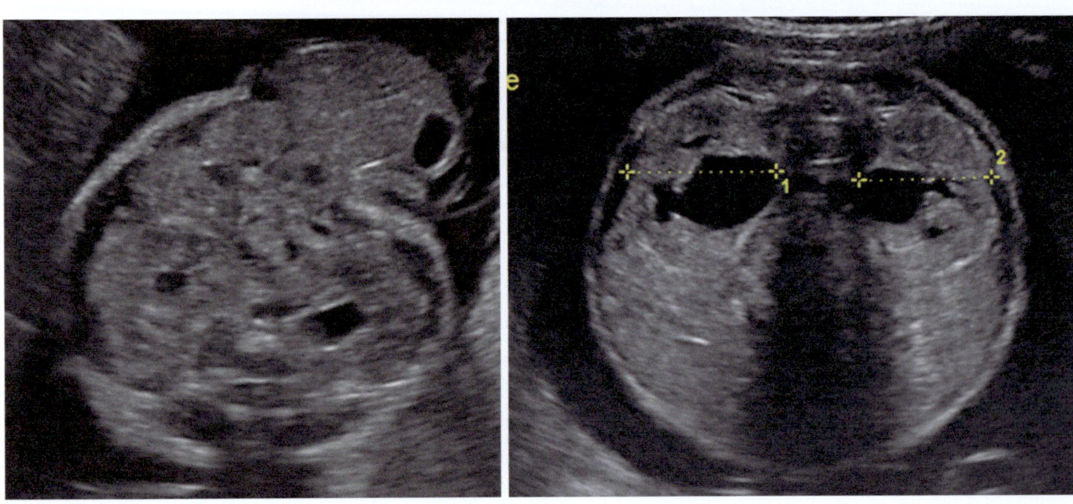

Fig. 4.84 Beckwith-Wiedemann syndrome: nephromegaly

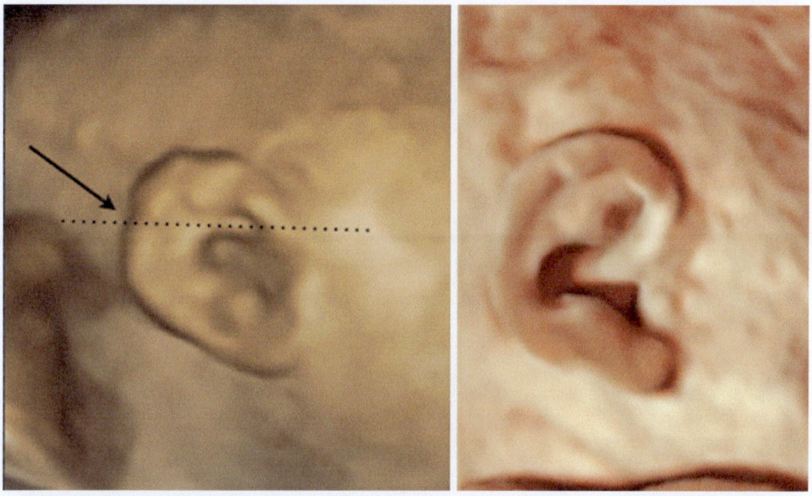

Fig. 4.85 Beckwith-Wiedemann syndrome: on the *right*, control ear, on the *left*, ear of affected patient, with *dotted line* indicated the anthelix anomaly and the *full arrow* the auricular notch

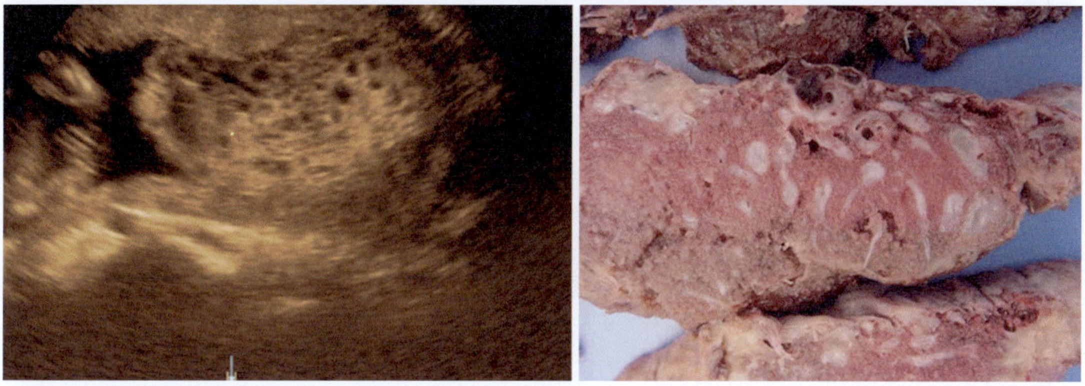

Fig. 4.86 Beckwith-Wiedemann syndrome: Mensenchymal dysplasia of the placenta, on ultrasound and fetal pathology examination

4.6.3 Fetal Alcohol Syndrome

The effects of alcohol exposure on the fetus have been known for several decades and are grouped together under the term "Fetal Alcohol Syndrome (FAS)". The criteria defined in 1989 to diagnose it are an association of markers:

- Prenatal and post-natal growth deficiency
- Microcephaly, biometric anomalies in the brain, in particular in the frontal lobes and the orbits, with late intellectual development and behavioral anomalies.
- Facial dysmorphism recognizable on ultrasound, combining short lower face, slight retrognathia, a long domed philtrum, and a thin upper lip (Figs. 4.87, 4.88, 4.89, and 4.90).

Revealing a context of fetal alcohol syndrome is often difficult and follow-up should be done with tact.

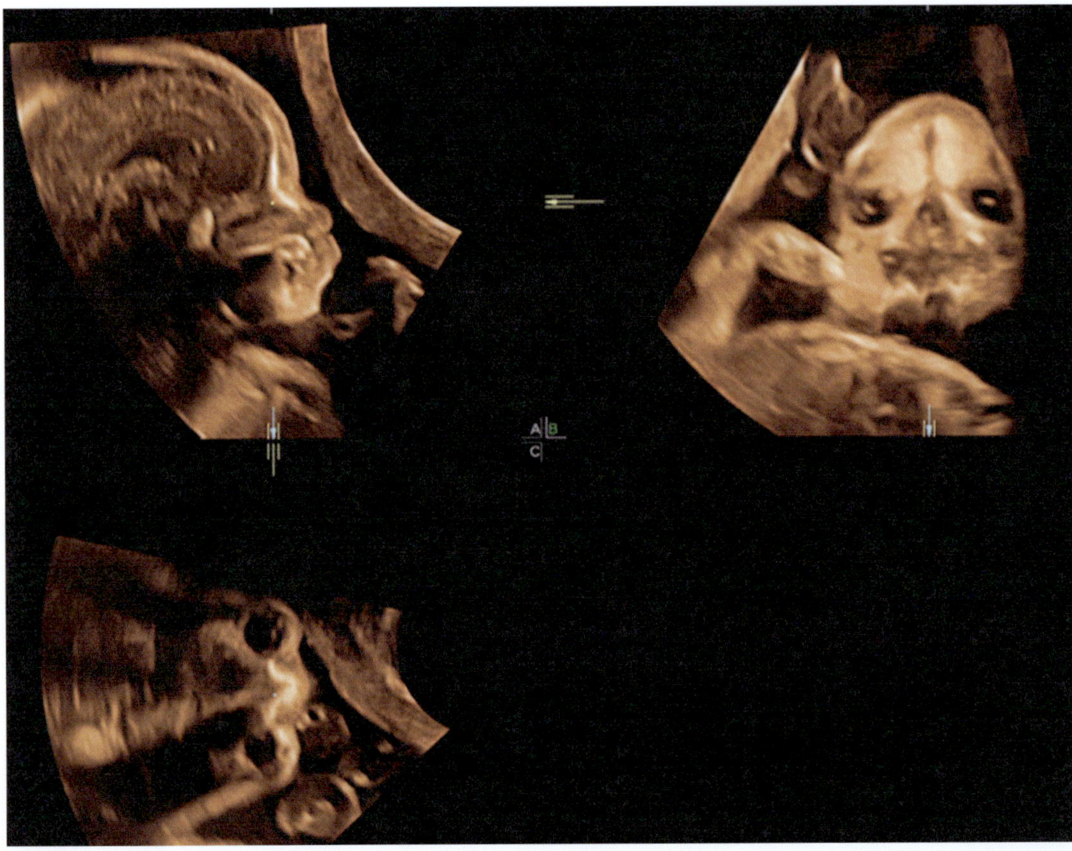

Fig. 4.87 FAS: multiplanar image

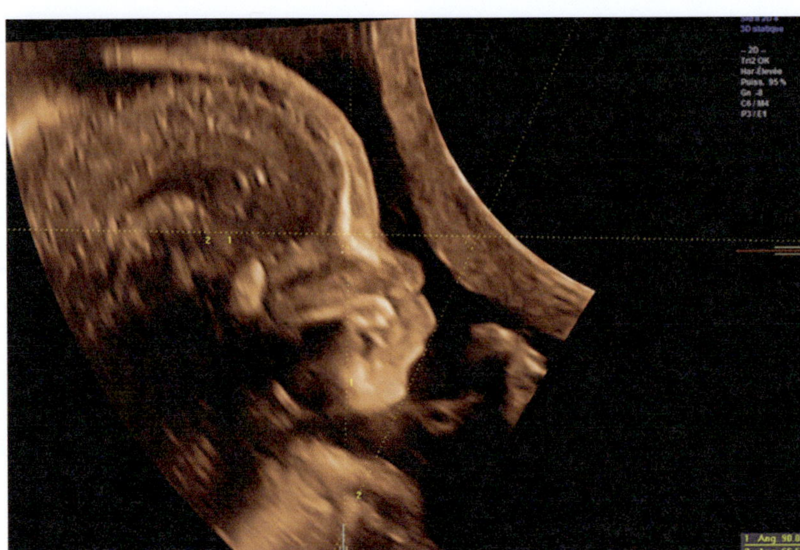

Fig. 4.88 FAS: practically normal inferior facial angle, long philtrum, short height of lower face

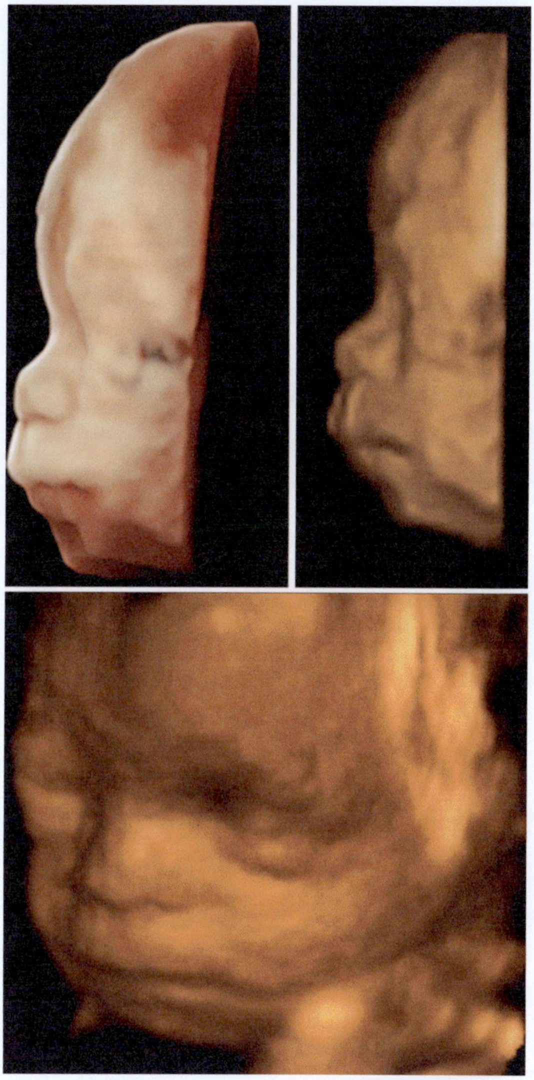

Fig. 4.89 FAS: surface rendering

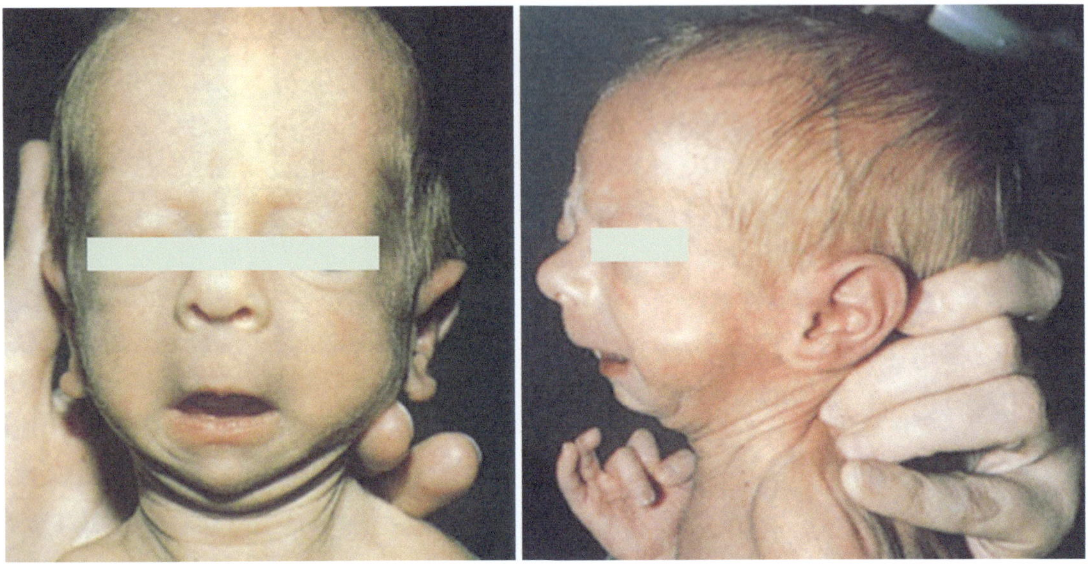

Fig. 4.90 FAS: post-natal appearance

4.7 Multiple-Level Pathologies

4.7.1 Cornelia de Lange Syndrome

This syndrome was described for the first time by Mrs. Cornelia de Lange in 1933. It has an estimated incidence of 1/45,000 to 1/62,500. It is often accompanied by severe mental retardation. Prenatal genetic diagnosis is difficult, which means that the diagnosing ultrasonography must know the markers well in order to analyze the dysmorphism and pinpoint cardinal markers and particularly evocative markers.

It should be noted that enlarged nuchal translucency could be seen in the first trimester. The classic ultrasound trio associates intrauterine growth retardation, limb defects and facial dysmorphism.

- Growth retardation appears as early as the second trimester, with its intensity seeming to correlate with the severity of mental retardation.
- Limb anomalies are variable. The most common are oligodactyly, ectrodactyly (Fig. 4.91), phocomelia (characteristic if present, Fig. 4.92), radial, cubital and ptergium anomalies.
- Facial dysmorphism is characterized by retrognathia that is more or less marked (Fig. 4.93), anteverted nares (Fig. 4.94), round nose, marked nasal bridge (Fig. 4.93), thin upper lip (Fig. 4.93), domed philtrum (Fig. 4.94), downturned corners of the mouth (Fig. 4.7), and short mandible (Fig. 4.98).

Discovery of all or any part of this trio of markers requires one to look for more subtle yet quite suggestive markers:

- Hypertrichosis (Fig. 4.99), with very long eyelashes (Fig. 4.100).
- A very small first metacarpal, which appears round (Fig. 4.101).
- The presence of a bony spine under the mandible (Fig. 4.102).

The genetics of this syndrome are complex: its expression is variable, it is most frequently transmitted sporadically, although a few cases of autosomal dominant family transmission have been described.

Mutations on three genes involved in chromosome cohesion (cohesion complex) are involved in this syndrome.

- Major form: the *NIPBL* gene (5p13.2) is mutated in approximately 50 % of patients.
- Minor form: the *SMC1A* gene (SMC1L1, Wp11.22-p11.21) is associated with a form of Cornelia de Lange Syndrome linked to the X, and the *SMC3* gene (10q25).

The *NIPBL* and *SMC1A* genes are involved in 65 % of cases. If this mutation is identified, a prenatal diagnosis can be suggested.

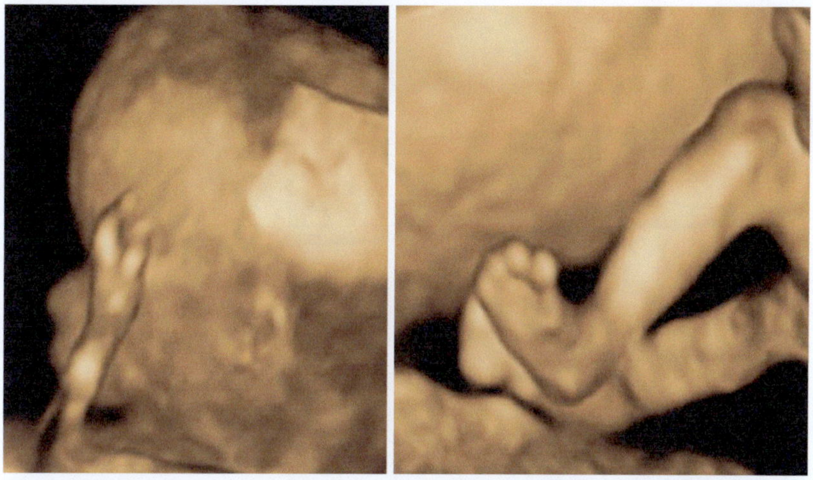

Fig. 4.91 Cornelia de Lange Syndrome: oligodactyly

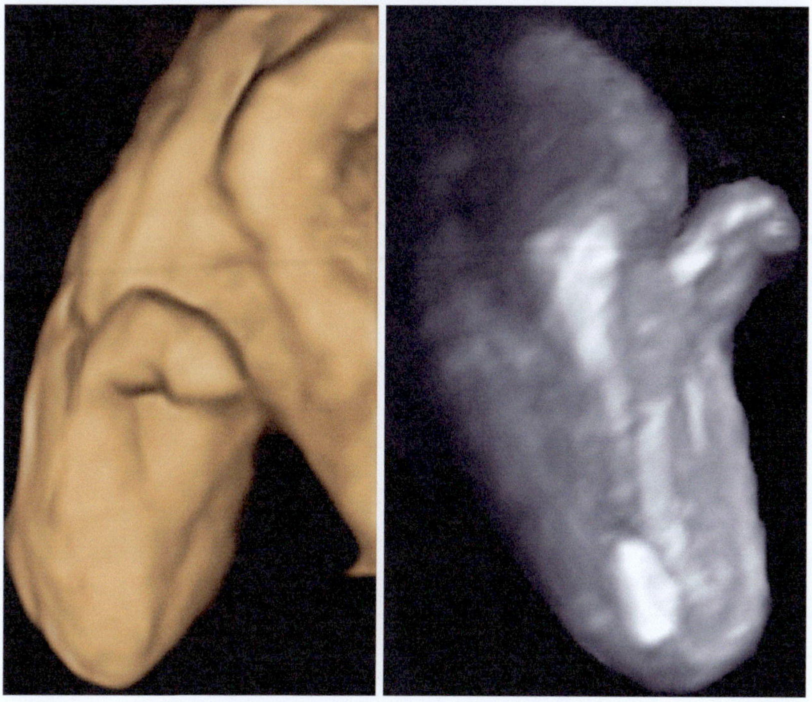

Fig. 4.92 Cornelia de Lange Syndrome: phocomelia

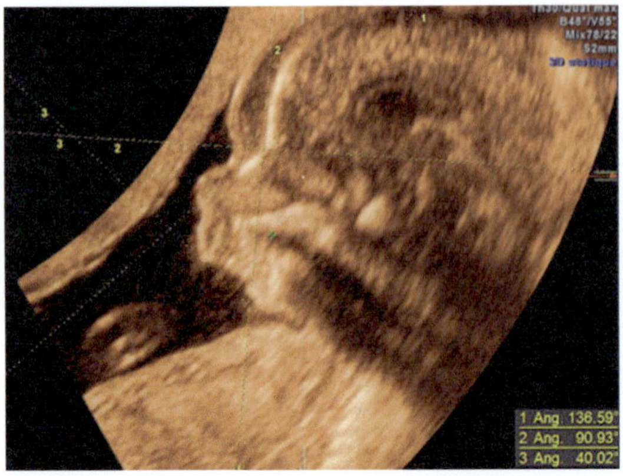

Fig. 4.93 Cornelia de Lange syndrome: retrognathia

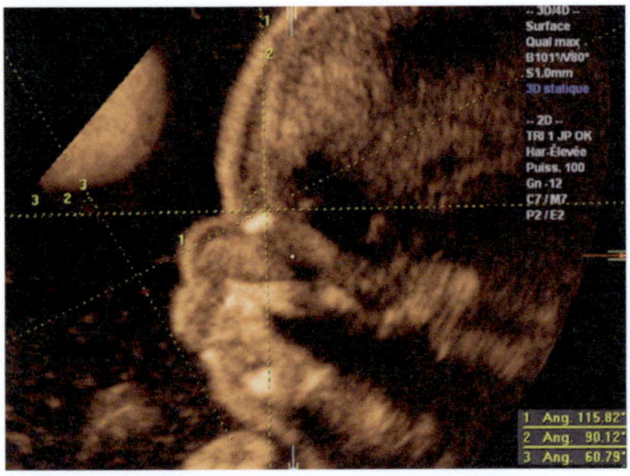

Fig. 4.94 Cornelia de Lange syndrome, 17 GW: marked nasal bridge, domed philtrum

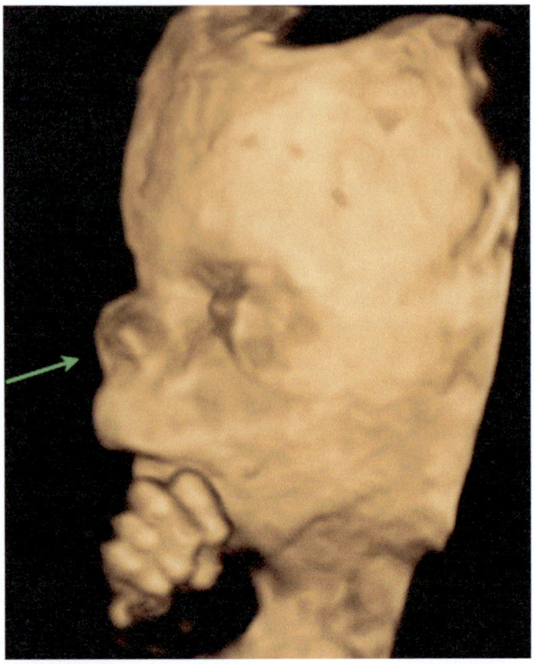

Fig. 4.95 Cornelia de Lange syndrome, 17 GW: anteverted nostrils, round nose

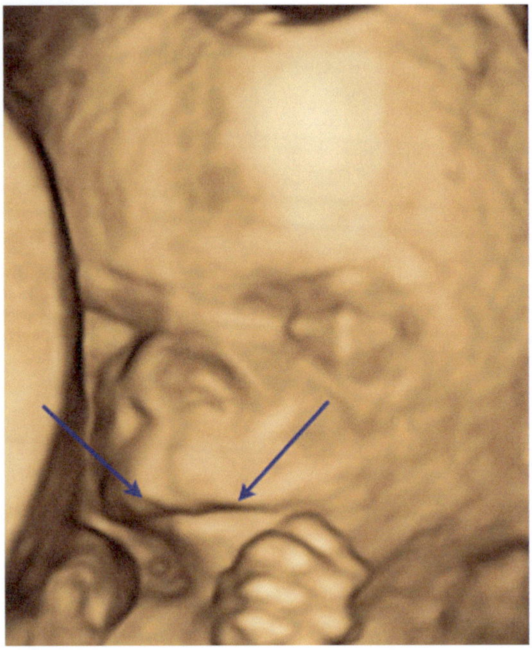

Fig. 4.96 Cornelia de Lange syndrome, 17 GW: thin upper lip

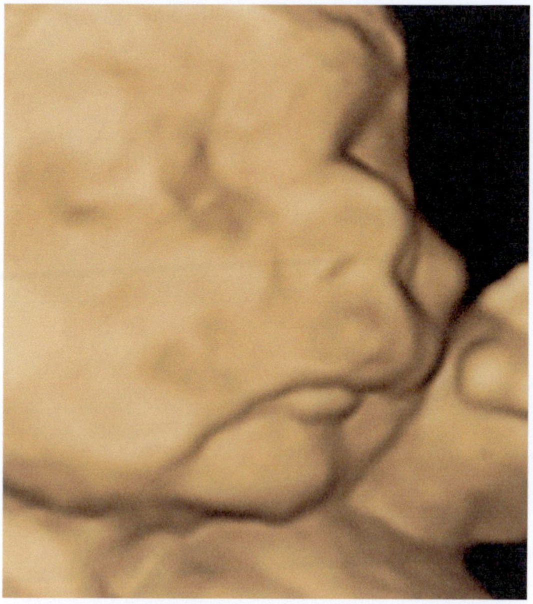

Fig. 4.97 Cornelia de Lange syndrome: downturned corners of the mouth

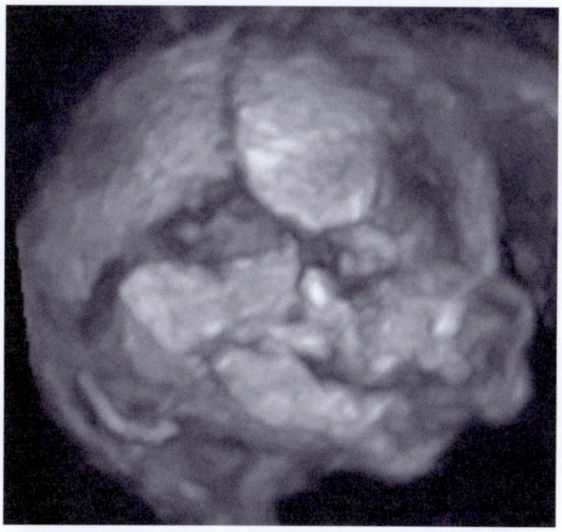

Fig. 4.98 Cornelia de Lange syndrome: short mandible

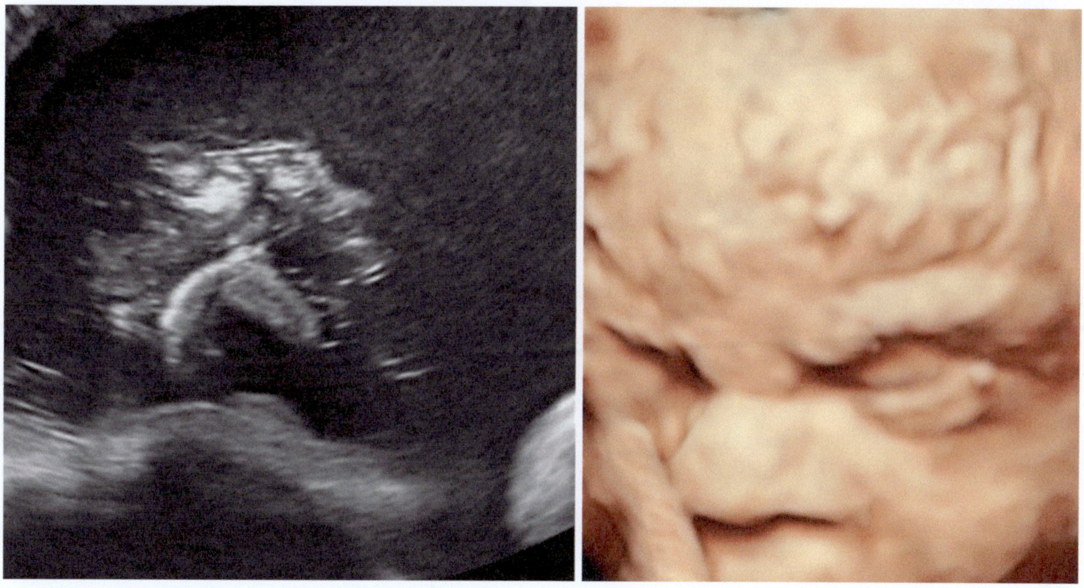

Fig. 4.99 Cornelia de Lange syndrome: hypertrichosis

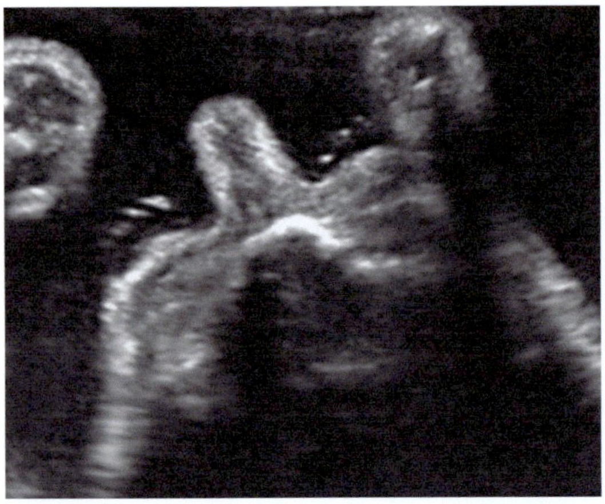

Fig. 4.100 Cornelia de Lange syndrome, 29 GW: very long eyelashes

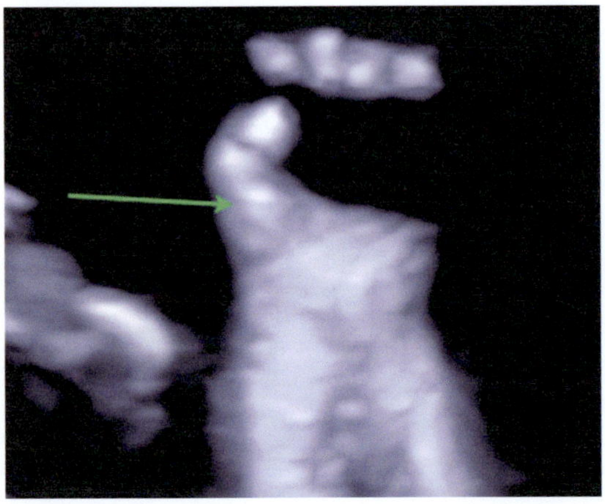

Fig. 4.101 Cornelia de Lange syndrome, 29 GW: round first metacarpal (*green arrow*)

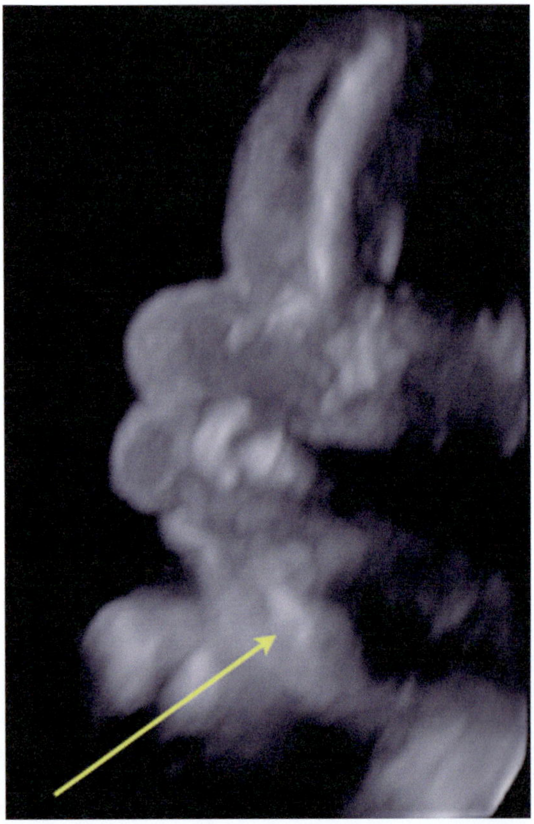

Fig. 4.102 Cornelia de Lange syndrome, 29 GW: bony spine under the mandible (*yellow arrow*)

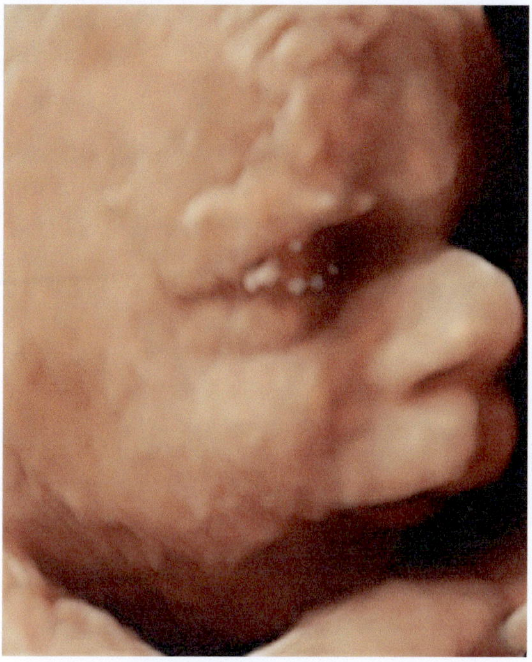

Fig. 4.103 Cornelia de Lange syndrome, 29 GW

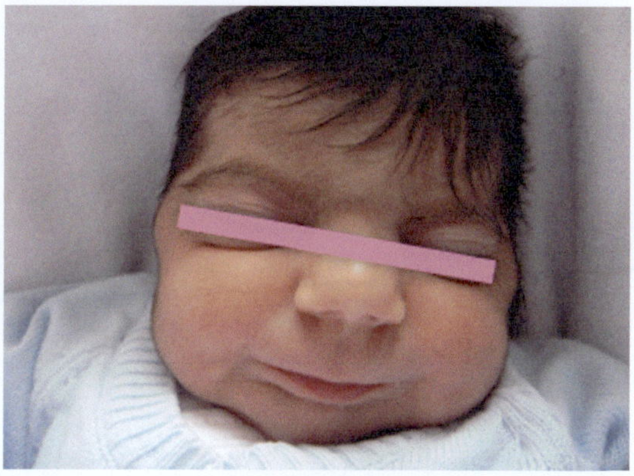

Fig. 4.104 Cornelia de Lange syndrome: post-natal appearance

4.7.2 Williams-Beuren Syndrome

Williams-Beauren syndrome was first described in 1961 and 1962 by J.C.P. Williams, who described the supravalvular aortic stenosis and A.J. Beuren who pinpointed the association with intellectual disabilities.

Today, this syndrome is still diagnosed during young childhood based on associated difficulties in school:

- Mental retardation (IQ<60)
- Cognitive deficiencies: disturbed perceptual functions in discord with proper language acquisition.
- Supravalvular aortic stenosis heart disease (in around 70% of cases) or peripheral pulmonary stenosis.
- Facial dysmorphism that combines a broad forehead, malar hypoplasia, full cheeks, long philtrum, and a wide mouth with an everted lower lip, giving an "elfin" appearance (Figs. 4.105, 4.107 and 4.108).

Progress in cytogenetics, and in particular techniques that enable targeted testing for lost genetic material that corresponds to mutations known to be responsible for syndromes associated with mental retardation, has led ultrasonographers to develop precise studies of subtle markers visible on ultrasound and to study in an objective manner dysmorphic anomalies that had previously gone unseen. Williams-Beuren syndrome is a typical example benefiting from these advancements: the dysmorphism is, indeed, analyzable on prenatal ultrasound and the subravalvular aortic stenosis can also be recognized (Figs. 4.106, 4.107, and 4.108).

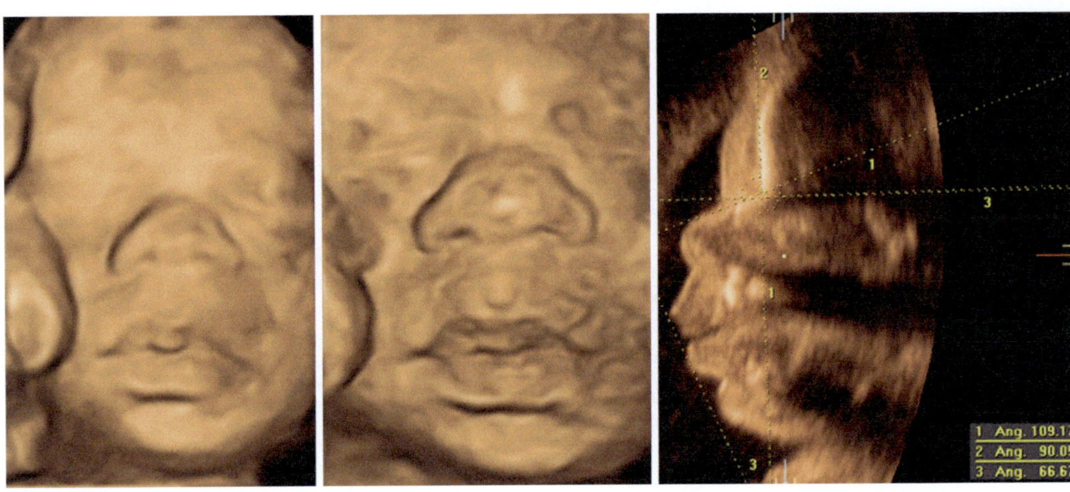

Fig. 4.105 Williams-Beuren syndrome, 30GW: small round nose, wide mouth with everted lower lip

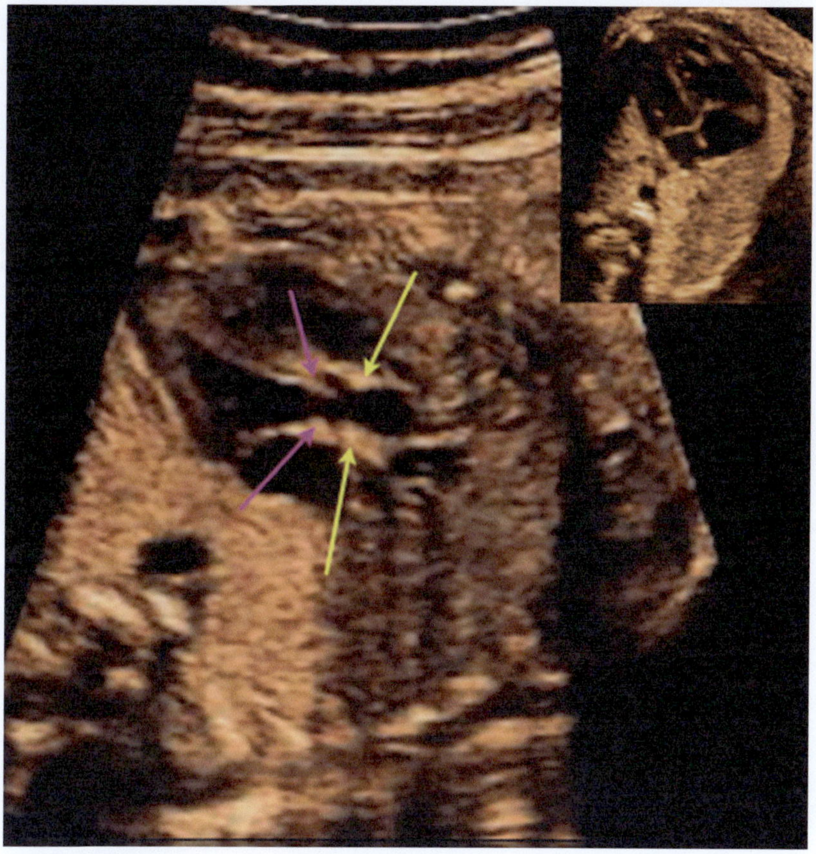

Fig. 4.106 Williams-Beuren syndrome, 30GW: supravalvular aortic stenosis (*pink arrows*, aortic valves; *yellow arrows*, stenosis)

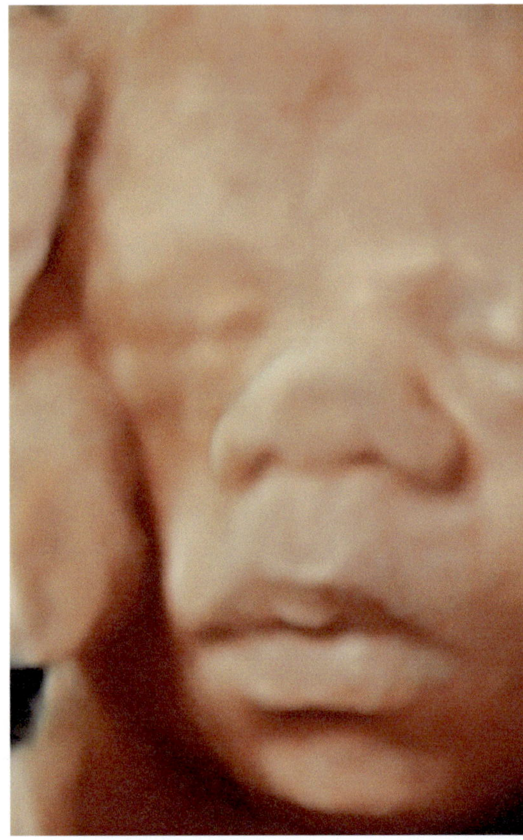

Fig. 4.107 Williams-Beuren syndrome, 30GW: HD Live surface rendering

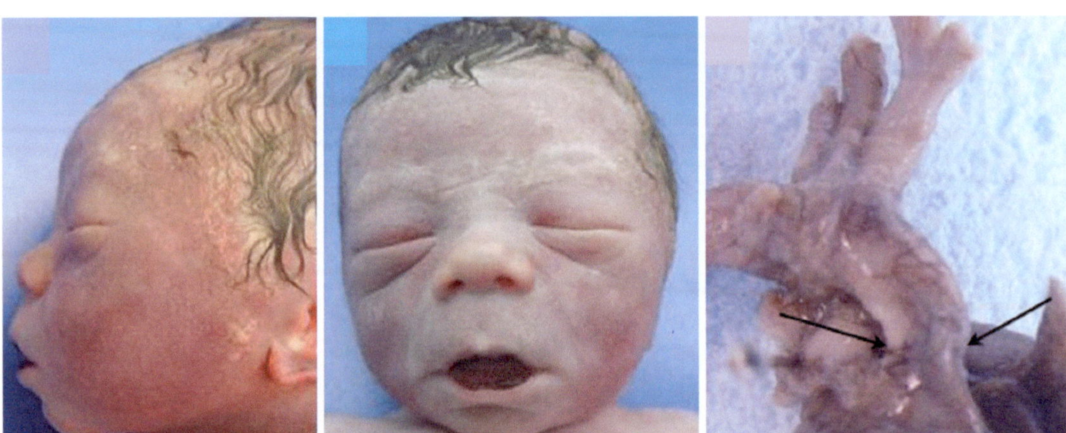

Fig. 4.108 Williams-Beuren syndrome, 30GW: fetopathological appearance. Left and central images show facial dysmorphism, while the right image shows the aortic stenosis

4.8 A Few Additional Syndromes

4.8.1 CHARGE Association

It may seem odd in a book focused on the fetal face to discuss the CHARGE association of birth defects, yet the presence of external ear and eye anomalies fit in the framework of facial dysmorphism and we considered that pertinent enough to include this association in this chapter.

The acronym CHARGE stands for a set of birth defects combining Coloboma of the eye (Fig. 4.109), Heart defects (tetralogy of Fallot, double outlet ventricle, atrioventricular canal defect, etc.), Atresia of the choanae, Retardation of growth and/or development, Genital hypoplasia, and Ear abnormalities.

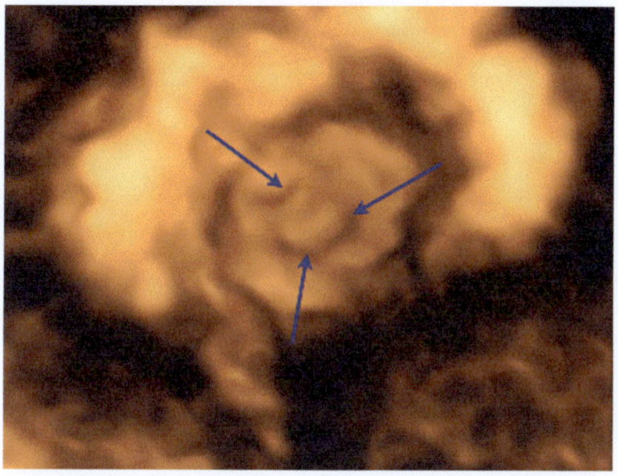

Fig. 4.109 CHARGE association: coloboma of the retina (virtual fundus exam technique)

Incidence is 1/8500 to 1/12,000, with mutations on the CDH7 gene being responsible in 75 % of cases.

Discovery of a heart defect associated with an anomaly of the posterior fossa (vermian hypoplasia Fig. 4.110) should lead to a search for dysmorphism such as auricular anomalies (they are often small with a rather rectilinear upper edge (Fig. 4.111), microphthalmia (Fig. 4.112), colomba (Fig. 4.110). All cases present arhinencephaly (Figs. 4.113 and 4.114) and anomalies of the semicircular canals—hypoplasia or more frequently agenesis (Fig. 4.115). One can use ultrasound to look for these two latter anomalies, and then confirm with MRI.

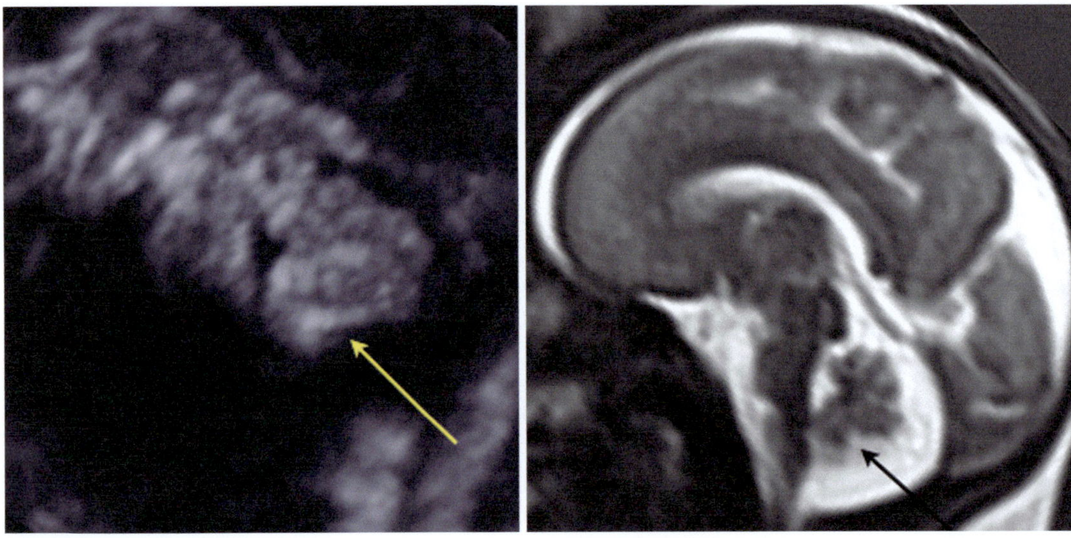

Fig. 4.110 CHARGE association: vermian anomaly (defect in lower region)

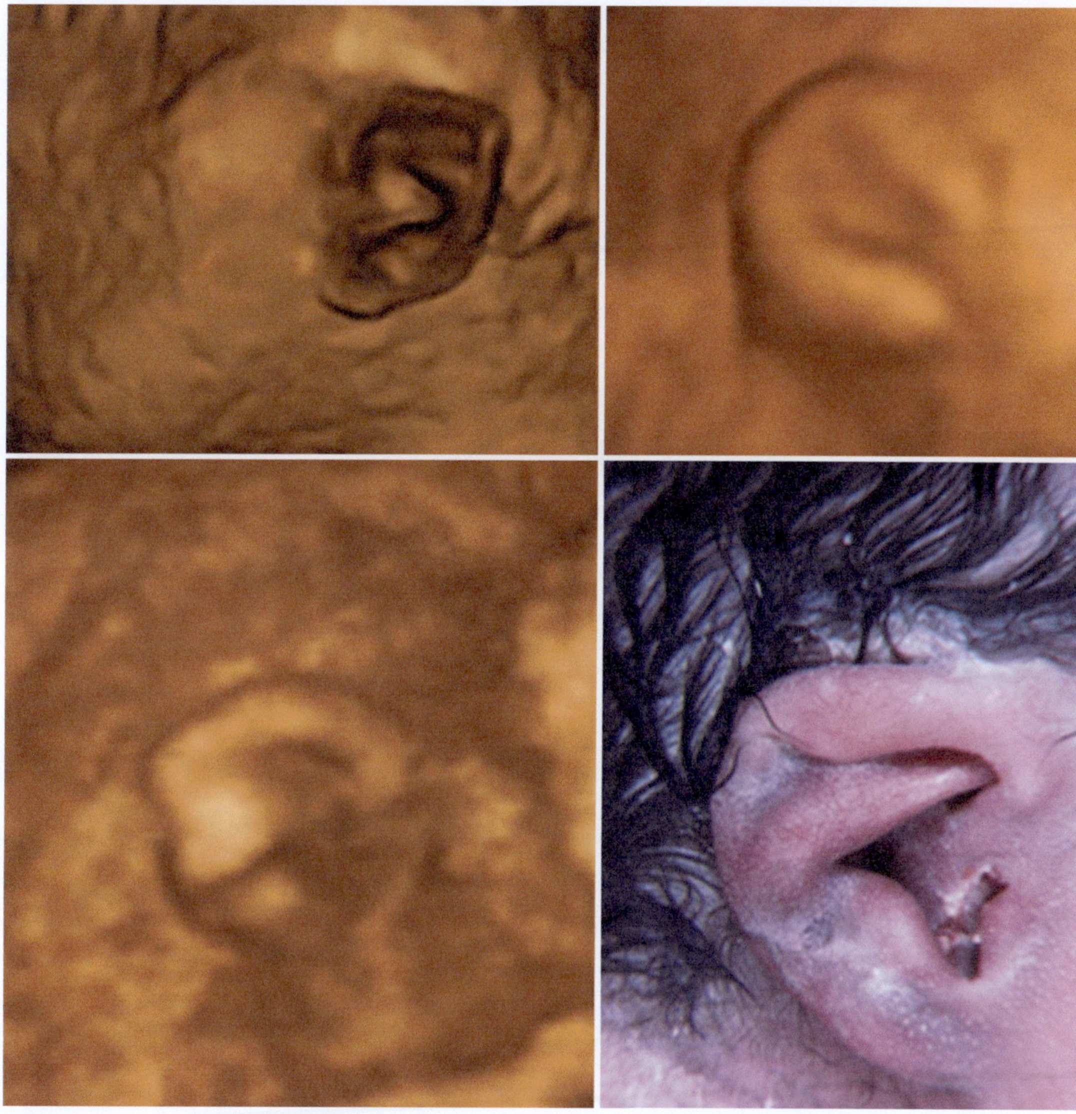

Fig. 4.111 CHARGE association: ear anomalies

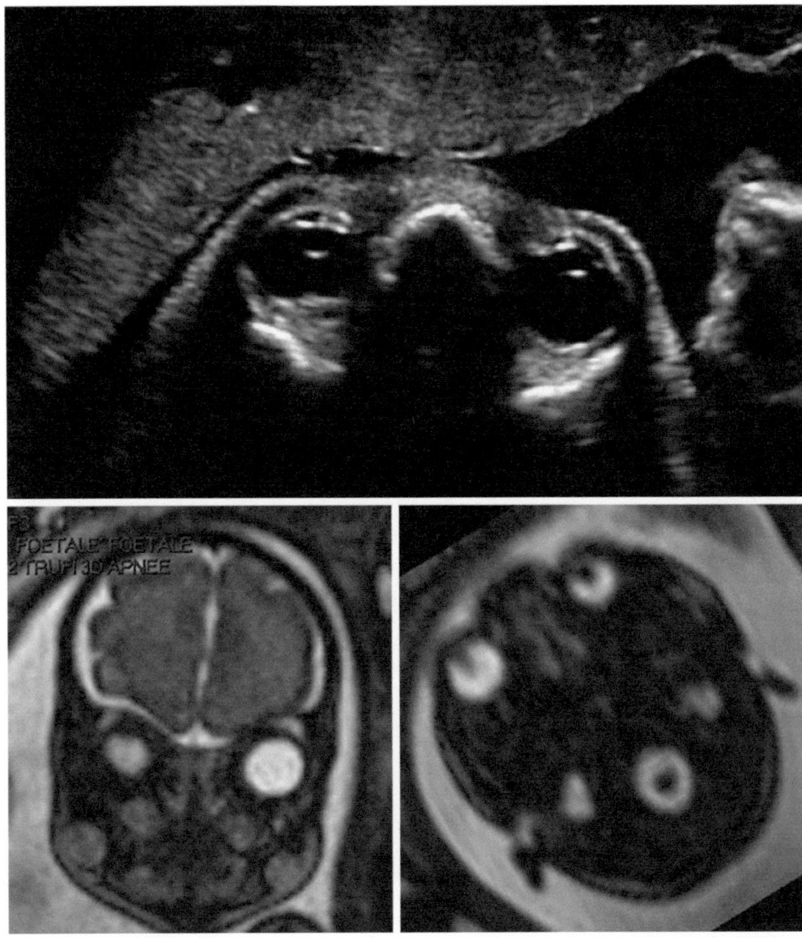

Fig. 4.112 CHARGE association: microphthalmia

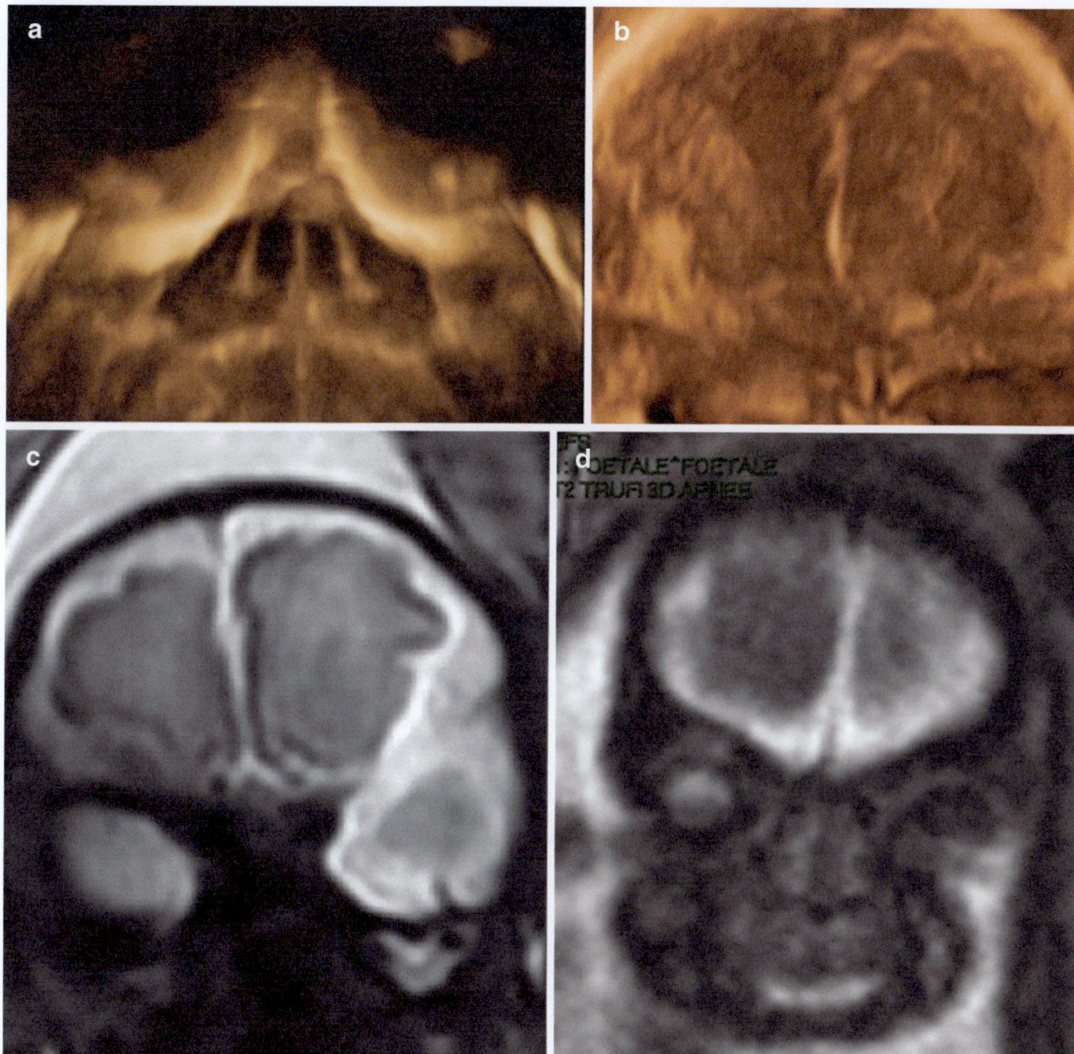

Fig. 4.113 CHARGE association: (**a**). olfactive grooves present on ultrasound; (**b**) absence of grooves (CHARGE), (**c**) bulbs present on MRI, (**d**) absence of bulbs (CHARGE)

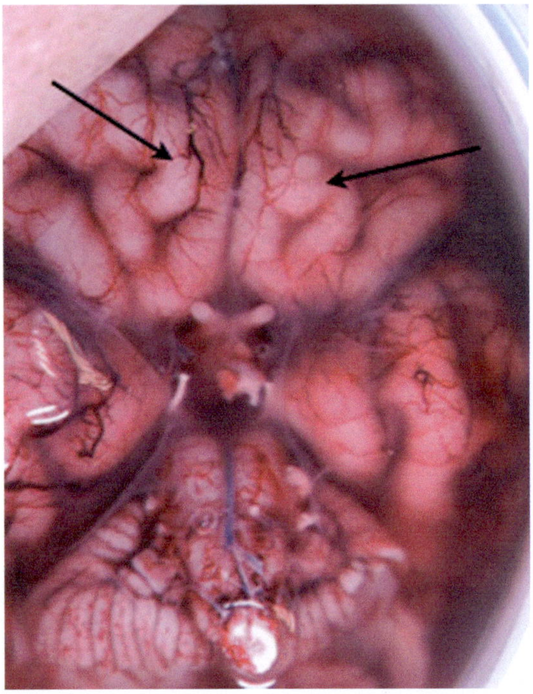

Fig. 4.114 CHARGE association: fetopathological appearance of arhinencephaly (olfactive bulbs are not visible, *arrows*)

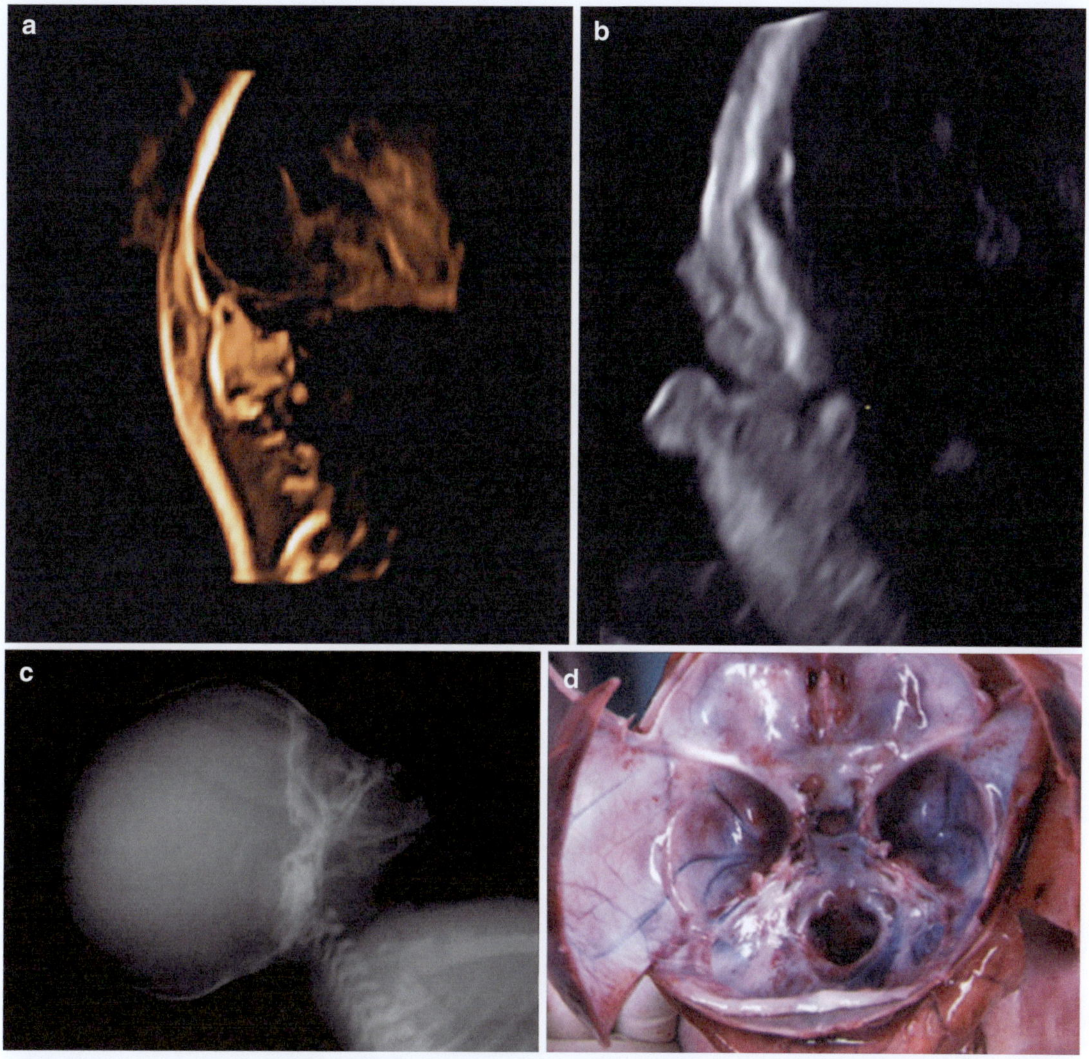

Fig. 4.115 CHARGE association: (**a**) normal semicircular canals (SCC), (**b**) very hypoplastic SCC in case of CHARGE, (**c**). absence of SCC (RX CHARGE), (**d**). fetopathological image, absence of SCC

4.8.2 Noonan Syndrome

This syndrome has an estimated incidence of between 1/1000 and 1/2500. This diagnosis should be considered when a fetus presents increased nuchal translucency thickness (especially if persistent, Fig. 4.115) with a normal karyotype, but also if a fetus presents hydamnios, pleural effusion or edemas if not ordinary cause can be found.

The characteristics of this syndrome are:

- Cardiovascular anomalies: pulmonary stenosis, hypertrophic cardiomyopathy
- Scoliosis-like vertebral anomalies
- Umbilical vein anomalies such as persistent right umbilical vein
- Cryptorchidism
- Facial dysmorphism associated with hypertelorism (Fig. 4.116), downwardly slanting palpebral fissures (Fig. 4.119), a deep vertical groove in the middle of the philtrum (Figs. 4.121 and 4.122), low-set ears with a thick, backward rotating helix (Figs. 4.118, 4.119, and 4.120), and a short, thick neck (Fig. 4.123).

This syndrome is inherited in an autosomal dominant pattern. For half the patients, there is a missense mutation in gene *PTPN11* localized on chromosome 12 leading to a gain in phsophotyrosine function. Other genes involved have recently been uncovered.

This chapter highlights the progress that has been made in the prenatal study of fetal dysmorphism thanks to the innumerable technical advances that have been made not only in imagery, but also in genetic medicine, cytogenetics and molecular genetics.

Working in a pluridisciplinary team allows the ultrasonographer to develop his or her knowledge and to sharpen his or her curiosity looking for markers that may be subtle but that are present. This work must be accompanied with an extremely rigorous methodology, as I hope we have demonstrated.

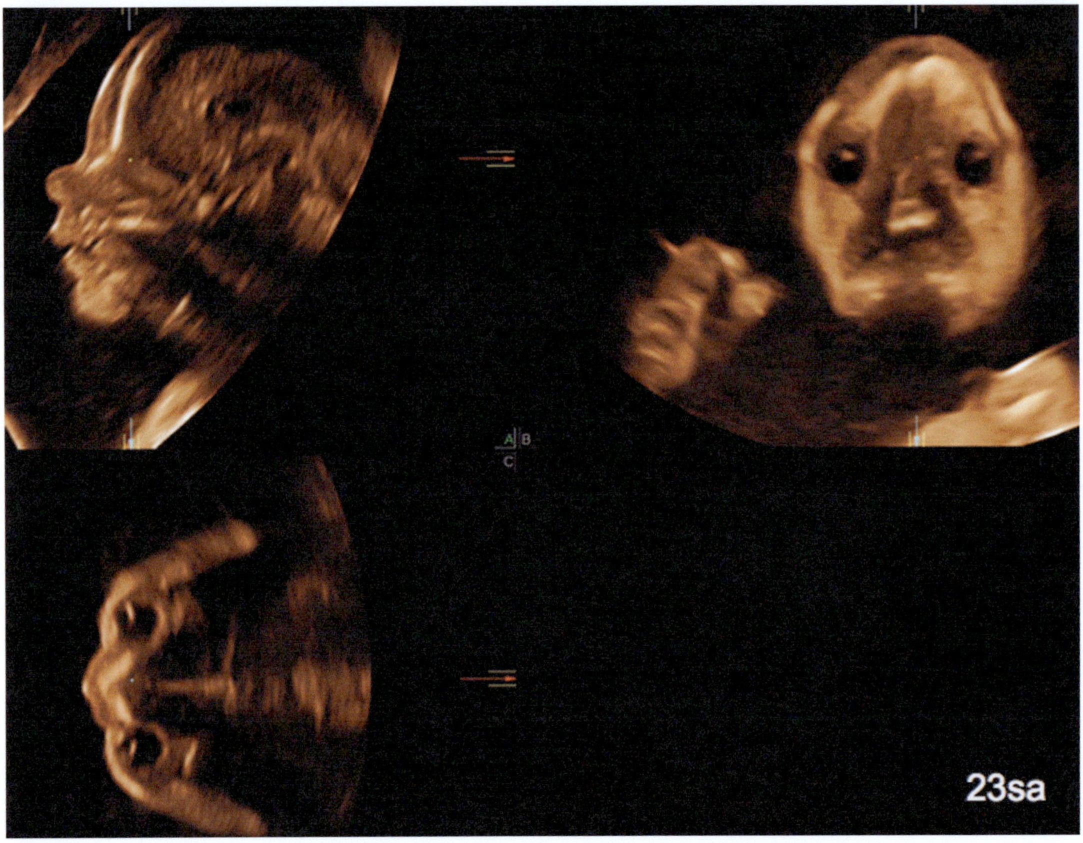

Fig. 4.116 Noonan syndrome: multiplanar view, hypertelorism

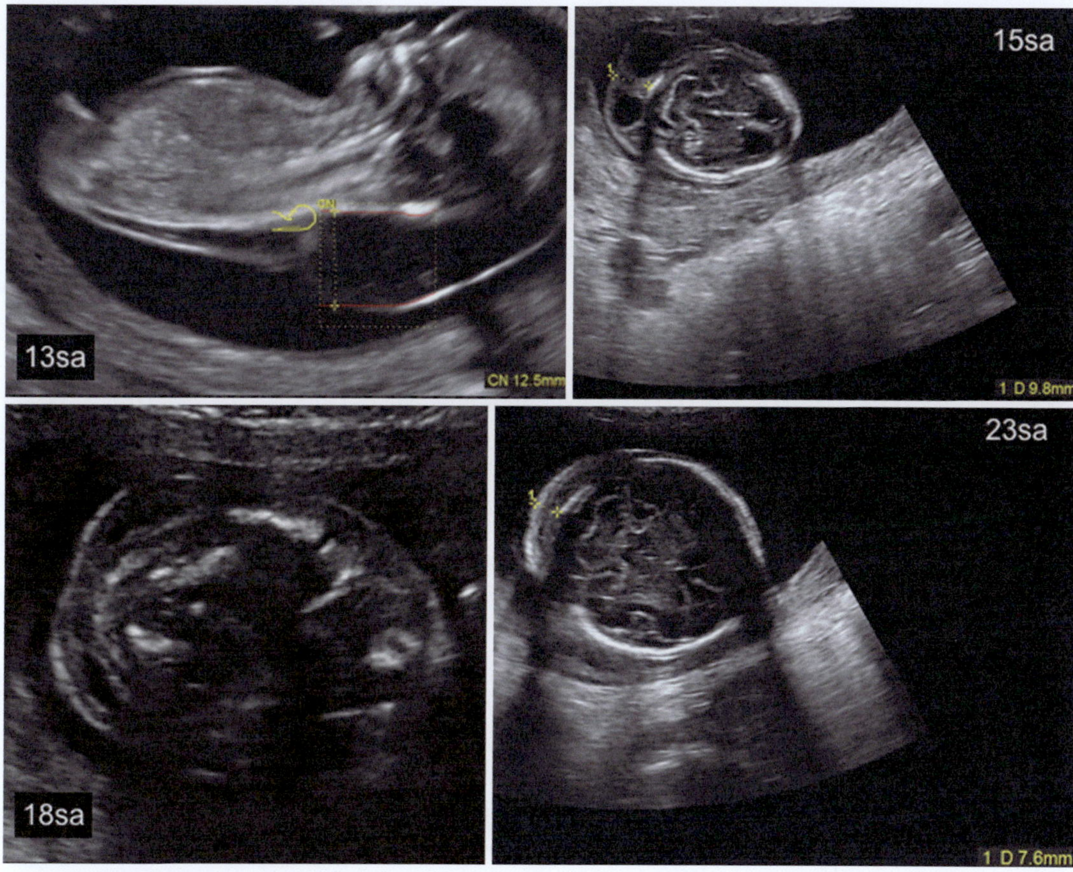

Fig. 4.117 Noonan syndrome: evolution of neck thickness

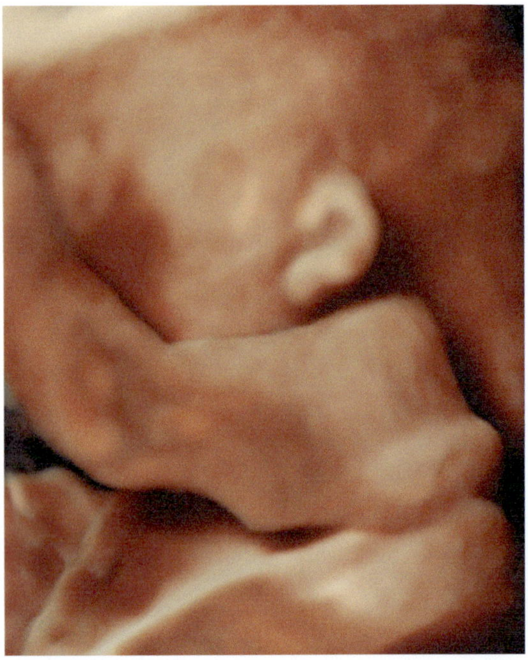

Fig. 4.118 Noonan syndrome: ears 23GW, poorly folded, low-set and rotated backward

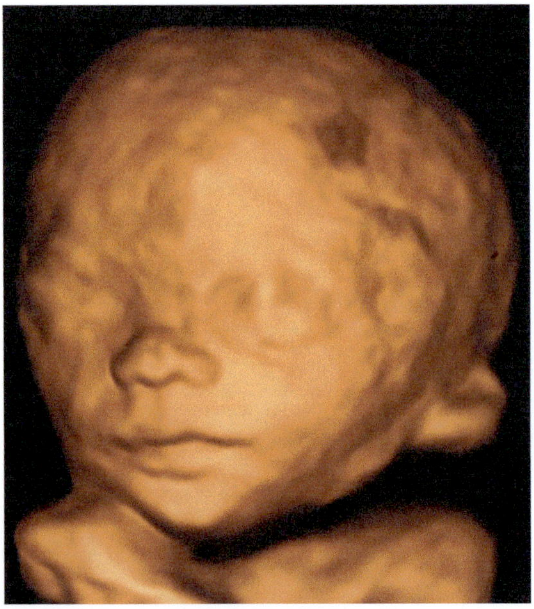

Fig. 4.119 Noonan syndrome: ears at 23GW and postnatal

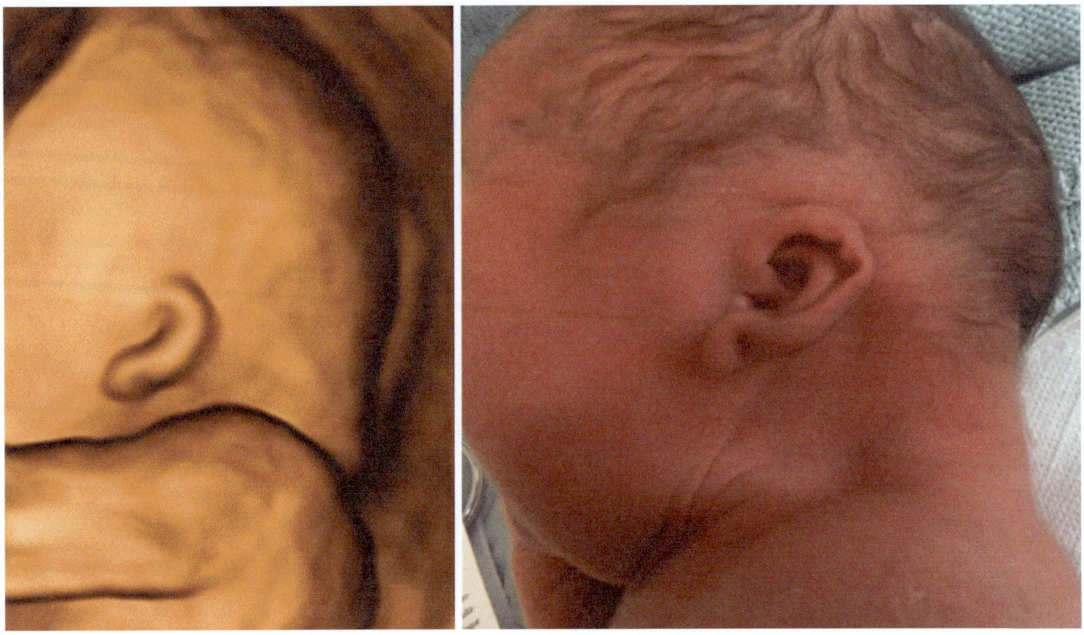

Fig. 4.120 Noonan syndrome, 28GW: dysmorphism, hypertelorism, downward slanting palpebral fissures

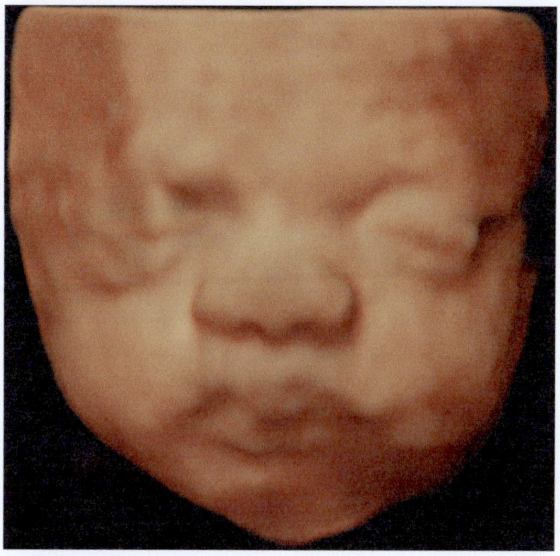

Fig. 4.121 Noonan syndrome, 23GW: Deep grooves in Cupid's arc

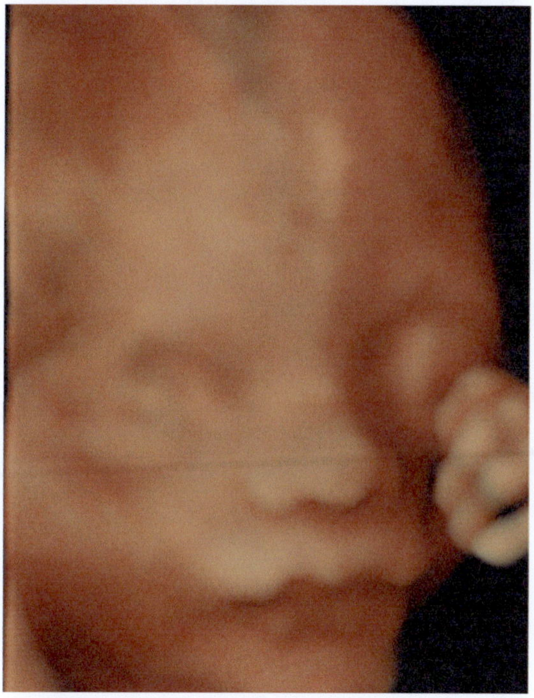

Fig. 4.122 Noonan syndrome: 21 SA: other deep grooves in Cupid's arc

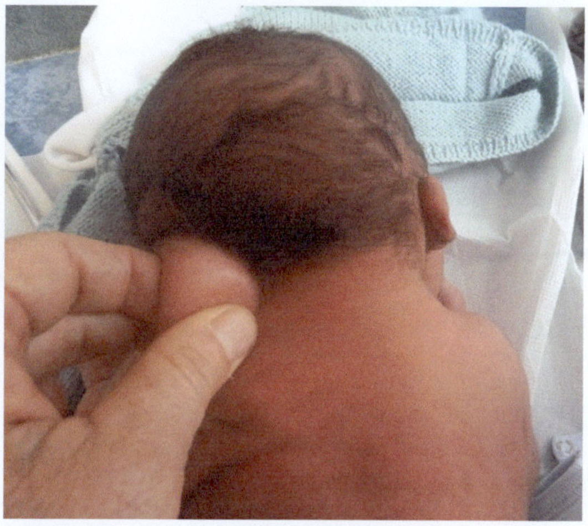

Fig. 4.123 Noonan syndrome: postnatal, thick neck

References

How to Define Objective Facial Dysmorphism

1. Cassidy SB, Allanson JE. Management of genetic syndromes. Hoboken: Wiley; 2010. TRI 21: 309–335, TRI 18, TRI 13: 807–823.
2. Smiths, Kenneth LJ. Recognizable paterns of human malformation. Elsevier Saunders. TRI 21:7–9, Tri 18 :13–17, Tri 13:18–21.
3. Baraitzer M, Winter RM. Color atlas of congenital malformation syndromes. Baltimore/London: Mosby-Wolfe Editors; 1996. Tri 21:1–2, Tri 18:2–3, Tri 13: 3.
4. Leon E, Jamal SM, Zou YS, Milunsky JM. Partial trisomy 8: mosaicism due to a pseudoisodicentric chromosome 8. Am J Med Genet A. 2011;155A(7):1740–4. doi:10.1002/ajmg.a.34073. Epub 2011 Jun 10.
5. Morichon-Delvallez N. Orphanet. 2010.

Upper Face Pathologies

6. Xu J, Freeman V, Carter RF, Paes B, Heshka T, Nowaczyk JM. Wolf-Hirschorn syndrome resulting from partial monosomy 4p/trisomy 9p. Am J Med Genet. 2000;93(4):285–9.
7. Suzanne BC, Judith EA. Deletion 4p: Wolf-Hirschhorn syndrome. In: Management of Genetic Syndromes. Hoboken: Wiley; 2010. p. 249–61.
8. Alain V. Orphanet: wolf-hirschhorn syndrome Dagmar Wieczorek. 2003.
9. Winter RM, Baraitzer M, Baraitzer M. Color atlas of congenital malformation syndromes. London/Baltimore: Mosby-Wolfe Editors; 1996. p. 6–7.
10. Dikkeboom CM, Roelfsema NM, Van Adrichem LNA, Wladimiro JW. The role of three-dimensional ultrasound in visualizing the fetal cranial sutures and fontanels during the second half of pregnancy. Ultrasound Obstet Gynecol. 2004;24(4):412–6.
11. Delahaye S, Bernard JP, Rénier D, Ville Y. Prenatal ultrasound diagnosis of fetal craniosynostosis. Ultrasound Obstet Gynecol. 2003;21(4):347–53.
12. Bernard JP, Levaillant JM. Diagnostic prénatal des craniosténoses. Neurochirurgie. 2006;52(2-3):246–58.
13. Baraitzer M, Winter RM. Color atlas of congenital malformation syndromes. London/Baltimore: Mosby-Wolfe Editors; 1996. Crouzon: 42–43, Apert: 43–44.
14. Suzanne B, Cassidy J E. Management of gene c syndromes. Wiley/Allanson; 2010. 227–37.
15. Chaoui R, Kalache KR, Heling KS, Tennsted T, Bommer C, Korner H. Absent or hypoplastic thymus on ultrasound: a marker for dele on 22q11.2 in fetal cardiac defects. Ultrasound Obstet Gynecol. 2002;20:546–52.
16. Chaoui R, Heling KS, Sarut Lopez A, Thiel G, Karl K. The thymic–thoracic ratio in fetal heart defects: a simple way to identify fetuses at high risk for microdeletion 22q11. Ultrasound Obstet Gynecol. 2011;37(4):397–403.
17. Paladini D. How to identify the thymus in the fetus: the thy-boxn. Ultrasound Obstet Gynecol. 2011;37(4):488–92.
18. Stoll C, Clemen M. Prenatal diagnosis of dysmorphic syndromes by routine fetal ultrasound examination across Europe. Ultrasound Obstet Gynecol. 2003;21(6):543–51. The Philadelphia story: the 22 .q.1.1 Dele on Report on 250 patients. Mc Donald et al Genet Couns 10:11–24.
19. Suzanne B, Cassidy Judith E. Management of genetic syndromes. Hoboken: Wiley; 2010. p. 263–84.
20. Kenneth LJ, editor. Smiths recognizable patterns of human malformation. Elsevier Saunders; p. 298–9.
21. Den Hollander NS, Wessels MW, Los FJ, Ursem NTC, Niermeijer MF, Wladimiro JW. Congenital microcephaly detected by prenatal ultrasound: genetic aspects and clinical significance. Ultrasound Obstet Gynecol. 2000;15(4):282–7.
22. Le Ray C, Viot G, Tsatsaris V, Adamsbaum C, Grangé G. Late onset microcephaly: failure of prenatal diagnosis. Ultrasound Obstet Gynecol. 2004;24(4):481–3.
23. Schwärzler P, Homfray T, Bernard JP, Bland JM, Ville Y. Late onset microcephaly: failure of prenatal diagnosis. Ultrasound Obstet Gynecol. 2003;22(6):640–2.
24. Pilu G, Falco P, Milano V, Perolo A, Bovicelli L. Prenatal diagnosis of microcephaly assisted by vaginal sonography and power Doppler. Ultrasound Obstet Gynecol. 1998;11(5):357–60.
25. Kenneth LJ, editor. Smiths recognizable patterns of human malformation. Elsevier Saunders; p. 870–1.

Mid-Face Pathologies

26. Ruano R, Molho M, Roume J, Ville Y. Prenatal diagnosis of fetal skeletal dysplasias by combining two-dimensional and three-dimensional ultrasound and intrauterine three-dimensional helical computer tomography. Ultrasound Obstet Gynecol. 2004;24(2):134–40.
27. Moeglin D, Benoit B. Three-dimensional sonographic aspects in the antenatal diagnosis of achondroplasia. Ultrasound Obstet Gynecol. 2001;18(1):81–3.

28. Macé G, Sonigo P, Cormier-Daire V, Aubry MC, Martinovic J, Elie C, Gonzales M, Carbonne B, Dumez Y, Le Merrer M, Brunelle F, Benachi A. Use of 3D helical computed tomography in the prenatal diagnosis of fetal skeletal dysplasia. Ultrasound Obstet Gynecol. 2013;42(2):161–8.
29. Althuser M, Bault JP, Fontanges M, Fries N. Epiphysis and metaphysis prenatal ultrasound examination: normal aspect and a new marker for achondroplasia. Ultrasound Obstet Gynecol. 2005;26(4):393.
30. Kenneth LJ, editor. Smiths recognizable patterns of human malformation. Elsevier Saunders. p. 390–1.
31. Suzanne B, Cassidy JE. Management of gene c syndromes. Wiley; 2010, p. 17–37.
32. Benaicha A, Dommergues M, Jouannic JM, Jacquette A, Alexandre M, Le Merrer M, Ducou Le Pointe H, Garel C. Prenatal diagnosis of brachytelephalangic chondrodysplasia punctata: case report. Ultrasound Obstet Gynecol. 2009;34(6):724–6.
33. Cook K, Prefumo F, Pres F, Homfray T, Campbell S. The prenatal diagnosis of Binder syndrome before 24 weeks of gestation: case report. Ultrasound Obstet Gynecol. 2000;16(6):578–81.
34. Vasquez MP. Orphanet. Dysplasie maxillo-faciale. 2006.
35. Kölble N, Sobetzko D, Ersch J, Stallmach T, Eich G, Huch R, Huch A, Superti-Furga A, Wisser J. Diagnosis of skeletal dysplasia by multidisciplinary assessment: a report of two cases of thanatophoric dysplasia. Ultrasound Obstet Gynecol. 2002;19(1):92–8.
36. Schild RL, Hunt GH, Moore J, Davies H, Horwell DH. Antenatal sonographic diagnosis of thanatophoric dysplasia: a report of three cases and a review of the literature with special emphasis on the differential diagnosis. Ultrasound Obstet Gynecol. 1996;8(1):62–7.
37. Machado LE, Bonilla-Musoles F, Osborne NG. Thanatophoric dysplasia. Ultrasound Obstet Gynecol. 2001;18(1):85–6.
38. Ngo C, Viot G, Aubry MC, Tsatsaris V, Grange G, Cabrol D, Pannier E. First-trimester ultrasound diagnosis of skeletal dysplasia associated with increased nuchal translucency thickness. Ultrasound Obstet Gynecol. 2007;30(2):221–6.
39. Baraitzer M, Winter RM. Color atlas of congenital malformation syndromes. Mosby-Wolfe Editors. 1996;99–100.
40. Fong BF, de Vries JIP. Obstetric aspects of the Prader–Willi syndrome. Ultrasound Obstet Gynecol. 2003;21(4):389–92.
41. Muñoz A, Scazzochio E, Devesa M, Prats P, del Campo M, Torrents M, Serra B. Prenatal features in Prader Willi syndrome with two and three dimensional sonographic assessment. Ultrasound Obstet Gynecol. 2006;28(4):580.
42. Suzanne B, Cassidy JEA. Management of genetic syndromes. Wiley; p. 625–650.
43. Diene G, Pinto G, Postel-Vinay A, Tauber M. ORPHANET. Syndrome de Prader-Willi. 2007.
44. Baraitzer M, Winter RM. Color atlas of congenital malforma on syndromes. Mosby-Wolfe Editors. 1996;160–161.
45. Kenneth LJ, editor. Smiths recognizable patterns of human malformation. Elsevier Saunders; p. 223–7.
46. Verloes A, Lesenfants S, Barr M, Grange DK, Journel H, Lombet J, Mortier G, Roeder E. Fronto-otopalatodigital osteodysplasia: clinical evidence for a single entity encompassing Melnick-Needles syndrome, otopalatodigital syndrome types 1 and 2, and frontometaphyseal dysplasia. Am J Med Genet CS. 2000;90(5):407–22.
47. Orphanet. Syndrome oto-palato-digital. 2005.
48. Robertson S. Otopalatodigital spectrum disorders, GeneReviews™.

Lower Face Pathologies

49. Burglen L, Soupre V, Diner PA, Gonzalès M, Vazquez M-P. Dysplasies oto-mandibulaires: génétique et nomenclature des formes syndromiques. Ann Chir Plast Esthet. 2001;46:400–9.
50. Thomson A. Notice of several cases of malformation of the external ear, together with experiments on the state of hearing in such persons. Mont J Med Sci. 1846–1847;7:420.
51. Collins ET. Cases with symmetrical congenital notches in the outer part of each lower lid and defective development of the malar bones. Trans Ophthal Soc UK. 1933;20:190–2.
52. Franceschetti A, Klein D. Mandibulo-facial dysostosis: new hereditary syndrome. Acta Ophthalmol. 1949;27:143–224.
53. Cannistra C, Barbet JR, Houette A, Loro Marchese J, Ianne G. Mandibulo-facial dysostosis: comparison study of a neonate with mandibulo-facial dysostosis and a normal neonate. J Craniomaxillofac Surg . 1998;26:92–7.
54. The Treacher Collins Syndrome Collaborative Group. Positional cloning of a gene involved in the pathogenesis of Treacher Collins syndrome. Nat Genet. 1996;12:130–6.
55. Trainor PA, Dixon J, Dixon MJ. Treacher Collins syndrome: etiology, pathogenesis and prevention. Eur J Hum genet CS. 2008.
56. Kaga K, Takegoshi H, Yamasoba T, Nakamura M, Kaneko M, Ino K. Aplasia of zygomatic arch and dislocation of temporomandibular joint capsule in Treacher-Collins syndrome: three-dimensional reconstruction of computed tomographic scans. Int J Pediatr Otorhinolaryngol. 2003;67(11):1189–94.
57. Edwards SJ, Fowlie A, Cust MP, Liu DTY, Young ID, Dixon MJ. Prenatal diagnosis in Treacher Collins syndrome using combined linkage analysis and ultrasound imaging. J Med Genet. 1996;33:603–6.

58. Hsu TY, Hsu JJ, Chang SY, Chang MS. Prenatal three-dimensional sonographic images associated with Treacher Collins syndrome. Ultrasound Obstet Gynecol. 2002;19(4):413–22.
59. Rotten D, Levaillant JM, Marnez H, Ducoule Le pointe H, Vicaut E. The fetal mandible: a 2D and 3D sonographic approach to the diagnosis of retrognathia and micrognathia. Ultrasound Obstet Gynecol. 2002;19(2):122–30.
60. Rotten D, Levaillant JM. Two- and three-dimensional sonographic assessment of the fetal face. 1. A systematic analysis of the normal face. Ultrasound Obstet Gynecol. 2004;23(3):224–31.
61. Rotten D, Levaillant JM. Two- and three-dimensional sonographic assessment of the fetal face. 2. Analysis of cleft lip, alveolus and palate. Ultrasound Obstet Gynecol. 2004;24(4):402–11.
62. Faure JM, Captier G, Bäumler M, Boulot P. Sonographic assessment of normal fetal palate using three dimensional imaging: a new technique. Ultrasound Obstet Gynecol. 2007;29(2):159–65.
63. Centre de référence des maladies endocriniennes rares de la croissance. APHP.
64. Gicquel C. Orphanet: syndrome de Beckwith-Wiedemann. 2005.
65. Gicquel C, Rossignol S, Le Bouc Y. Orphanet: encyclopédia Mars 2005 Beckwith-Wiedemann syndrome. D. Lacombe, Scientific Editor.
66. Baraitzer M, Winter RM. Color atlas of congenital malformation syndromes. Mosby-Wolfe Editors. 1996;164–165.
67. Suzanne B, Cassidy AJE. Management of gene c syndromes. Wiley; 139–148.
68. Nyberg D, et al. Diagnostic of fetal anomalies. Philadelphia: Lippinco Williams & Wilkins; 2003. p. 338.
69. Kfir M, Yevtushok L, Onishchenko S, Wertelecki W, Bakhireva L, Chambers CD, Jones KL, Hull AD. Can prenatal ultrasound detect the effects of in-utero alcohol exposure? A pilot study. Ultrasound Obstet Gynecol. 2009;33(6):683–9.
70. Diagnostic; Grossesse; Prise en charge; Syndrome d'alcoolisation foetale sub l D. Hôpital Jeanne de Flandre, Lille France Thésaurus TOXIBASE. http://bdoc.ofdt.fr/pmb/opac_css/index.php?lvl=no ce_display&id=21585.
71. Larroque B, Kaminski M, Dehaene P, Sub ID, Querleu D. Prenatal alcohol exposure and markers of minor neurological dysfunction at preschool age. Dev Med Child Neurol. 2000;42(8):508–14.
72. Larroque B, Kaminski M, Lelong N, Sub ID, Dehaene P. Effects on birthweight of alcohol and caffeine consumption during pregnancy. Am J Epidemiol. 1993;9:941–50.
73. Larroque B, Dehaene P, Sub ID, Delfosse MJ, Querleu D. Prenatal alcohol exposure and psychomotor development at preschool age American. J Public Health. 1995;85:1654–61.
74. Larroque B, Kaminski M. Prenatal alcohol exposure and development at preschool age: main results of a French study. Alcohol Clin Exp Res. 1998;22:295–303.
75. Kaminski M, Lelong N, Bean K, Chwalow J, Sub l D. Exposition prénatale à l'alcool: données … – iPubli-Inserm. www.ipubli.inserm.fr/bitstream/handle/10608/168/?sequence=13.

Multiple-Level Pathologies

76. Spaggiari E, Vuillard E, Khung-Savatovsky S, Muller F, Oury JF, Delezoide AL, Guimiot F. Ultrasound detection of eyelashes: a clue for prenatal diagnosis of Cornelia de Lange Syndrome. Ultrasound Obstet Gynecol. Accepted manuscript online: 20 Aug 2012.
77. David RF, Antonie D. Cornelia de Lange Syndrome. Management of genetic syndromes. Kline; 2010, p. 195–209.
78. Kline AD, Krantz ID, Sommer A, Kliewer M, Jackson LG, FitzPatrick DR, Levin AV, Selicorni A. Cornelia de Lange syndrome: clinical review, diagnostic and scoring systems, and anticipatory guidance. Am J Med Genet A. 2007;143A(12):1287–96.
79. Clark DM, Sherer I, Deardor MA, Byrne JLB, Loomes KM, Nowaczyk MJM, Jackson LG, Krantz ID. Identification of a prenatal profile of Cornelia de Lange syndrome (CdLS): a review of 53 CdLS pregnancies. Am J Med Genet A. 2012;158A(8):1848–56.
80. Pajkrt E, Grin DR, Chiy LS. Brachmann–de Lange syndrome: definition of prenatal sonographic features to facilitate definitive prenatal diagnosis. Prena Diagn. 2010;30(9):865–72.
81. Chong K, Keang S, Hurst S, Summers A, Berger H, Seaward G, Marn N, Friedberg T, Chitayat D. Cornelia de Lange syndrome (CdLS): prenatal and autopsy findings. Prenat Diagn. 2009;29(5):489–94.
82. Williams JC, Barra -Boyes BG, Lowe JB. Supravalvular aortic stenosis. Circulation. 1961;24:1311–8.
83. Beuren AJ, Apitz J, Harmjanz D. Supravalvular aortic stenosis in association with mental retardatoon and a certain facial appearance. Circulation. 1962;26:1235–40.
84. Morris C, Suzanne B. Cassidy JEA. Management of gene c syndromes. 2005; (vol Second Edi on): 909–20.
85. Hoffman JIE. Supravalvar aortic stenosis: the natural and unnatural history of congenital heart disease. Chichester/Hoboken: Wiley-Blackwell; 2009, p. 277–81.
86. Ewart AK, Morris CA, Atkinson D, et al. Hemizygosity at the elastin locus in a developmental disorder, Williams syndrome. Nat Genet. 1993;5:11–6.
87. Kontos H, Manolakos E, Malligiannis P, Plachouras N. Prenatal diagnosis of a fetus with 7q11.23 deletion detected by multiplex ligation-dependent probe

amplification (MLPA) screening. Prenat Diagn. 2008;28:556–8.

88. Krzeminska D, Steinfeld C, Cloez JL, et al. Prenatal diagnosis of Williams syndrome based on ultrasound signs. Prenat Diagn. 2009;29:710–2.

A Few Additional Syndromes

89. Alby-Averseng C, Benzina N, Bault JP, Stirnemann J, Russell NE, Sonigo P, Millischer A, Attie Bitach T, Salomon LJ, Ville Y. The role of prenatal ultrasound and MRI in the prenatal diagnosis of CHARGE: a single centre prenatal series of 10 cases of CHD7 mutated fetuses. Ultrasound Obstet Gynecol. 2012;40(S1):294–5.

90. Chaoui R, Thiel G, Sarut Lopez A, Bergann A, Heling KS. Prenatal diagnosis of a fetus with a CHARGE syndrome. Ultrasound Obstet Gynecol. 2010;36(S1):201.

91. Bault JP. Fetal olfactory sulci and bulbs: feasibility of 3D US assessment. Ultrasound Obstet Gynecol. 2012;40(S1):221.

92. Bault J-P, Quarello E. Retinal coloboma: prenatal diagnosis using a new technique, the 'virtual fetal eyeground'. Ultrasound Obstet Gynecol. 2009;33(4):495–6.

93. Tellier AL, Cormier-Daire V, Abadie V, Amiel J, Sigaudy S, et al. CHARGE syndrome: report of 47 cases and review. Am J Med Genet CS. 1998;76(5):402–9.

94. Amiel J, Attieé-Bitach T, Marianowski R, Cormier-Daire V, Abadie V, Bonnet D, Gonzales M, Chemouny S, Brunelle F, Munnich A, Manach Y, Lyonnet S. Temporal bone anomaly proposed as a major criteria for diagnosis of CHARGE syndrome. Am J Med Genet CS. 2001;99(2):124–7.

95. Menashe M, Arbel R, Raveh D, Achiron R, Yagel S. Poor prenatal detection rate of cardiac anomalies in Noonan syndrome. Ultrasound Obstet Gynecol. 2002;19(1):51–5.

96. Bradley E, Kean L, Twining P, James D. Persistent right umbilical vein in a fetus with Noonan's syndrome: a case report. Ultrasound Obstet Gynecol. 2001;17(1):76–8.

97. van Huizen ME, Pighe M, Bijlsma EK, Knegt AC, Bilardo CM. Increased nuchal translucency thickness: a marker for chromosomal and gene c disorders in both o spring and parents. Ultrasound Obstet Gynecol. 2005;26(7):793–4.

98. Timmerman E, Pajkrt E, Maas SM, Bilardo CM. Enlarged nuchal translucency in chromosomally normal fetuses: strong association with orofacial clefts. Ultrasound Obstet Gynecol. 2010;36(4):427–32.

99. Hiippala A, Eronen M, Taipale P, Salonen R, Hiilesmaa V. Fetal nuchal translucency and normal chromosomes: a long-term follow-up study. Ultrasound Obstet Gynecol. 2001;18(1):18–22.

100. Mesquita MM, Barreiro M, Dias C, Ferreira L, Taborda F, Pestana A, Vasconcelos E, Ferreira A. Prenatal diagnosis of Noonan syndrome: a case report. Ultrasound Obstet Gynecol. 2004;24(3):353. 7. Van der Burgt I. Orphanet : Syndrome de Noonan. Juin 2008.

101. Suzanne B. Cassidy JEA. Noonan syndrome. Management of genetic syndromes. Wiley; p. 569–86.

J.-M. Levaillant et al., *Normal and Abnormal Fetal Face Atlas*, DOI 10.1007/978-3-319-43769-9_5

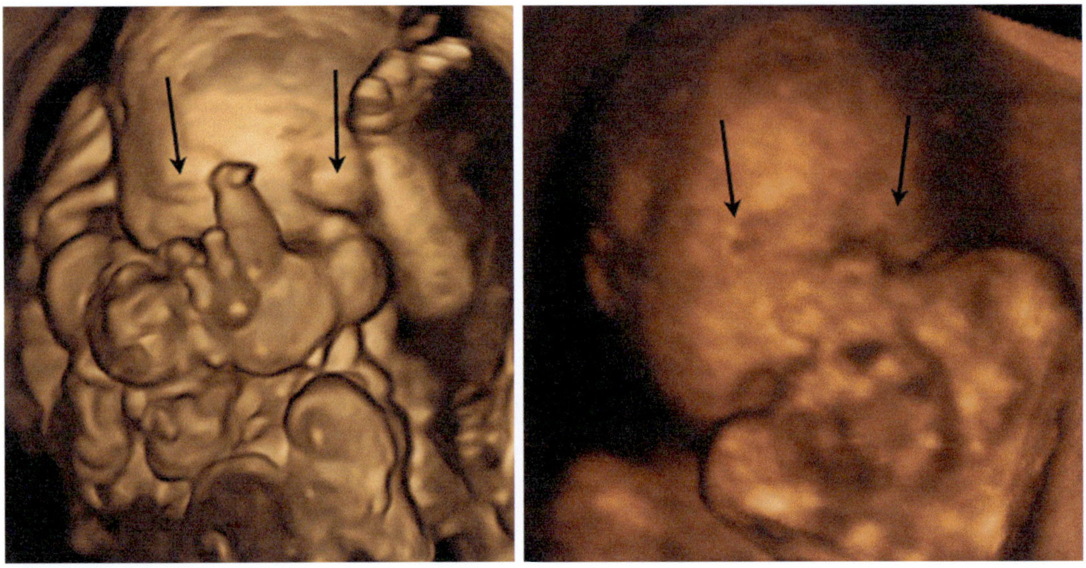

Fig. 5.1 Giant epignathus (*arrows*: ocular structures)

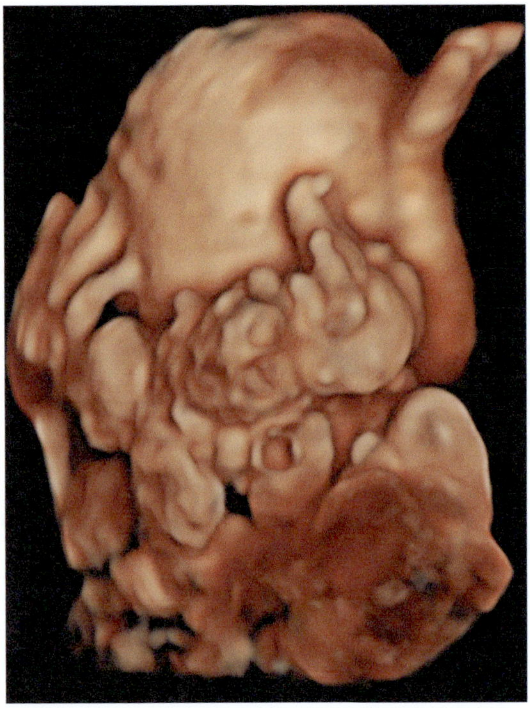

Fig. 5.2 Giant epignathus: HD live rendering highlighting the heterogeneous aspect of the tumor

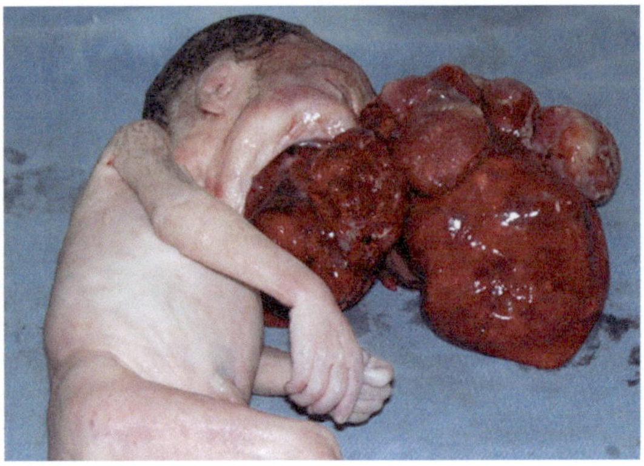

Fig. 5.3 Giant epignathus: post-natal appearance

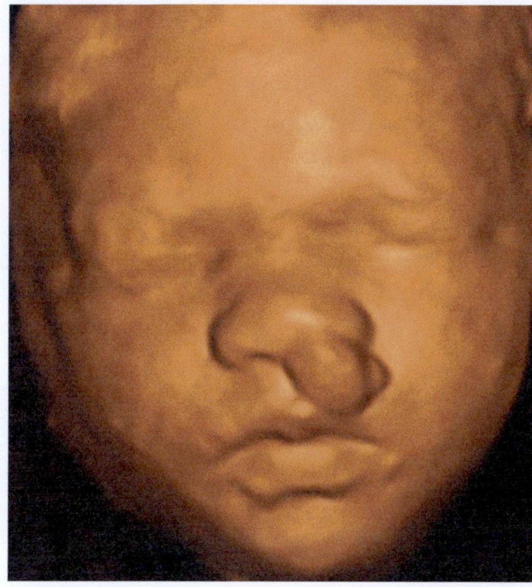

Fig. 5.4 Small facial teratoma outside of one nostril

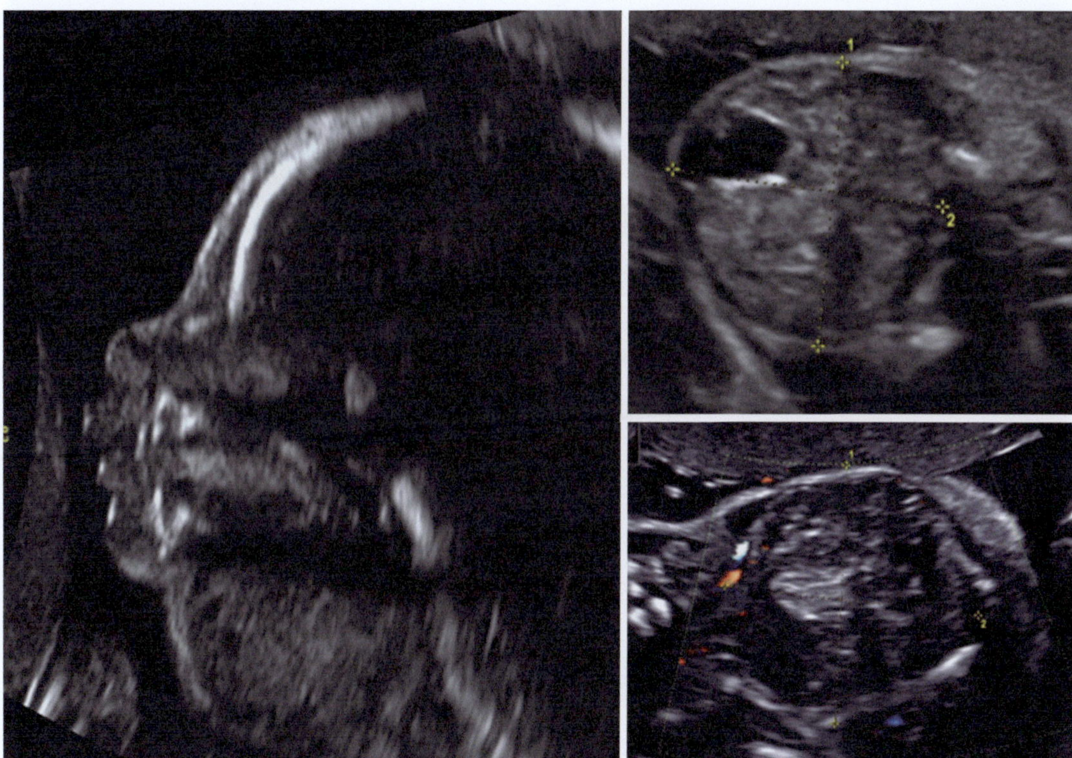

Fig. 5.5 Cervical teratoma: heterogenous aspect, vascularization

5.2 Hemangiomas

Vascular anomalies of the face can be divided into vascular malformations and vascular tumors. The most frequent kind of vascular tumor is the hemangioma. These are further divided into those that spontaneously disappear (regression is usually complete by the age of 14 months) and those for which involution is not spontaneous.

The aspect is most often echogenic, usually homogenous, although sometimes heterogeneous (Fig. 5.7). The Doppler highlights hypervascularization at the level of the tumor (Figs. 5.6 and 5.7). The opposite bone base could be the seat of a slight depression related to the subperiostal process origin (Fig. 5.8).

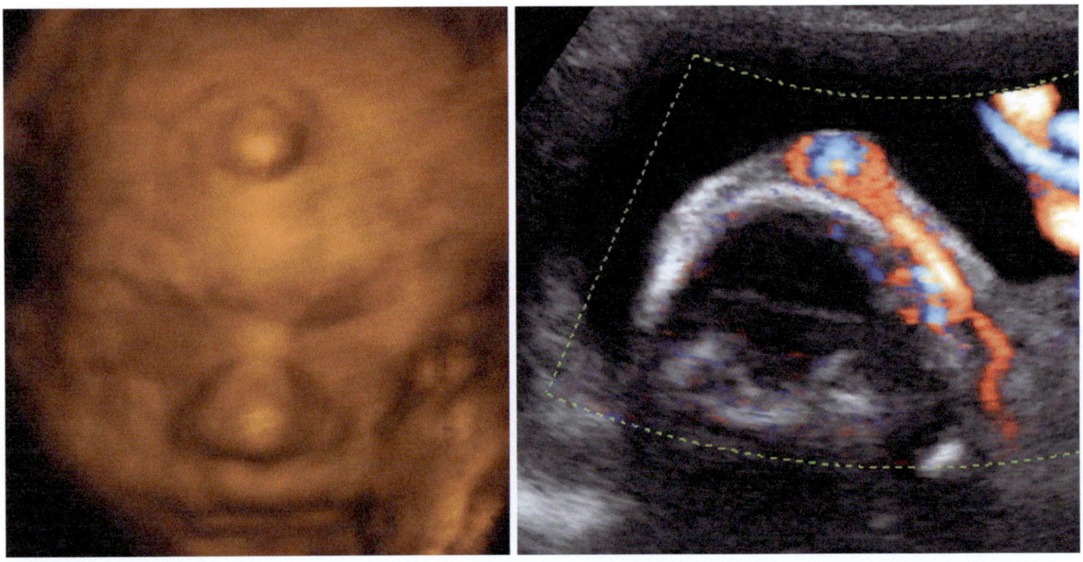

Fig. 5.6 Frontal hemangioma: *left*, rendered aspect, *right* vascularization and slight bone depression

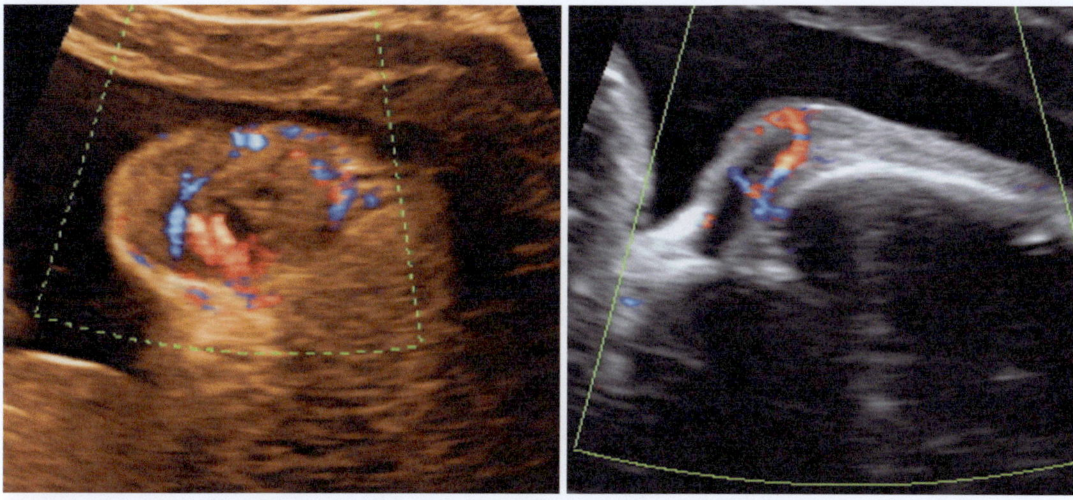

Fig. 5.7 Extraorbital hemangioma: slightly heterogeneous tumor aspect and vascularization

Fig. 5.8 Left extraorbital hemangioma in HD live rendered mode

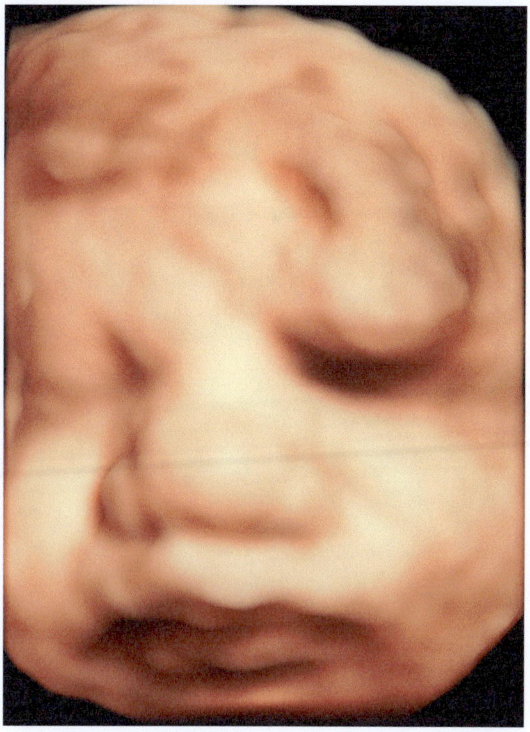

5.3 Lymphangiomas

Frequently found at the cervical level, they are much less frequent on the face. Surgery is very difficult and could leave major damage, particularly from an esthetic point of view. Their appearance on the ultrasound is that of a multicystic formations (the cysts can be of variable sizes Figs. 5.9, 5.10, 5.11, 5.12 and 5.13).

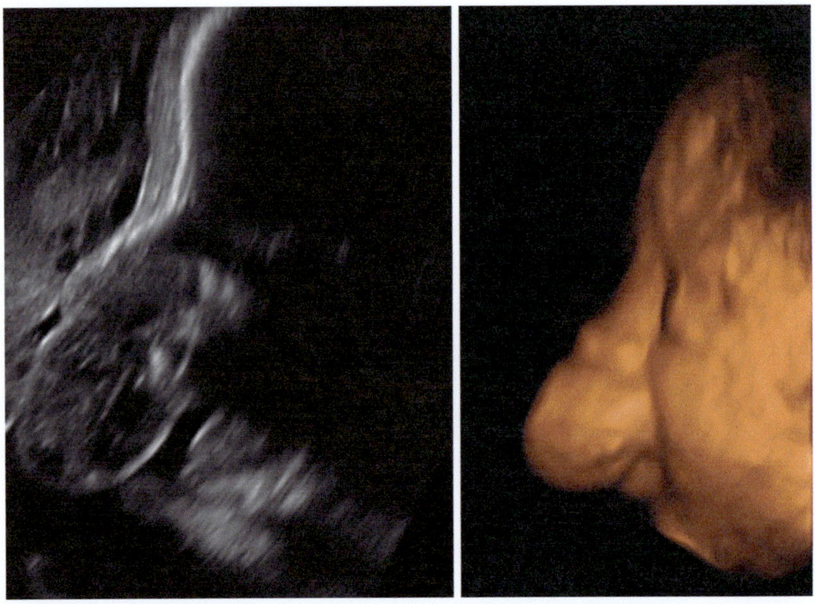

Fig. 5.9 Cystic lymphangioma of the upper lip: left, cystic aspect of the tumor; right, rendered mode

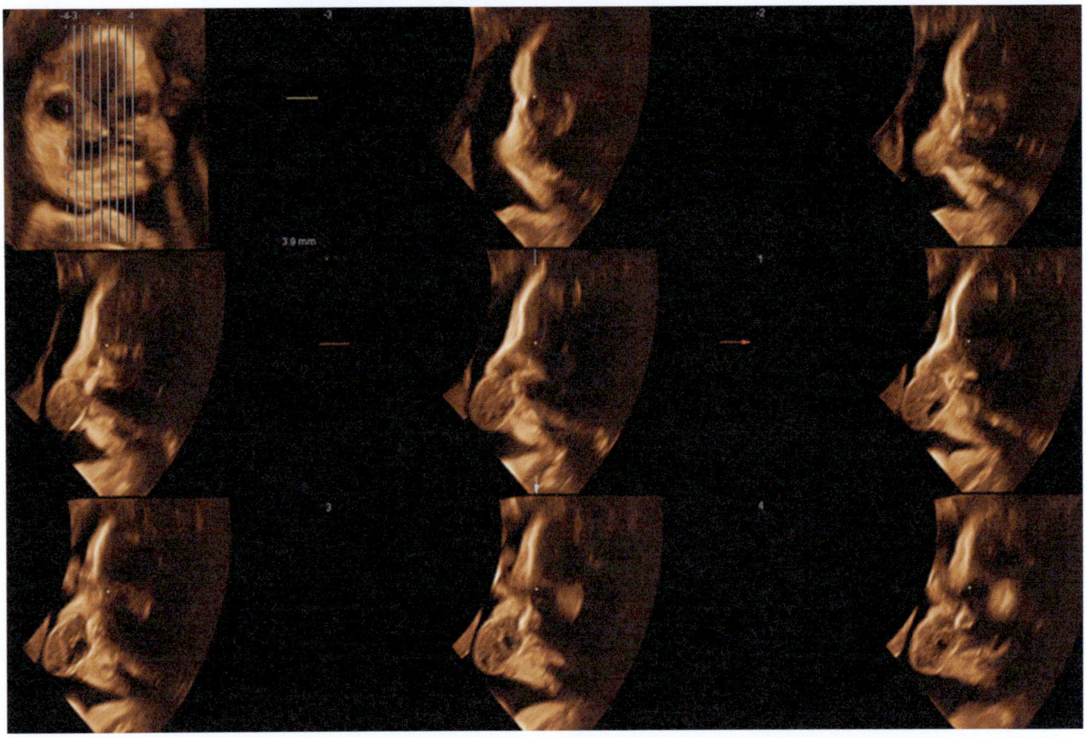

Fig. 5.10 Cystic lymphangioma of the upper lip: TUI mode

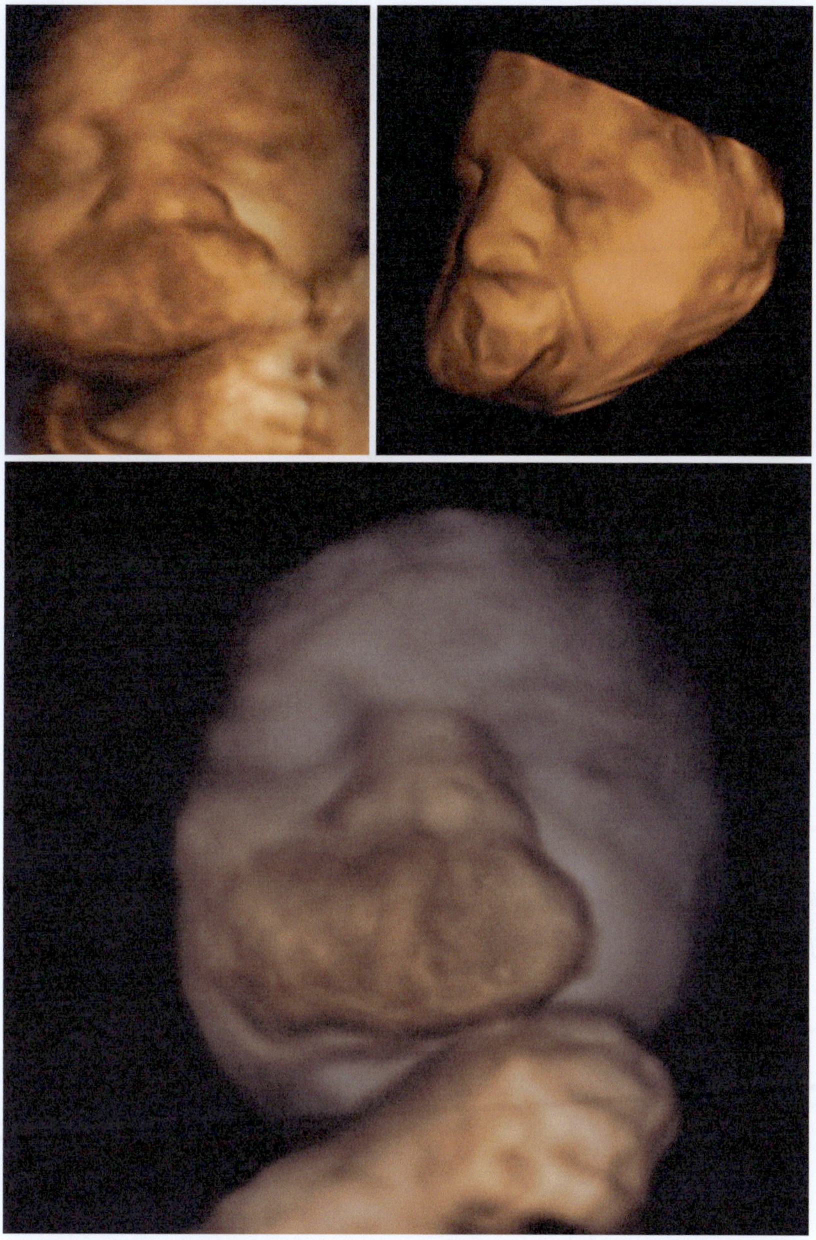

Fig. 5.11 Cystic lymphangioma of the upper lip: classic rendering and HD live (*lower image*)

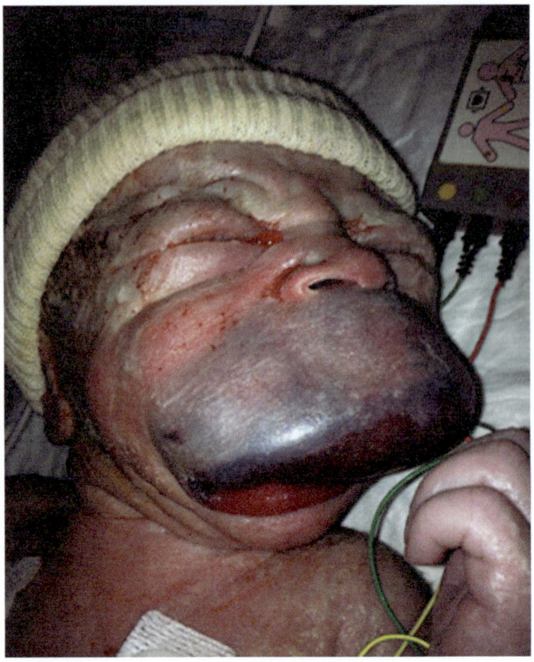

Fig. 5.12 Lymphangioma of the upper lip (post-natal appearance)

5.4 Expansive Process of Brain Tumors

Brain tumors can develop to be big and "fuse" through the skull bone structures such as the orbits. The histological nature of these tumors is variable. We will illustrate with a cerebral hamartoma that spread, destroying the left orbit and invading the mouth (Figs. 5.13, 5.14, 5.15, and 5.16).

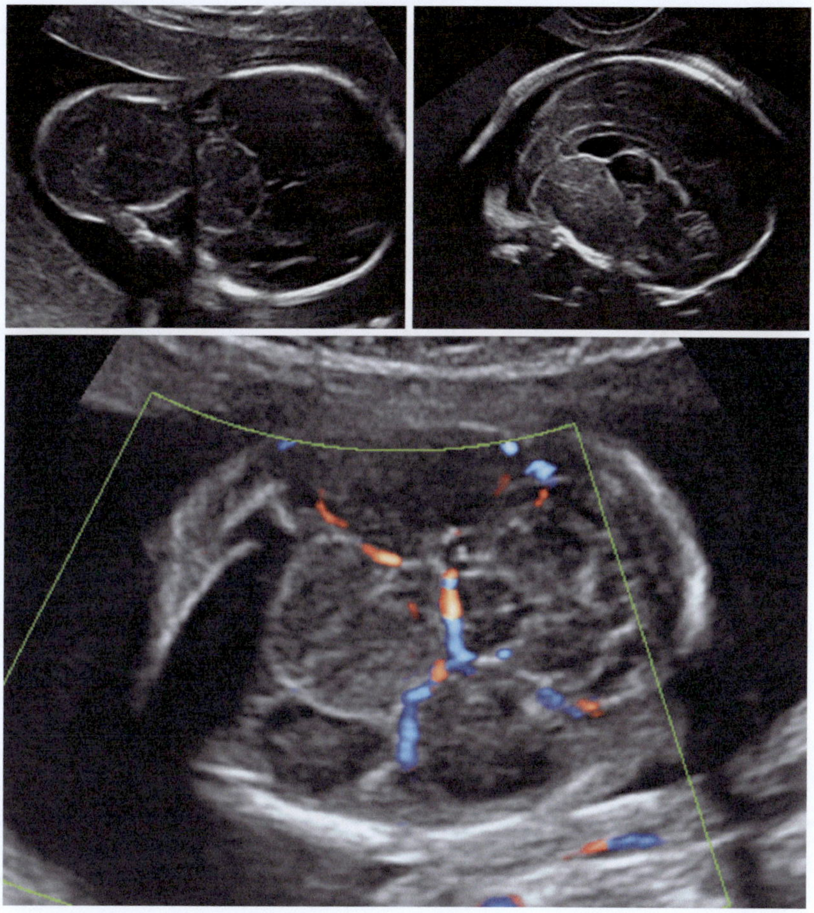

Fig. 5.13 Cerebral hamartoma: medio-cerebral origin

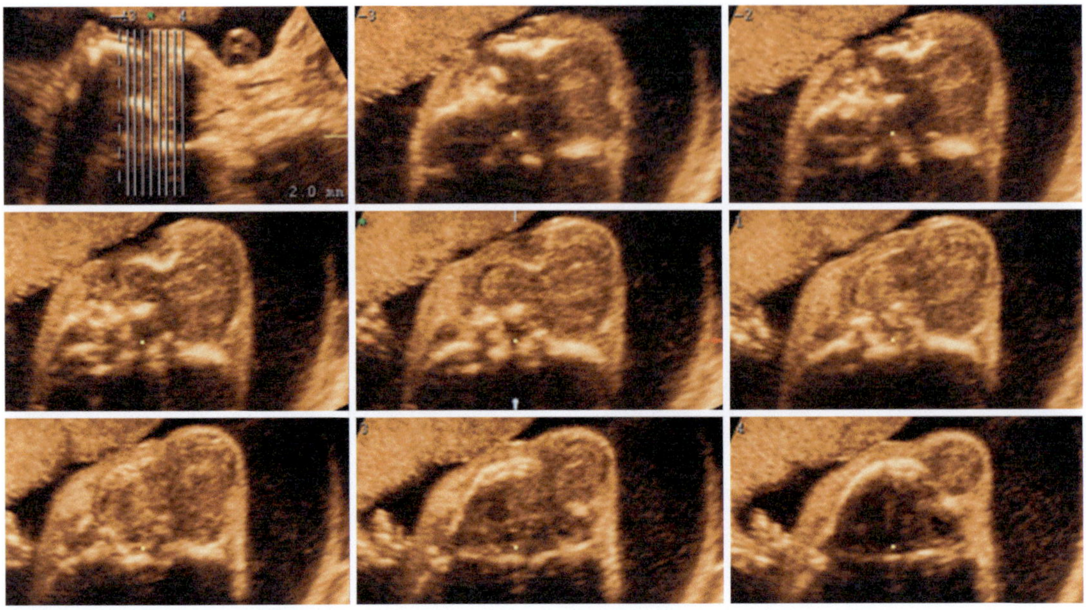

Fig. 5.14 Cerebral hamartoma: TUI extension of tumor into the orbit and mouth

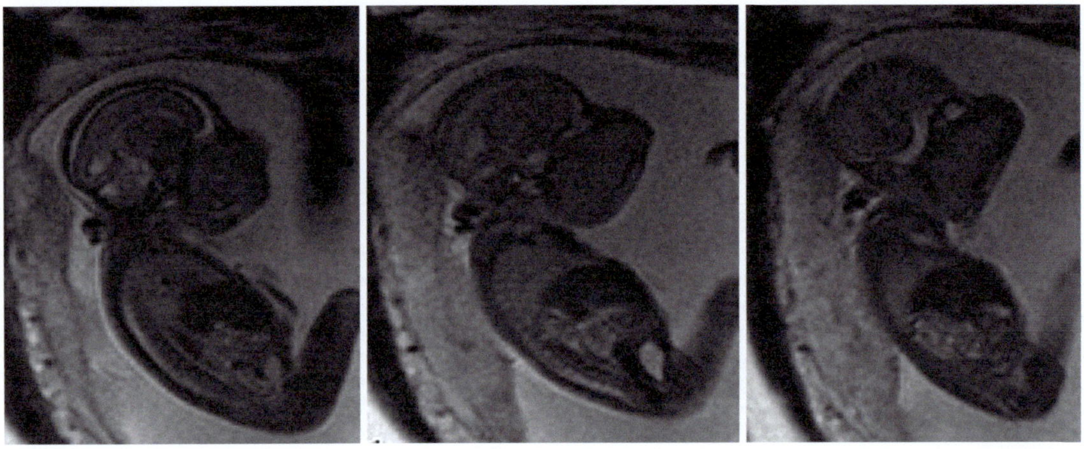

Fig. 5.15 MRE of cerebral hamartoma

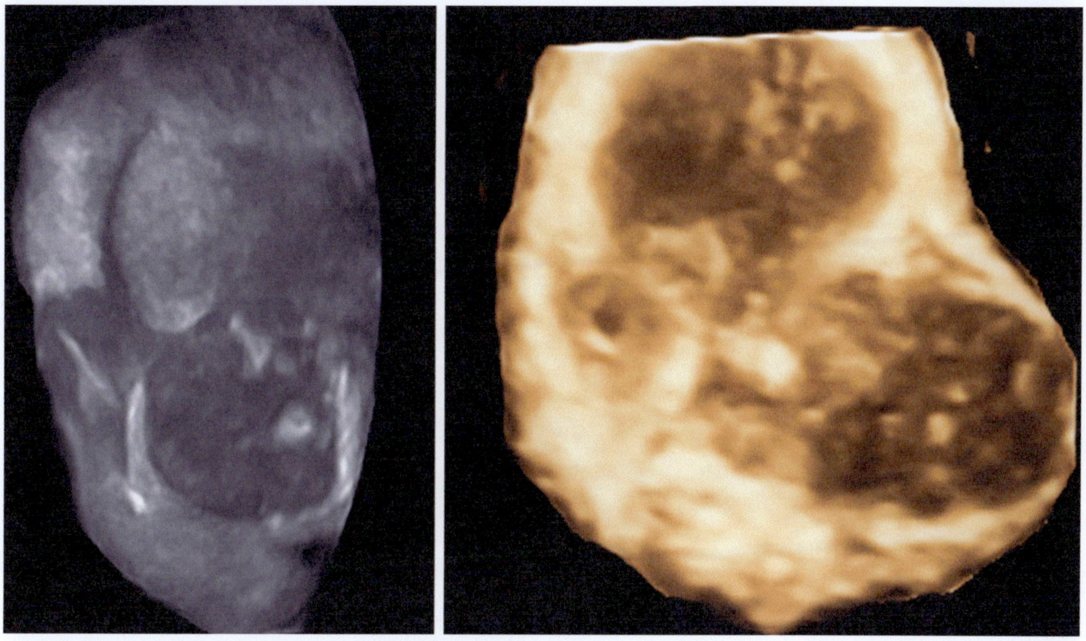

Fig. 5.16 Cerebral hamartoma: destruction of the orbit by the tumor (maximum mode)

5.5 Pai Syndrome

This rare syndrome is characterized by the association of pericallosal lipomas, median cleft of the palate and upper lip, and by the presence of skin polyps on the face. It can include bifid nose. Psychomotor development is normal.

The discovery of a callous lipoma should, evidently, lead to the attentive examination of the fetal face and the secondary hard palate (Fig. 5.17).

The few examples found in this chapter illustrate the difficulty of delivering a precise etiological diagnosis and highlight the importance of pluridisciplinary work to best counsel couples facing these difficult situations.

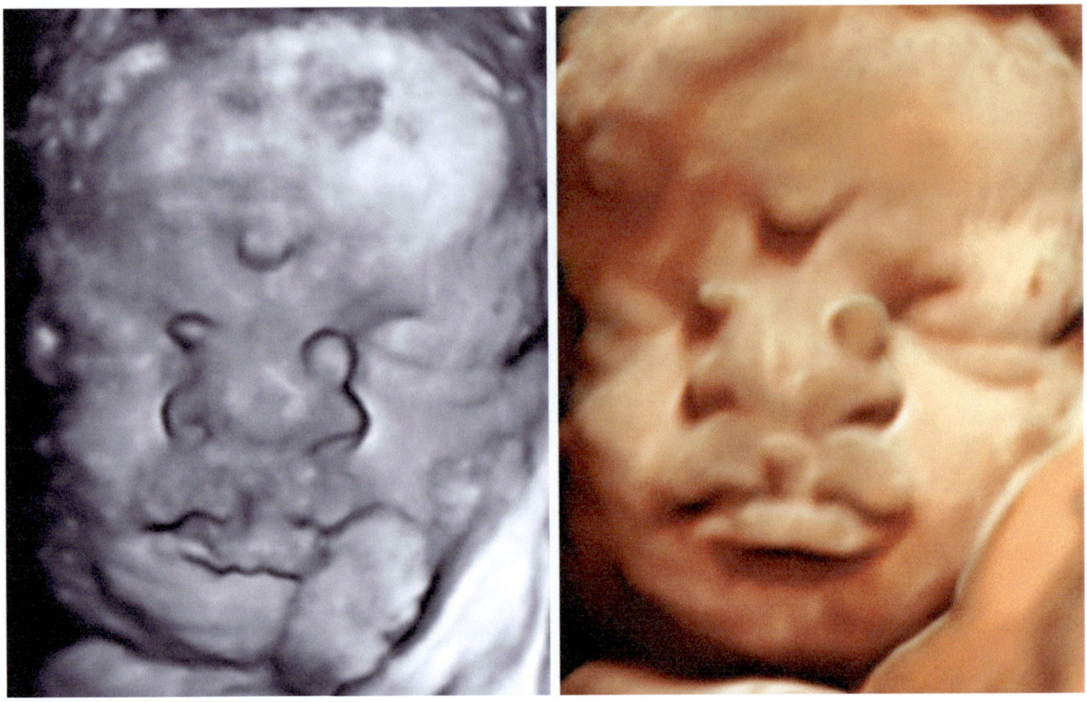

Fig. 5.17 PAI syndrome in rendered mode: multiple skin polyps, bifid nose, median cleft

References

1. Ruano R, Benachi A, Aubry MC, Parat S, Dumez Y, Mannach Y, Dommergues M. Three-dimensional prenatal diagnosis of an epignathus teratoma. Ultrasound Obstet Gynecol. 2003;22(S1):116.
2. Chen CP, Shih JC, Huang JK, Chin DTH, Tzen CY, Lin YH, Wang W. Two- and three-dimensional ultrasound demonstration of a giant epignathus. Ultrasound Obstet Gynecol. 2003;21(4):407–9.
3. Clement K, Chamberlain P, Boyd P, Molyneux A. Prenatal diagnosis of an epignathus: a case report and review of the literature. Ultrasound Obstet Gynecol. 2001;18(2):178–81.
4. Elia D, Garel C, Enjolras O, Vermouneix L, Soupre V, Oury JF, Guibaud L. Prenatal imaging findings in rapidly involuting congenital hemangioma of the skull. Ultrasound Obstet Gynecol. 2008;31(5):572–5.
5. Kashima H, Unno N, Hyodo H, Hyodo HM, Takagi K, Nakamura T, Kondoh Y, Noguchi M, Konishi I. Antenatal sonographic and magnetic resonance images of a giant hemangioma of the fetal skull. Ultrasound Obstet Gynecol. 2005;25(5):522–3.
6. Paladini D, Vassallo M, Sglavo G, Lapadula C, Longo M, Nappi C. Cavernous lymphangioma of the face and neck: prenatal diagnosis by three-dimensional ultrasound. Ultrasound Obstet Gynecol. 2005;26(3): 300–2.
7. Schlembach D, Bornemann A, Rupprecht T, Beinder E. Fetal intracranial tumors detected by ultrasound: a report of two cases and review of the literature. Ultrasound Obstet Gynecol. 1999;14(6):407–18.
8. Sahinoglu Z, Aktug Ertekin A, Cerrah Celayir A, Gucluer B. Prenatal diagnosis of a huge facial tumor: report of a rare case and literature review. Congenit Anom. 2012;52(2):111–4.
9. Chousta A, Ville D, James I, Foray P, Bisch C, Depardon P, Rudigoz RC, Guibaud L. Pericallosal lipoma associated with Pai syndrome: prenatal imaging findings. Ultrasound Obstet Gynecol. 2008;32(5):708–10.

6 The Eye

The study of the fetal eye is an important moment in the examination of the face as numerous ocular anomalies are often found in syndromic disorders. Precise knowledge of the of normal semeiology is essential to better understand pathological aspects.

6.1 Embryology Review

Beginning on the 21st day (of pregnancy), neural folds in the diencephaly region invaginate to form gutters that transform into optic vesicles, which then rapidly invaginate again into optic cups. The ectoderm in contact with the vesicle thickens to constitute the lens placode, which in turn invaginations to become the lens vesicle. This then transforms into the lens.

The retina and lens are irrigated by the hyaloid artery (a branch of the ophthalmic artery) which reaches into the ocular globe by the coloboma fissure located on the ventral side of the optic pedicle. If this fissure does not close, there will be coloboma anomalies.

6.1.1 Ultrasound of a Normal Eye

6.1.1.1 First Trimester

The ocular structures are visible as early as the first trimester, in particular the ocular globes, the lens and the orbits. Note that at 13GW, it is possible to visualize a slight posterior thickening of the lens that corresponds to remnants of the primitive vitreous (Fig. 6.1).

J.-M. Levaillant et al., *Normal and Abnormal Fetal Face Atlas*, DOI 10.1007/978-3-319-43769-9_6

6.1.1.2 Second Trimester

The orbits are perfectly identifiable: in particular in maximum volume mode, which makes it possible to see the various bone components (Fig. 6.2). The ocular globes and the lenses are quite visible; the hyaloid artery appears as a thin line connection the back side of the lens to the retina (it does not persist beyond 28GW) (Fig. 6.3).

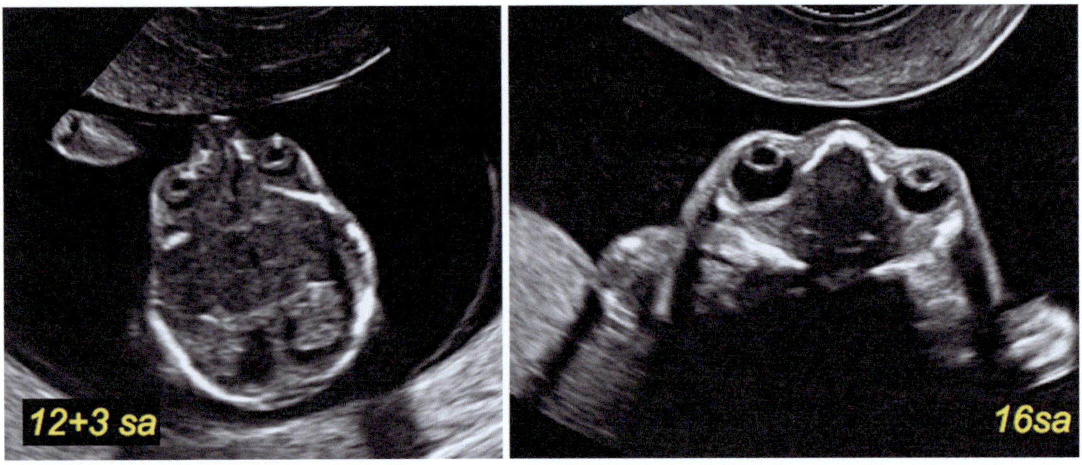

Fig. 6.1 The eye during the first trimester

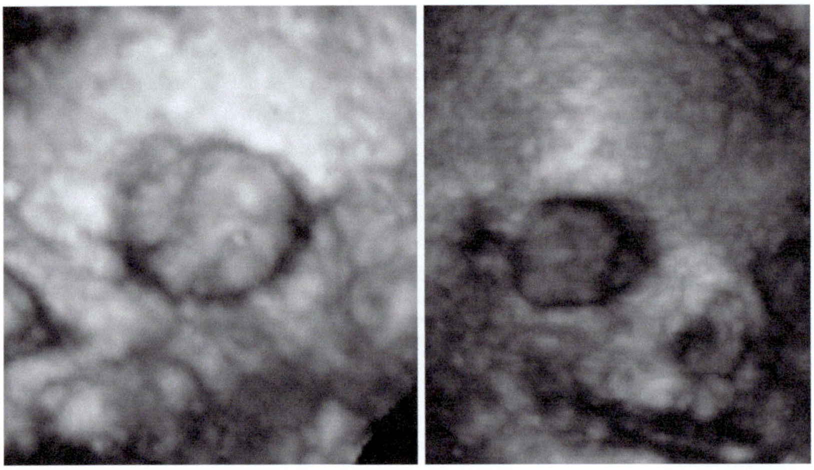

Fig. 6.2 Bony orbits (maximum volume mode)

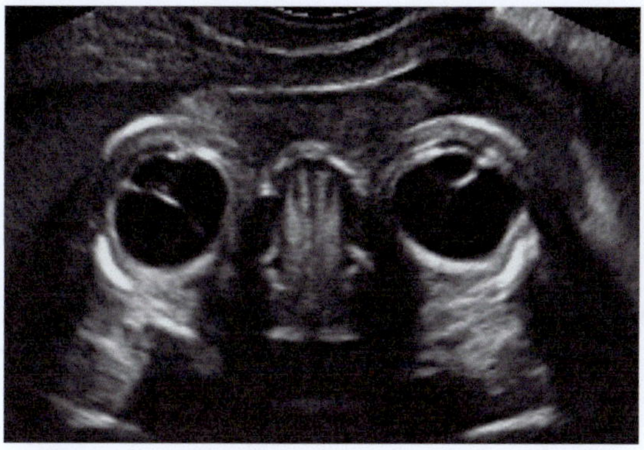

Fig. 6.3 Eye at 22 GW (note the presence of hyaloid arteries)

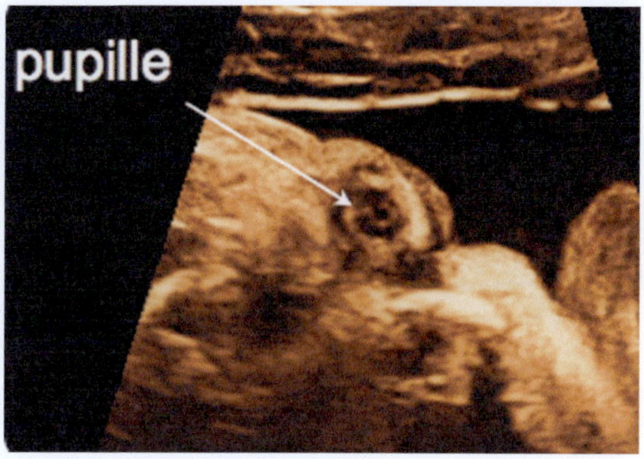

Fig. 6.5 Pupil on a frontal view
Pupil

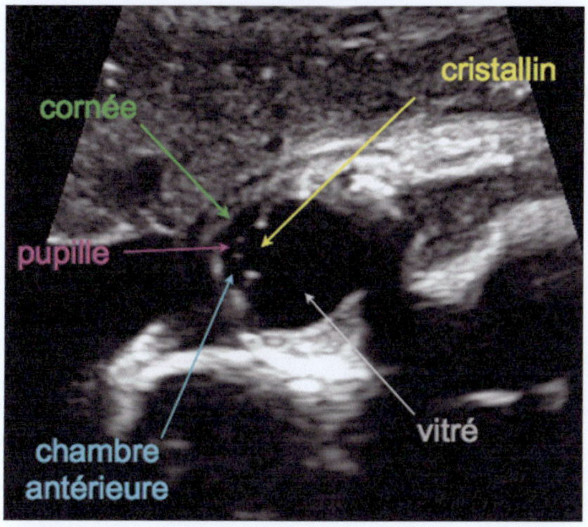

Fig. 6.4 Eye during the third trimester
Cornea
Pupil
Front chamber
Vitreous body
Lens

6.1.1.3 Third Trimester

The various eye structures are perfectly visible, as seen in Fig. 6.4, with the pupil being visible on a frontal view (Fig. 6.5).

Furthermore, one can see the optic nerve, the ophthalmic artery, the retrobulbar fat (which appears as clearly echogenic) (Fig. 6.6).

The chiasma opticum can be seen from 20GW using the volume mode. We described the capture in 2006 and the biometrics in 2007 in the journal *Ultrasound in Obstetrics and Gynecology* (cf. references) (Figs. 6.7, 6.8, and 6.9).

The eyelids and eyelashes are visible from the second trimester (Fig. 6.10).

It is possible to evaluator the fundus (virtual) using volumetric capture using the technique that we described in 2008 in the journal *Ultrasounds in Obstetrics and Gynecology* (cf. References) (Figs. 6.11 and 6.12).

It is possible to measure the inter-orbital distance. In France, we tend to use the measurement from the middle of one orbit to the middle of the other (Fig. 6.13), the relationship between the IOD/BPD is constant: 0.47 at 22 GW and 0.42 at 32 GW.

A simple technique is that of the "virtual orbit": if the two eyes are separated a normal distance, it is possible to put just one "virtual orbit" in between the two normal orbits (Fig. 6.14).

When making a diagnostic study, one will use various published tables, in particular those regarding measurements regarding the orbits, the ocular globes and the lenses.

In addition, an attentive study will make it possible to visualize ocular movements. It should be noted that strabismus, particularly when divergent, is often observed in the fetus.

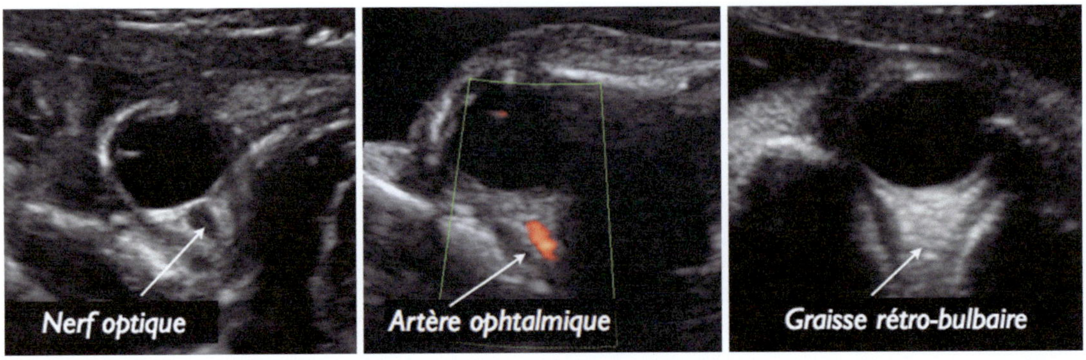

Fig. 6.6 Optic nerve, ophthalmic artery, retrobulbar fat
Optic nerve
Ophthalmic artery
Retrobulbar fat

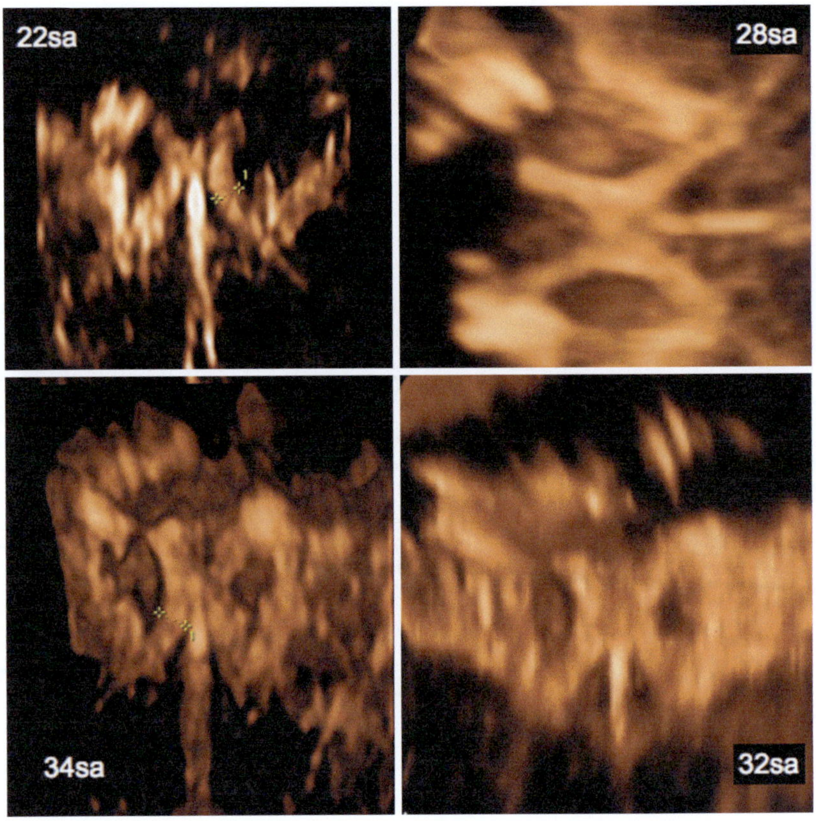

Fig. 6.7 Chiasma opticum at 22, 28, 32 and 34 GW

Fig. 6.8 Chiasma opticum at 22 GW: fetopathological examination (* olfactive bulbs, ° chiasma, *Sph* sphenoid wings)

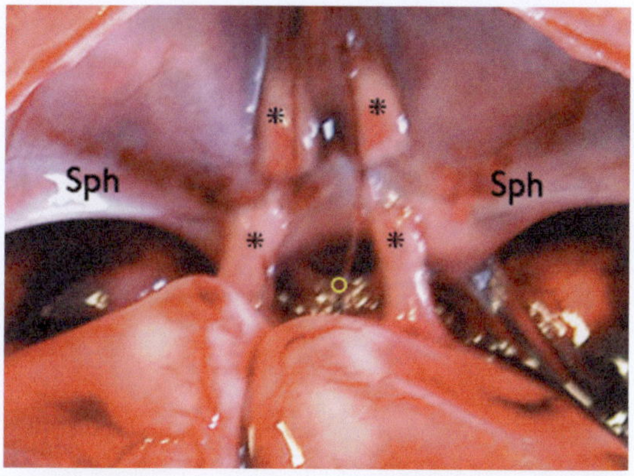

Fig. 6.9 Chiasma biometrics

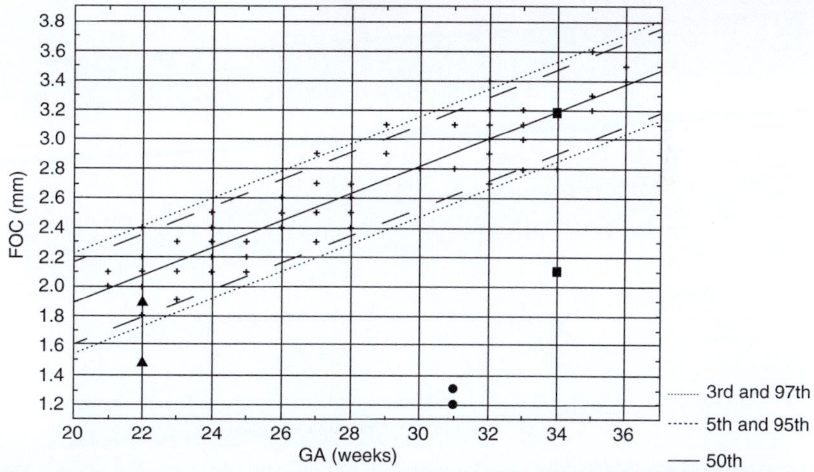

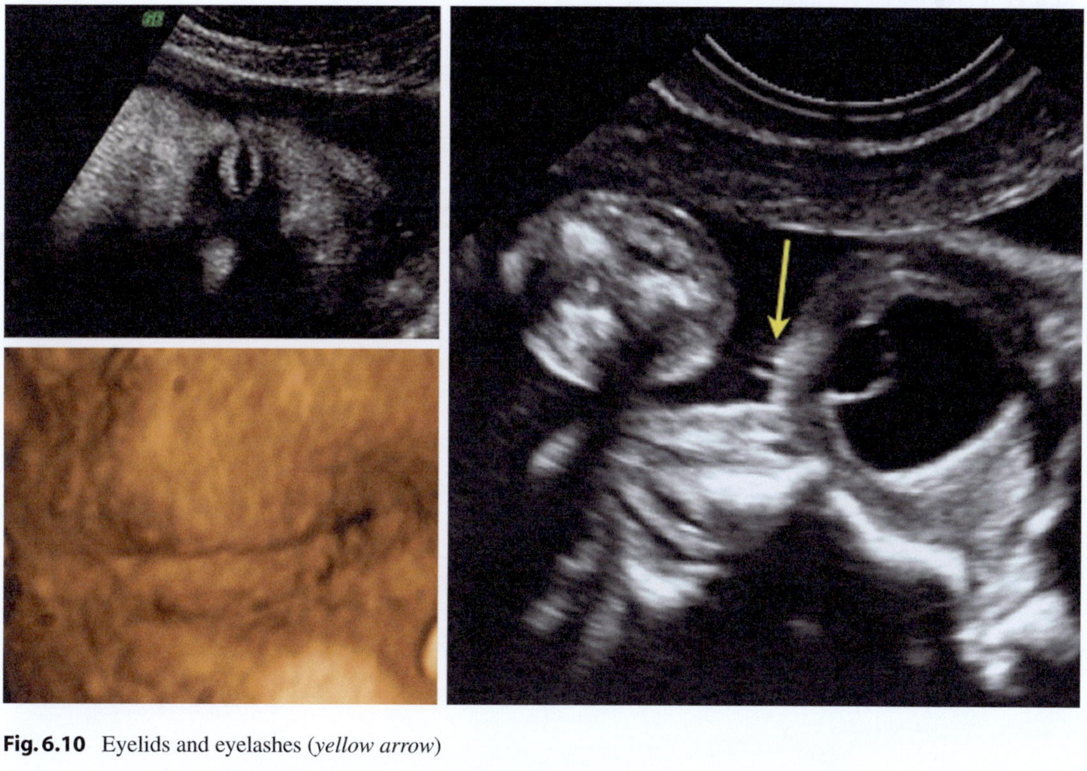

Fig. 6.10 Eyelids and eyelashes (*yellow arrow*)

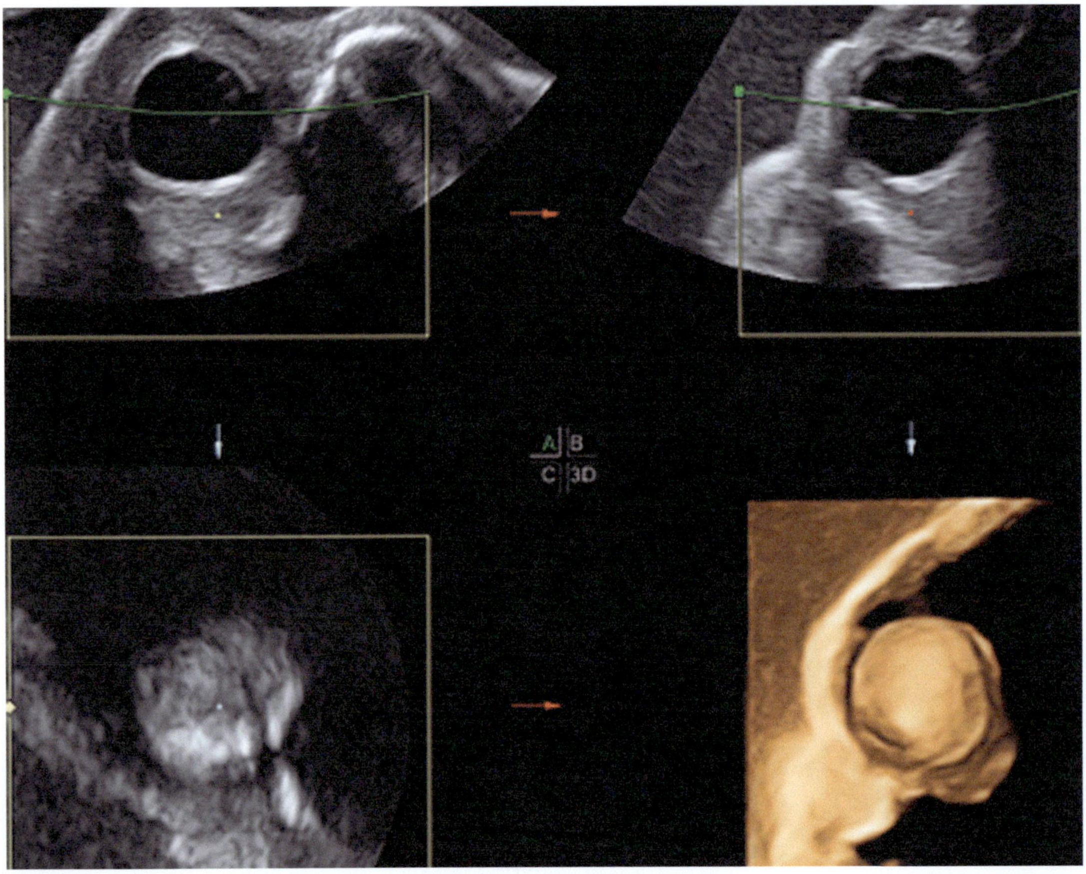

Fig. 6.11 Capturing the eye fundus

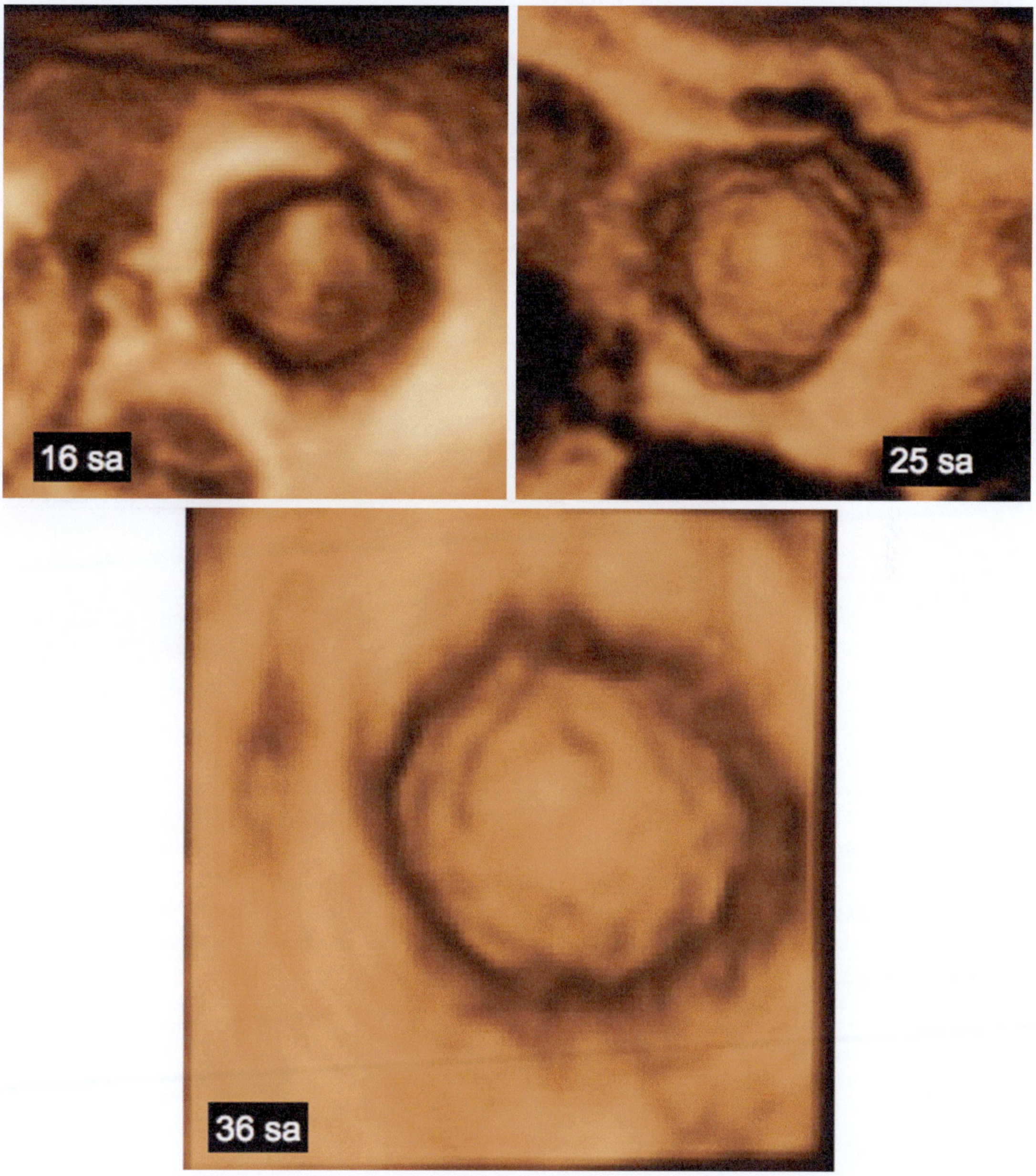

Fig. 6.12 Fundus at 16 GW (note the slight depression corresponding to the insertion of the hyaloid artery) at 25 GW and 36 GW

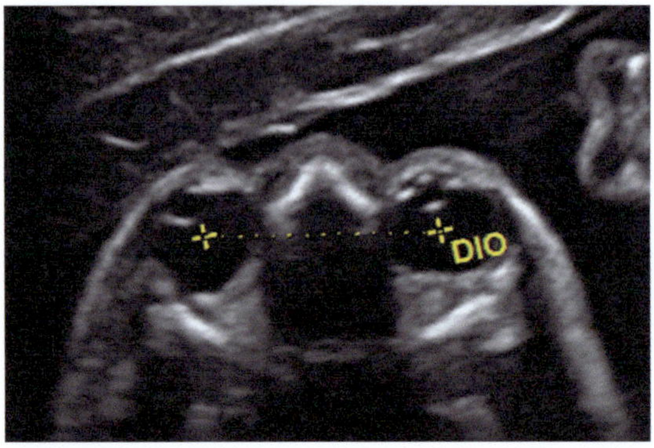

Fig. 6.13 Interorbital distance measurement IOD

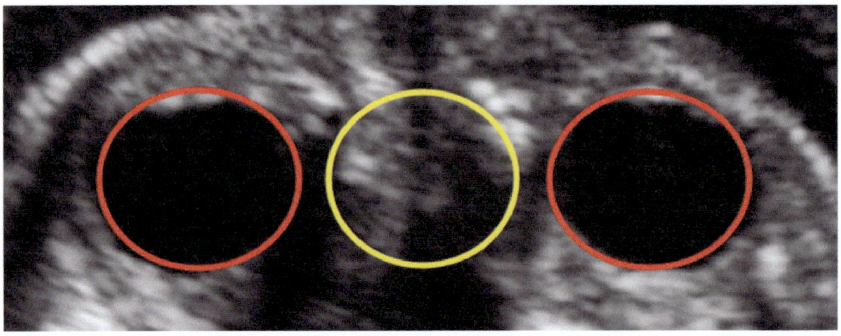

Fig. 6.14 Virtual orbit

6.1.2 Main Aspects of Ocular Pathologies

6.1.2.1 Dacryocystocele

This is a cyst in the lachrymal canal that can be seen on an ultrasound as a small swelling in the internal corner of the eye. On an axial view, this cyst appears as an anechogenic formation inside and to the front of the eye. It is sometimes possible to reveal a small lithiathis in the cyst (Fig. 6.15). In the majority of cases, these dacryocystoceles are easily resolvent after birth or even in utero (Fig. 6.16).

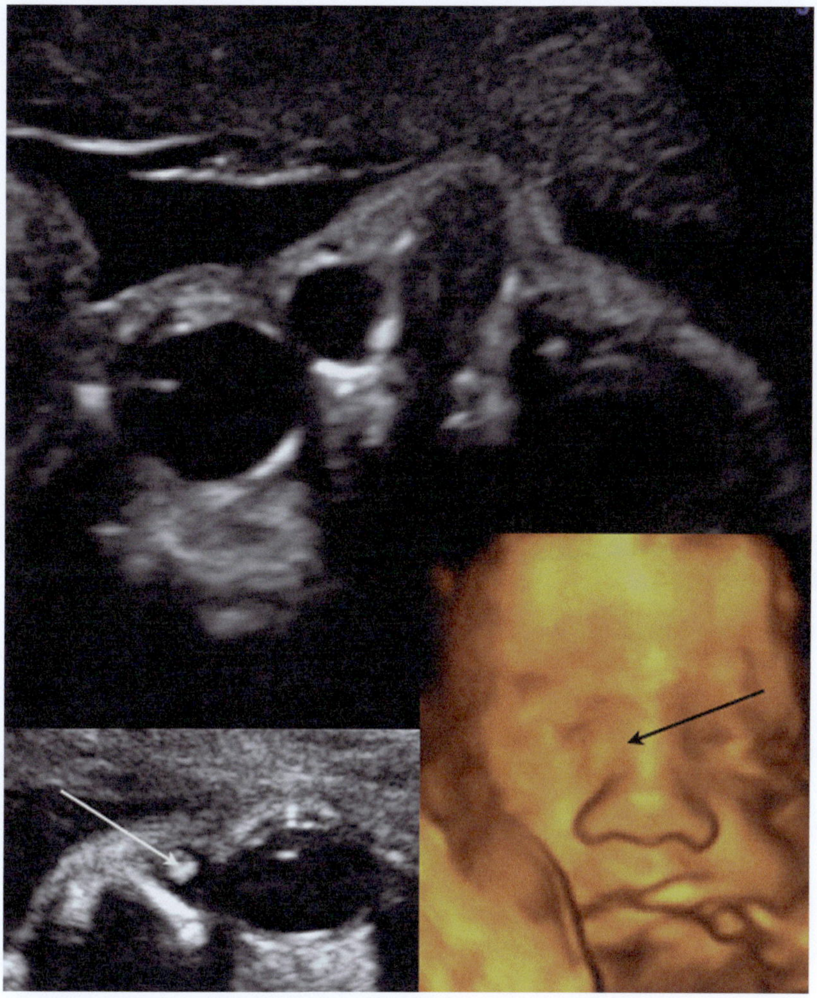

Fig. 6.15 Dacryocystoceles (*white arrow*: lithiathis, *black arrow*: volumetric aspect)

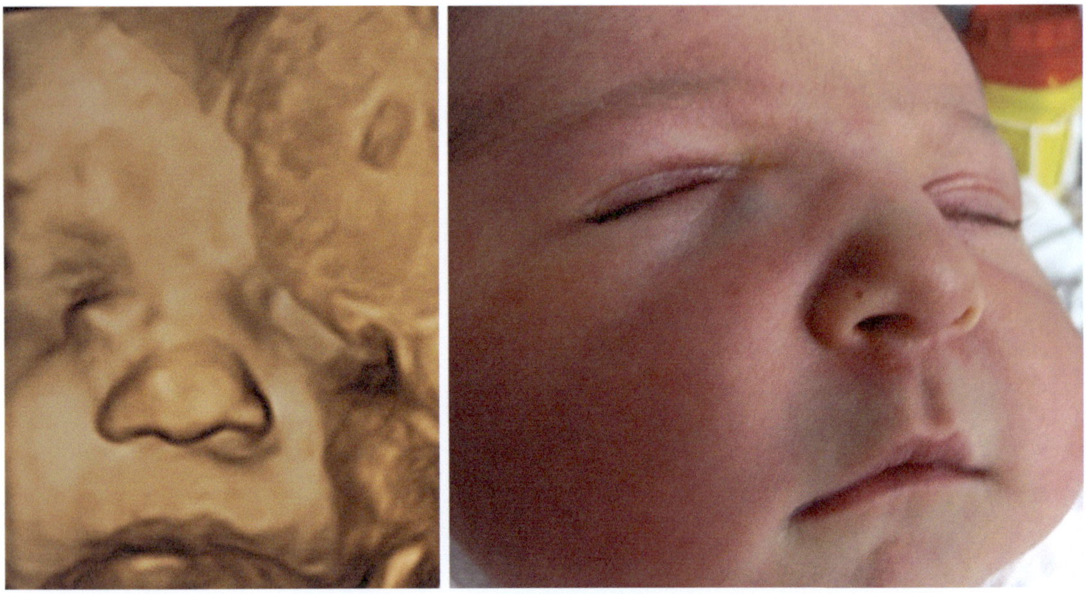

Fig. 6.16 Dacryocystoceles: pre and post natal

Fig. 6.17 Hypotelorism related to semi-lobar holoprosencephaly. BPD/OFD: 0.40 for 29 GW

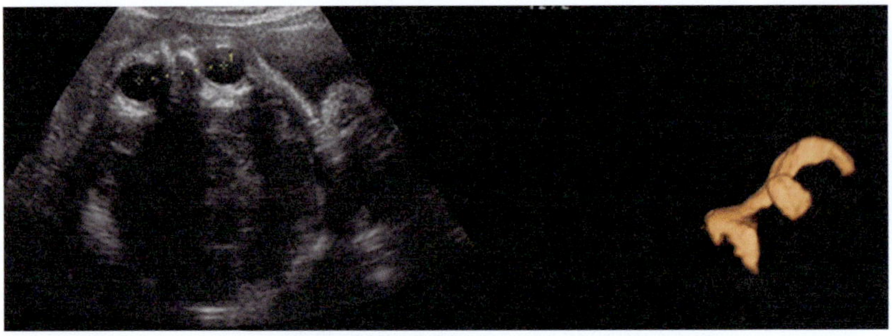

6.1.2.2 Hypoltelorism, Cyclopia

If hypotelorism is brought to light, it is imperative to look for holoprosencephaly, as illustrated in the below example (Fig. 6.17).

Extreme forms of hypotelerism result in cyclopia that can be seen as early as the first trimester (Fig. 6.18). Sometimes, the aspect is found in the second trimester, in particular if an ultrasound was not carried out during the first trimester (Fig. 6.19). These cyclopias can present with one or two lenses, the front bones are fused, and the metopic suture is not visible. These anomalies are most often seen as part of lobar holoprosencephaly, and in particular in trisomy 13-type anomalies (Fig. 6.19).

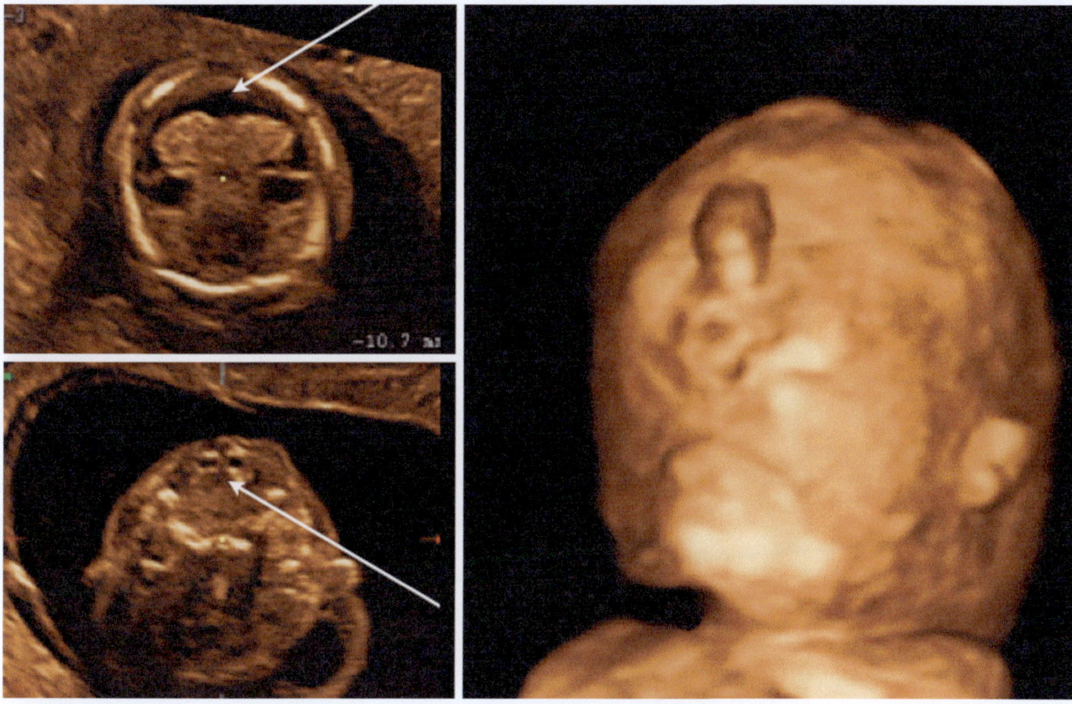

Fig. 6.18 Cyclopia in the first trimester: upper right, alobar holoprosencephaly (*arrow*); lower left: to lenses (*arrow*); right: volumetric rendered image (proboscis, single orbit, nose not visible)

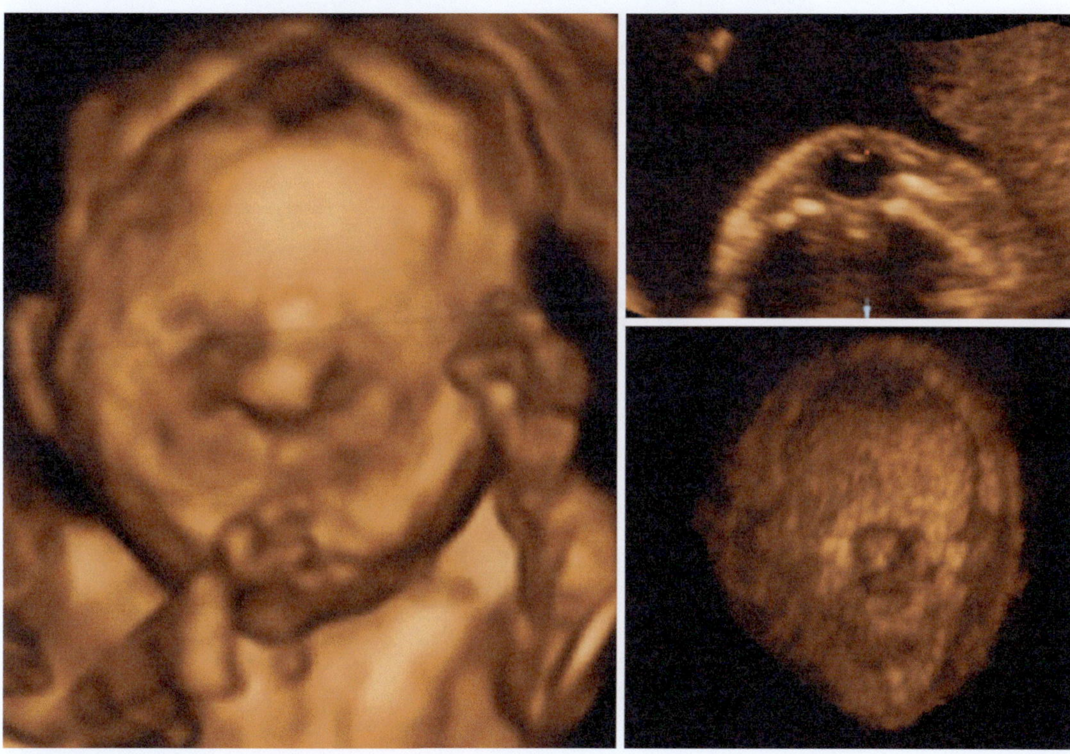

Fig. 6.19 Cyclopia in the second trimester: *left*, rendered volumetric image; *lower right*, single lens; *lower left*: fusion of frontal bones, no visible metopic suture

6.1.2.3 Cataracts

The lens is opaque, appearing echogenic. This aspect could be visible as early as the first trimester (Fig. 6.20), or during the morphological examination done during the second trimester (Fig. 6.21).

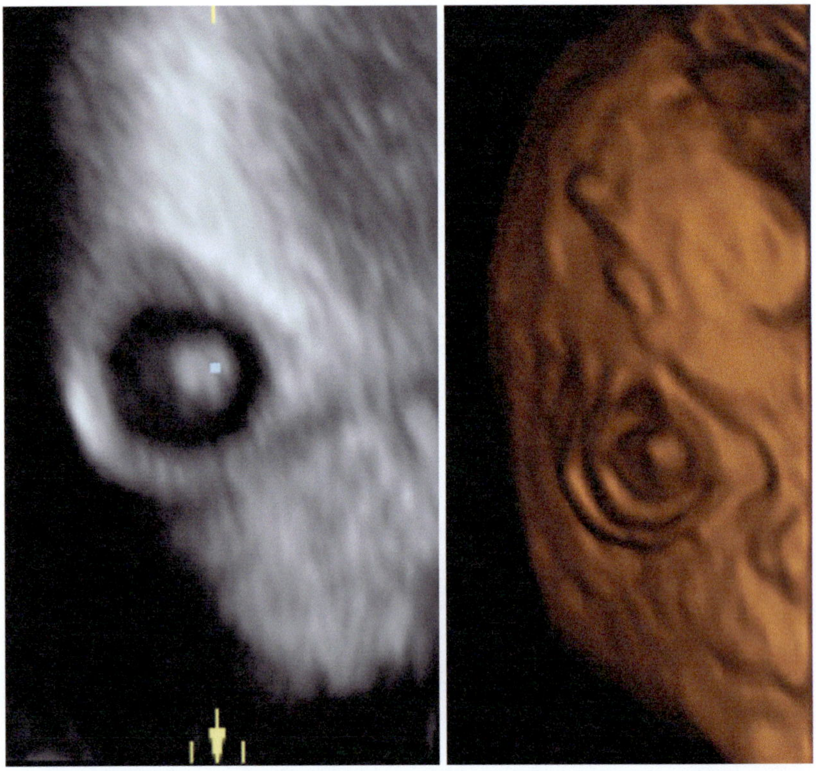

Fig. 6.20 Cataract during the first trimester

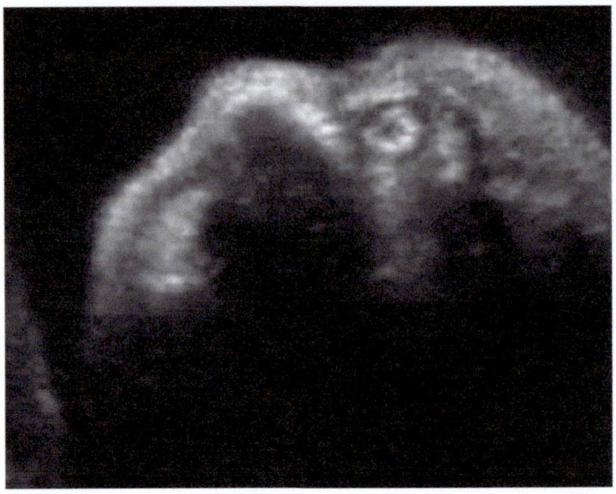

Fig. 6.21 Cataract during the second trimester

6.1.2.4 Microphthalmias

Microphthalmias can involve one eyeball (Figs. 6.22, 6.23, or 6.24) or be bilateral (Figs. 6.28 and 6.29).

The size of the eyes can have a very minimal difference, which means it is indispensable to refer to measurement tables (Goldstein 1998 OUG). And MRI can be complementary to ultrasound imagery (Fig. 6.25).

It is possible to use the reverse view technique. The volumetric image of the fetal face is captured from the front, and then the volume is visualized from behind the facial mass, looking for internal orbital contours (Fig. 6.26). This technical possibility can be assimilated to the reverse view technique described by S. Campbell to visualize palatal defects.

A colobomatous origin is frequent in microphthalmia, so one should be particularly attentive to looking for a retinal defect (Fig. 6.27).

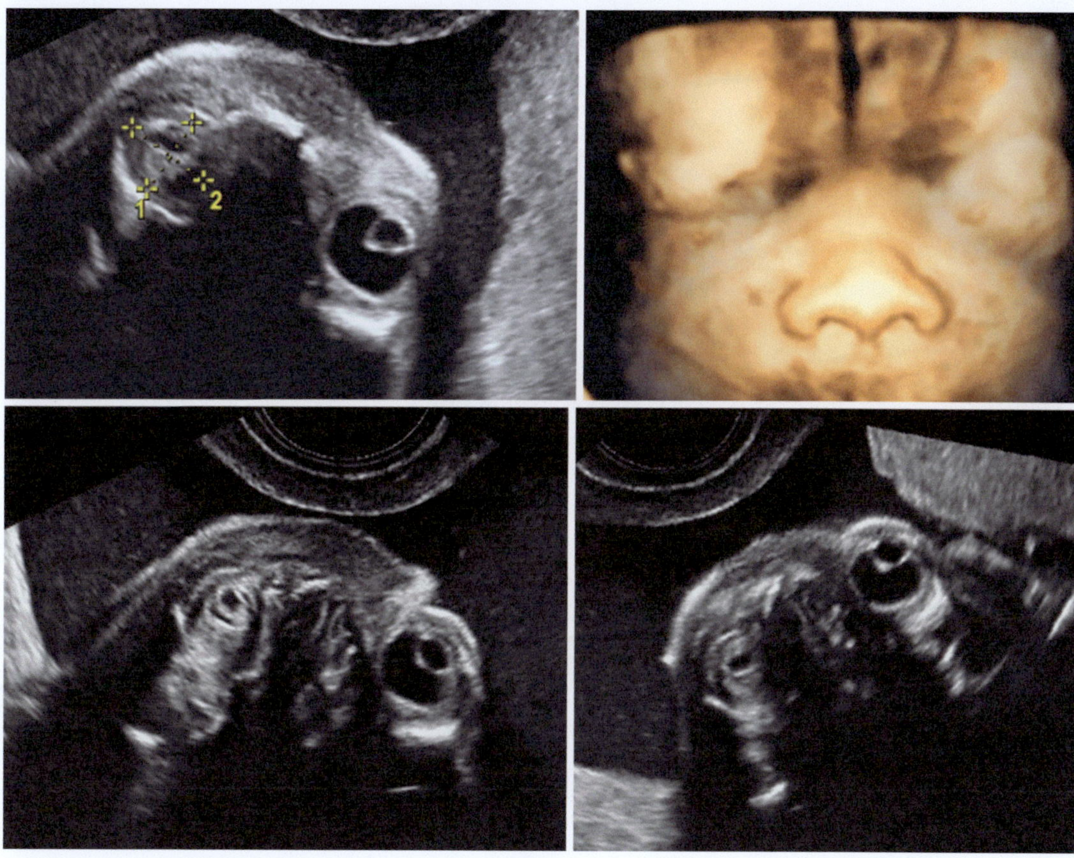

Fig. 6.22 Microphthalmia at 22 GW

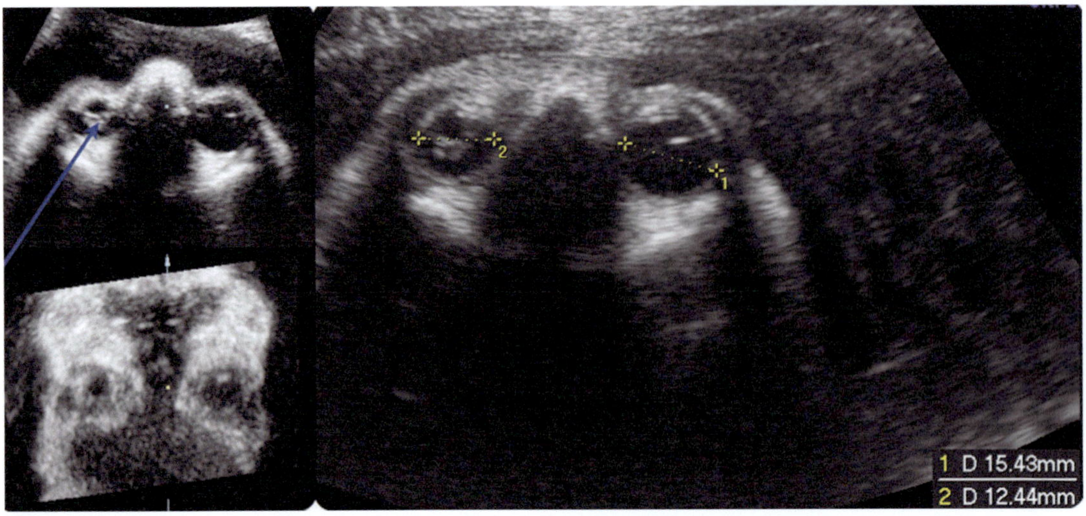

Fig. 6.23 Unilateral microphthalmia at 30 GW (*blue arrow*: persistence of primitive vitreous)

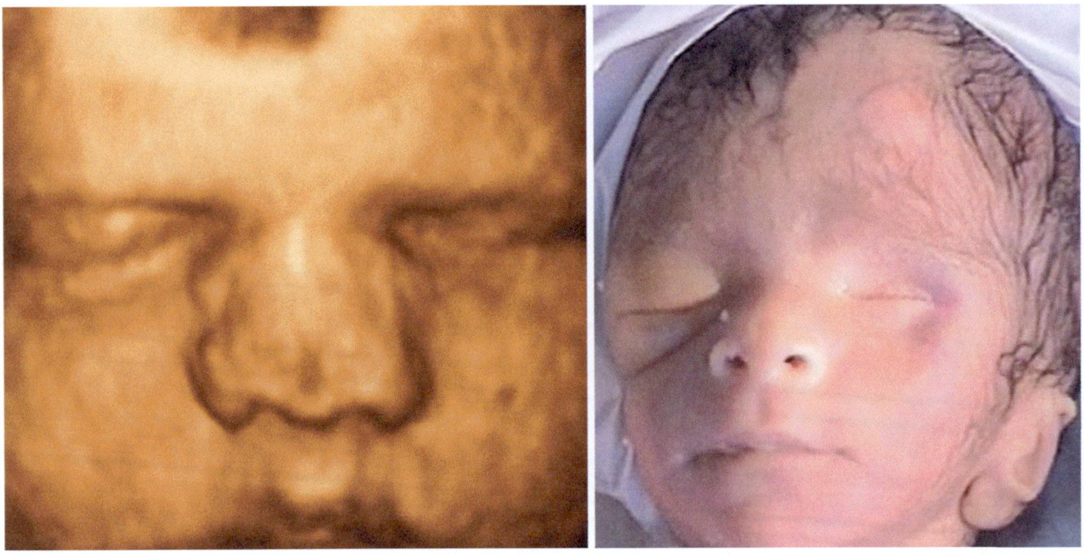

Fig. 6.24 Unilateral microphthalmia: prenatal rendered image and post-natal appearance

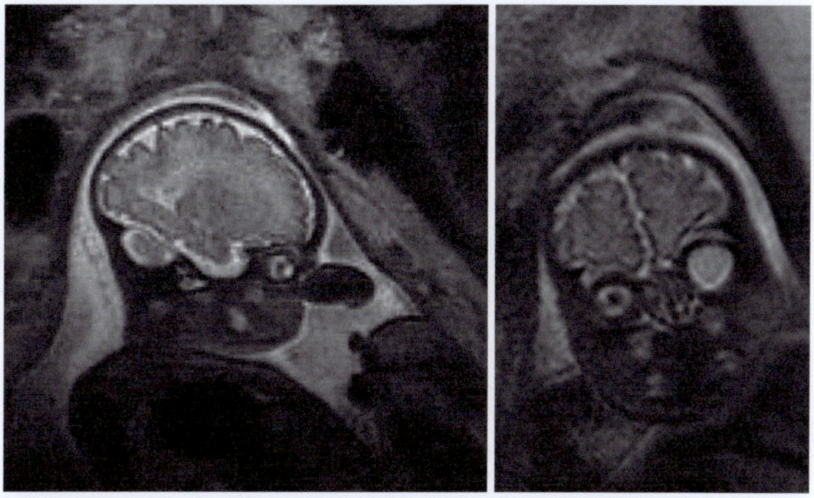

Fig. 6.25 Unilateral microphthalmia: MRI

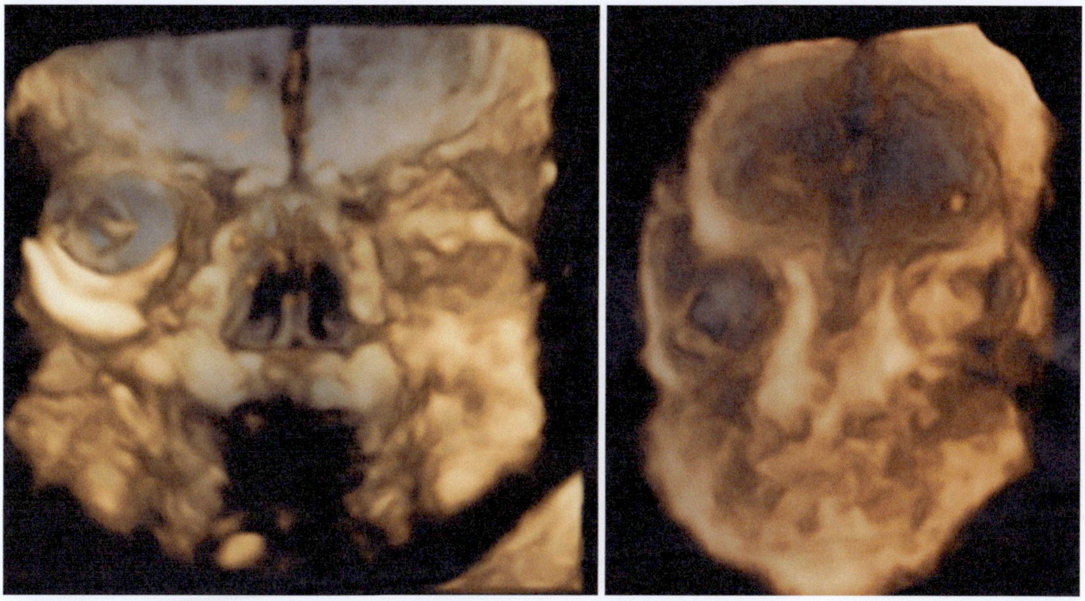

Fig. 6.26 Unilateral microphthalmia – reverse view: *right*, capture from the front; *left*: view of the facial mass from behind (not the small size of the left orbit, which is hard to visualizes from behind)

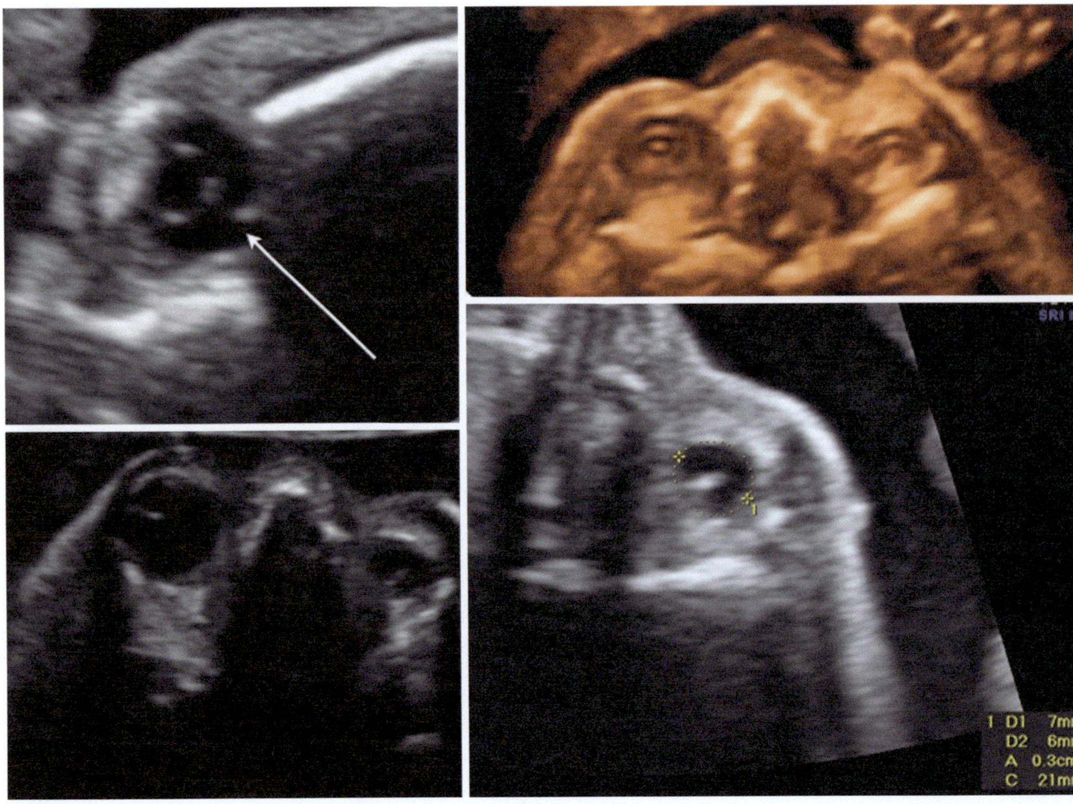

Fig. 6.27 Unilateral colobomatous microphthalmia (*arrow*: defect)

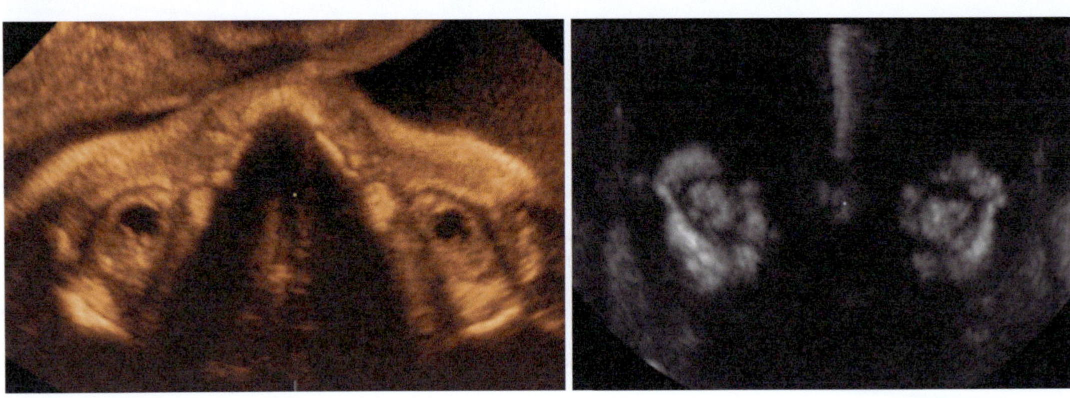

Fig. 6.28 Bilateral microphthalmia

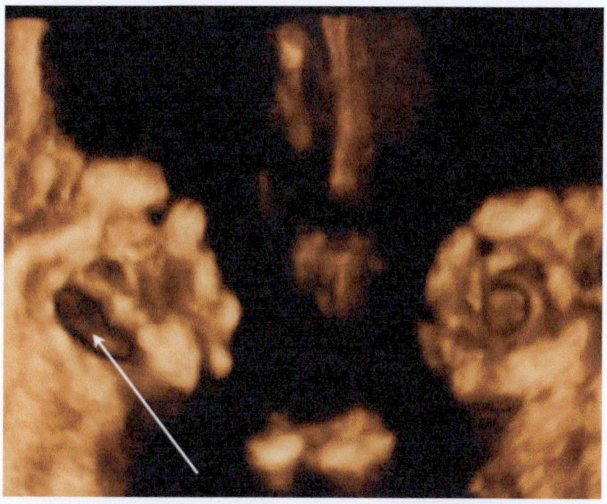

Fig. 6.29 Bilateral colobomatous microphthalmia (*arrow*: colobomatous cyst)

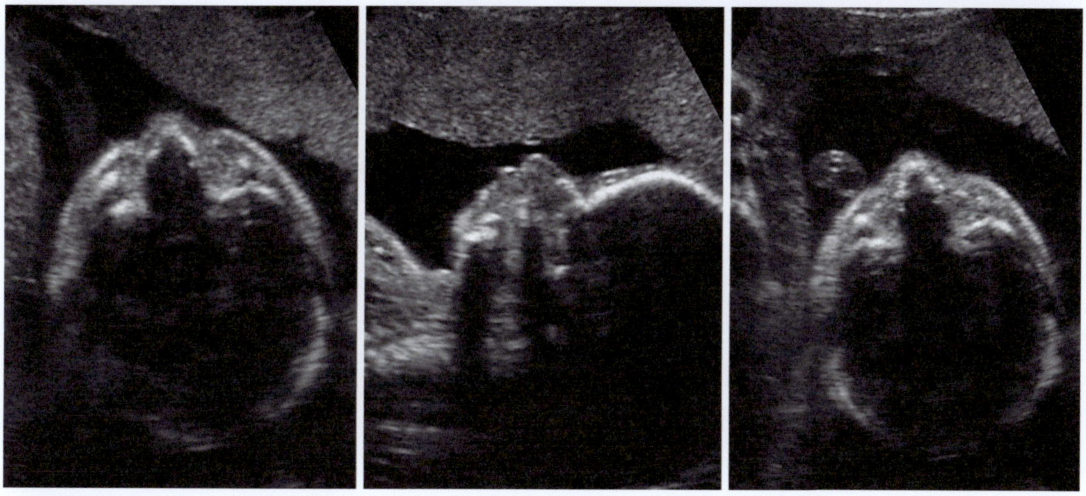

Fig. 6.30 Bilateral anophthalmia

6.1.2.5 Anophthalmia

Unilateral or bilateral: no orbital relief and not ocular structures are visible (Fig. 6.30).

Remember that to confirm there is an eye in an orbit, one needs to visualize the ocular structures. Normally, it is easiest to visualize the lenses.

6.1.2.6 Buphthalmia

This should not be confused with exophthalmia: the bony orbit is normal sized, but the ocular globe is too large, creating an impression of exophthalmia. Biometrics will help to determine the diagnosis (Fig. 6.31).

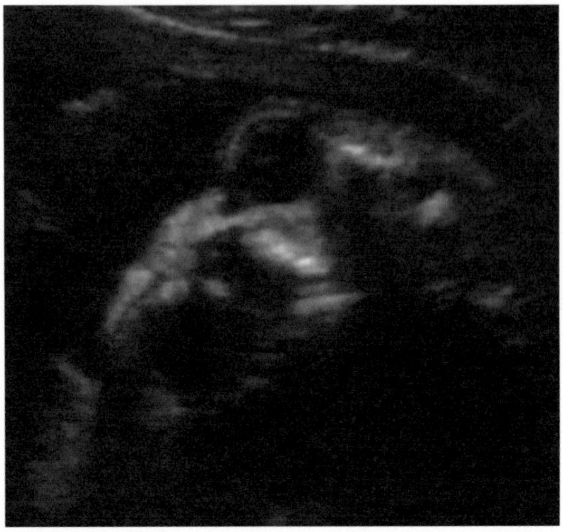

Fig. 6.31 Bilateral buphthalmia

6.1.2.7 Exophthalmia

Unlike buphthalmia, the ocular globe is normal sized, but the orbit is too small or deformed and cannot contain the ocular globe, leading to its exteriorization (Fig. 6.32).

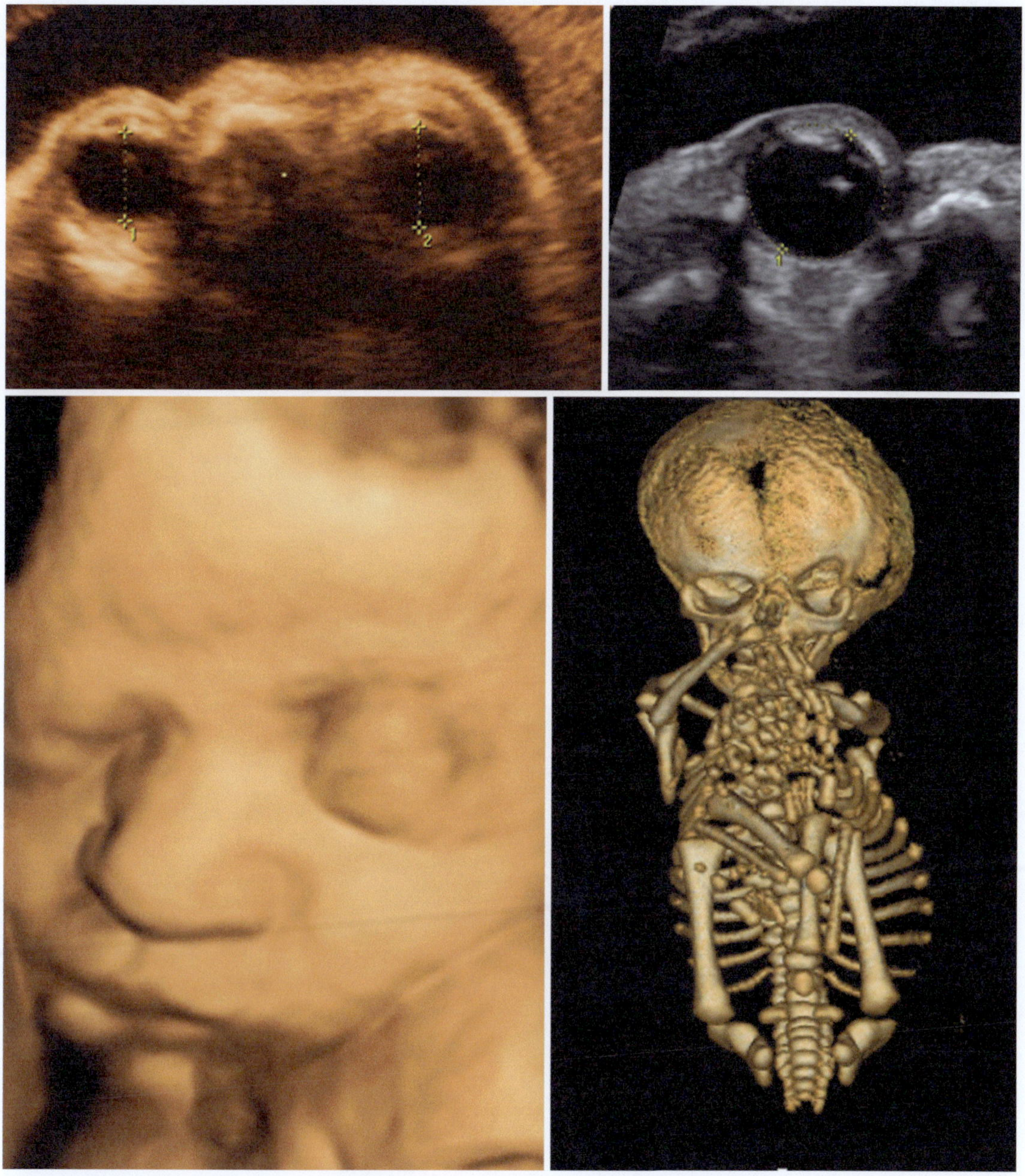

Fig. 6.32 Exophthalmia in a case of craniostenosis (the scanner clearly shows orbital asymmetry)

6.1.2.8 Persistent Primary Vitreous

This arises from the persistence of embryonic structures: the primary vitreous and the hyaloid vessels.

On an ultrasound, it can be seen by the present of an echogenic structure behind the lens. Most often the aspect is an arc of a circle or triangular with a posterior point. This persistence often accompanies microphthalmia (Figs. 6.23 and 6.33).

Buphthalmia, persistent primary vitreous (Fig. 6.35) and retinal dysplasia (Fig. 6.34) are often present in Walker-Warburg syndrome, with type 2 lissencephaly, associating, in particular, cerebral gyration, hydrocephaly, and a "Z"-shaped cerebral trunk.

(See the chapter on cerebral gyration troubles in the "Normal and Pathological Fetal Brain" by the same publisher).

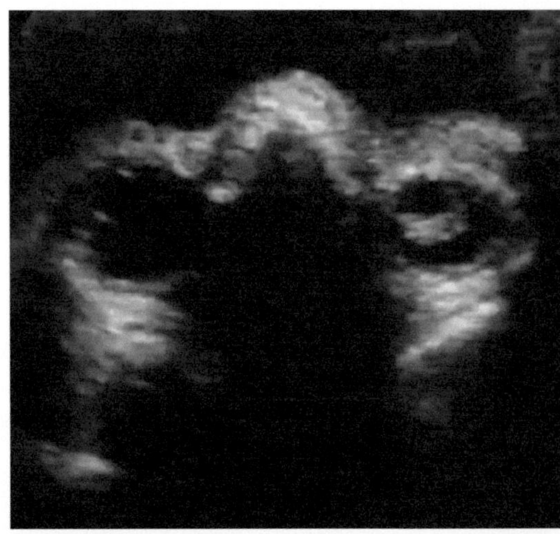

Fig. 6.33 Persistence of primary vitreous and microphthalmia

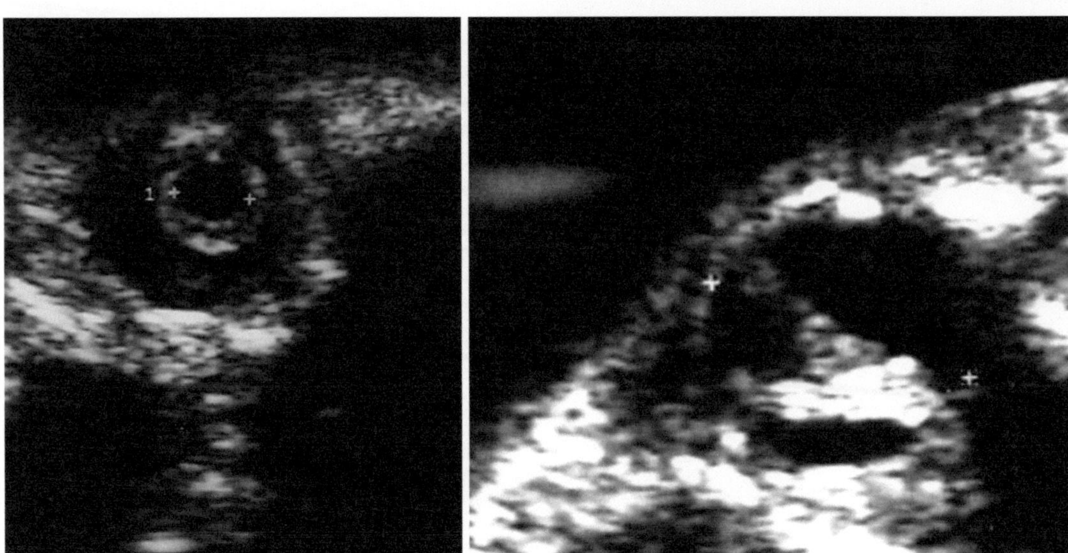

Fig. 6.34 Retinal dysplasia: Walker-Warburg syndrome

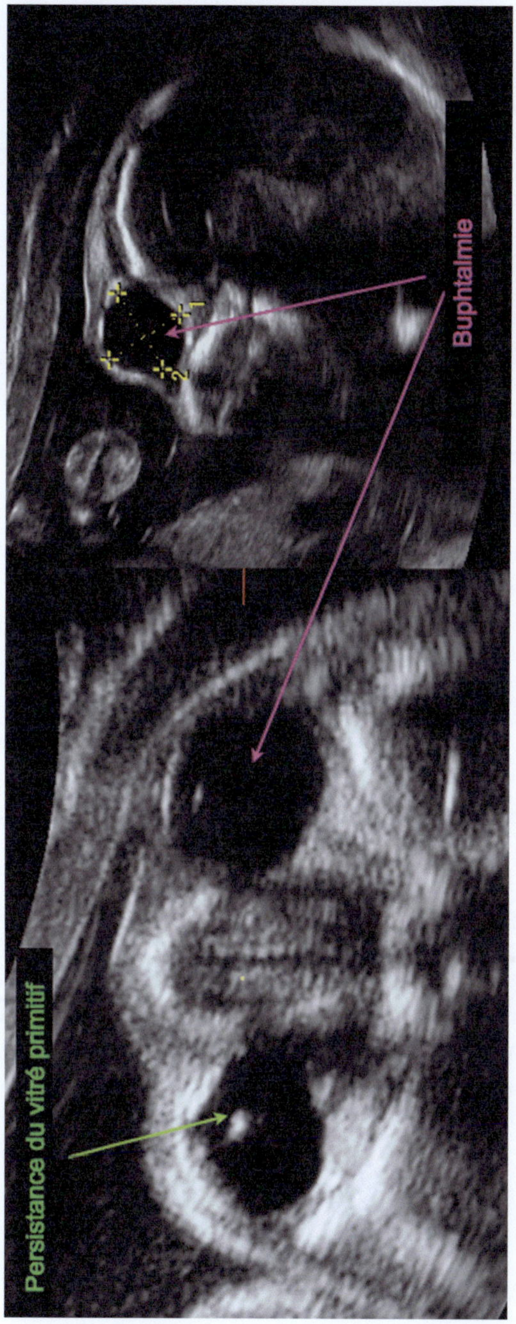

Fig. 6.35 Persistent primary vitreous and buphthalmia – Walker-Warburg syndrome

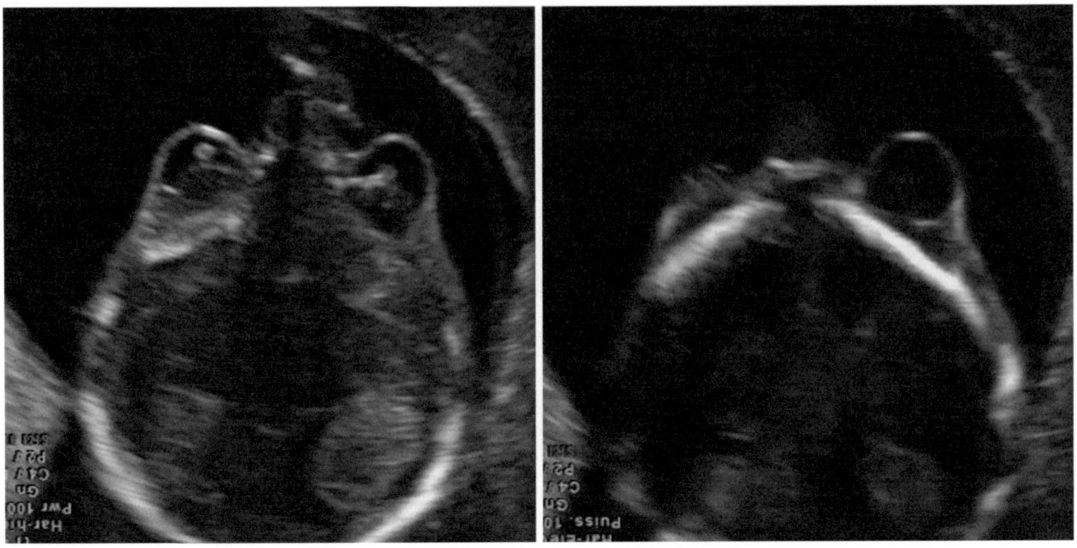

Fig. 6.36 Peters syndrome: glaucoma and cataract

6.1.2.9 Peters Syndrome

This syndrome is linked to an anomaly in the anterior eye chamber related to an anomaly in embryologic development that leads to glaucoma and opacification of the central part of the cornea (Fig. 6.36). This is a recessive autosomal anomaly. It can be associated with facial dysmorphism and hypospadias.

6.1.2.10 Septal and Chiasma Agenesis

When encountering a septel agenesis, the problem when it comes to advising the couple, is whether or not there is a septo-optical dysplasia, with is terrible consequences, which are endocrine troubles and blindness.

C. Lépinard and his team in Angers focused their research on measuring maternal plasma oestriol. When the rate collapses, there is a dysfunction of the hypothalamo-hypophyseal axis in the fetus.

Measurement of the posterior branches of the chiasma can currently be considered as a major prognostic element.

In a series that we recently published in ultrasound, the only two children who were affected by a septo-optic dysplasia had posterior chiasmatic branches presenting measurements very much below what was expected. Fig. 6.36 illustrates a case of serious hypoplasia of the posterior branches of the chiasma at 31 GW, in a fetus who will be affected by total blindness (Z-score: −5) (Fig. 6.37).

Several anomalies can be discovered in the eyes, including excessively long eyelids in Cornelia de Lange syndrome (see chapter on dysmorphism), cryptophthalmia in Frazer syndrome, and more.

As a result, the serious and complete examination of the fetus cannot neglect and attentive and detailed examination of the fetal eyes.

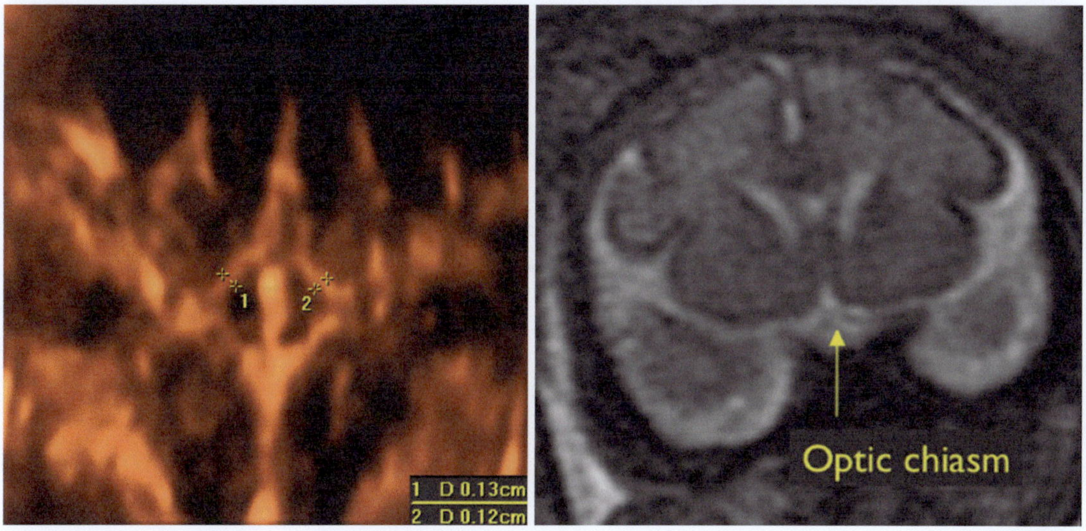

Fig. 6.37 Hypoplasia of the posterior branches of the chiasma in volumetric mode and MRI

References

1. Spaggiari E, Vuillard E, Baumann C, Dupont C, Belarbi N, Oury JF, Delezoide AL, Guimiot F. Ultrasound detection of hyaloid artery in the third trimester of pregnancy: a pathological finding. Ultrasound Obstet Gynecol. 2012;39(4):478–9.
2. Bault JP. Visualisation of the fetal optic chiasm. Ultrasound Obstet Gynecol. 2006;28(4):433–4.
3. Bault JP, Quarello E. Retinal coloboma: prenatal diagnosis using a new technique, the 'virtual fetal eyeground'. Ultrasound Obstet Gynecol. 2009;33(4):495–6.
4. Goldstein I, Tamir A, Zimmer EZ, Itskovitz-Eldor J. Growth of the fetal orbit and lens in normal pregnancies. Ultrasound Obstet Gynecol. 1998;12(3):175–9.
5. Horimoto N, Hepper PG, Shahidullah S, Koyanagi T. Fetal eye movements. Ultrasound Obstet Gynecol. 1993;3(5):362–9.
6. Mashiach R, Vardimon D, Kaplan B, Shalev J, Meizner I. Early sonographic detection of recurrent fetal eye anomalies. Ultrasound Obstet Gynecol. 2004;24(6):640–3.
7. Bault JP. Prognos c value of fetal op c chiasm measurements in fetuses with septal agenesis. Ultrasound Obstet Gynecol. 2007;30(4):390.
8. Bault JP. Re nal coloboma: a case report using "virtual eyeground": a new technique to visualize the fetal retina with 3D ultrasound. Ultrasound Obstet Gynecol. 2008;32(3):410.
9. Pietryga M, Karolczak-Kulesza M, Jankowiak A, Brazert J, Kociecki J. Prenatal diagnosis of eye abnormalities. Ultrasound Obstet Gynecol. 2007;30(4):421.
10. Wenling S, Manhua C, Zongxiang L, Hui L, Zongxiang L. Prenatal diagnosis of dacrocystocele. Ultrasound Obstet Gynecol. 2010;36(S1):239–40.
11. Wong HS, Parker S, Tait J, Pringle KC. Antenatal diagnosis of anophthalmia by three-dimensional ultrasound: a novel application of the reverse face view. Ultrasound Obstet Gynecol. 2008;32(1):103–5.
12. Yazicioğlu HF, Ocak Z. Walker–Warburg syndrome with persistent hyperplastic primary vitreous detected by prenatal ultrasonography. Ultrasound Obstet Gynecol. 2010;35(2):247–9.
13. Boog G, Le Vaillant C, Joubert M. Prenatal sonographic findings in Peters-plus syndrome. Ultrasound Obstet Gynecol. 2005;25(6):602–6.
14. Lepinard C, Coutant R, Boussion F, Loisel D, Delorme B, Biquard F, Bonneau D, Guichet A, Descamps P. Prenatal diagnosis of absence of the septum pellucidum associated with septo-optic dysplasia. Ultrasound Obstet Gynecol. 2005;25(1):73–5.
15. Bault JP, Salomon LJ, Guibaud L, Achiron R. Management of fetuses with septal agenesis. A role for 3D US measurement of the fetal optoc tract. Ultrasound Obstet Gynecol. 2011;37(5):570–5.

7 Biometric Parameters

7.1 The Eyes

Table 7.1 The fetal orbital diameter (mm)

				Centiles				
GA (weeks)	N	Mean	95 % CI	10	25	50	75	90
14	10	5.2	4.8–5.7	4.5	5.0	5.3	5.7	5.7
15	26	6.1	5.9–6.3	5.4	5.5	6.2	6.5	6.7
16	25	6.6	6.3–6.9	5.8	6.2	6.5	7.0	7.6
17–18	19	7.3	6.7–7.8	6.2	6.5	6.7	9.0	9.0
19–20	23	9.8	9.3–10.2	8.6	9.0	10.0	10.1	11.3
21	19	10.5	10.0–10.9	9.4	9.9	10.0	11.0	12.0
22	26	10.4	10.0–10.7	9.5	9.6	10.5	11.0	11.3
23	21	10.7	10.4–11.1	9.6	10.0	10.5	11.4	11.5
24	19	11.6	11.3–11.8	10.7	11.0	11.5	12.0	12.5
25	13	11.2	11.4–12.4	10.3	11.0	12.2	12.5	12.8
26	16	12.7	12.0–13.4	11.0	11.0	12.7	13.8	14.5
27	14	13.0	12.4–13.5	11.9	12.0	12.9	13.4	14.8
28	21	13.0	12.7–13.3	21.1	12.0	13.1	13.3	14.1
29	23	13.9	13.4–14.4	12.6	13.0	13.7	14.6	15.7
30–31	24	14.2	13.8–14.5	13.3	13.0	13.9	14.7	15.4
32–33	24	14.4	13.7–15.1	12.2	13.0	14.1	14.8	17.5
34–36	26	15.8	15.4–16.2	14.6	15.0	15.7	16.5	16.9

Growth of the fetal orbit and lens in normal pregnancies. *Ultrasound in Obstetrics and Gynecology*–Volume 12, issue 3, December 2002, 175–179 I. Goldstein, A. Tamir, E.Z. Zimmer, J. Itskowitz-Eldor
GA gestational age, *CI* confidence interval

J.-M. Levaillant et al., *Normal and Abnormal Fetal Face Atlas*, DOI 10.1007/978-3-319-43769-9_7

Table 7.2 The fetal orbital circumference (mm)

GA (weeks)	N	Mean	95 % CI	Centiles 10	25	50	75	90
14	10	16.4	15.1–17.8	14.1	15	16.5	17.9	17.9
15	26	19.1	18.4–19.8	17.0	17	19.5	20.4	21.0
16	25	20.7	19.9–21.6	18.1	19	20.4	21.9	23.8
17–18	19	22.9	21.2–24.6	19.3	20	20.9	28.3	28.3
19–20	23	30.6	29.3–31.9	26.9	28	31.4	31.7	35.5
21	19	32.9	31.5–34.3	29.4	31	31.4	34.6	37.7
22	26	32.6	31.5–33.6	29.7	30	33.0	34.5	35.4
23	21	33.7	32.6–34.8	30.2	31	33.0	35.8	36.1
24	19	36.3	35.4–37.2	33.5	35	36.1	37.7	39.3
25	13	37.5	35.8–39.1	32.4	36	38.3	39.3	40.3
26	16	40.0	37.8–42.1	34.7	36	39.8	43.4	45.5
27	14	40.7	39.0–42.4	37.4	38	40.5	42.0	46.4
28	21	40.8	39.9–41.7	38.0	39	41.0	41.6	44.4
29	23	43.7	42.2–45.1	39.6	40	42.9	45.9	49.4
30–31	24	44.58	43.5–45.6	41.9	42	43.8	46.0	48.2
32–33	24	45.3	43.0–47.5	38.3	41	44.4	46.4	54.9
34–36	26	49.7	48.5–50.8	46.0	47	49.4	51.7	52.9

Growth of the fetal orbit and lens in normal pregnancies. *Ultrasound in Obstetrics and Gynecology*–Volume 12, issue 3, December 2002, 175–179 I. Goldstein, A. Tamir, E.Z. Zimmer, J. Itskowitz-Eldor
GA gestational age, *CI* confidence interval

Table 7.3 The fetal orbital area (mm^2)

GA (weeks)	N	Mean	95 % CI	Centiles 10	25	50	75	90
14	10	21.6	18.0–25.2	15.9	19.3	21.7	25.5	25.5
15	26	29.2	27.1–31.3	23.0	23.8	30.2	32.9	35.1
16	25	34.4	31.6–37.3	26.2	30.2	32.9	38.2	44.9
17–18	19	42.7	36.2–49.3	29.7	33.2	34.7	63.6	63.6
19–20	23	75.3	68.7–81.8	57.4	63.6	78.6	80.2	100.5
21	19	86.5	79.0–94.1	68.7	76.2	78.6	95.1	113.1
22	26	84.9	79.4–90.4	70.2	72.8	86.6	94.4	99.5
23	21	90.5	84.6–96.5	72.4	79.8	86.6	102.1	103.9
24	19	105.2	100.2–100.3	89.1	100.0	103.0	113.1	122.8
25	13	112.2	102.8–121.5	84.2	103.0	116.0	122.8	129.2
26	16	128.2	114.4–142.0	95.8	106.0	126.0	149.7	165.2
27	14	132.4	120.8–144.0	111.0	116.0	130.0	140.4	170.9
28	21	132.6	127.0–138.3	114.0	124.0	133.0	137.9	157.0
29	23	152.5	142.0–162.9	124.0	131.0	146.0	167.5	194.3
30–31	24	158.3	150.5–166.2	139.0	143.0	152.0	168.8	185.1
32–33	24	165.1	148.2–182.1	117.0	137.0	156.0	170.9	240.0
34–36	26	196.6	187.9–205.3	167.00	181.0	194.0	212.6	222.9

Growth of the fetal orbit and lens in normal pregnancies. *Ultrasound in Obstetrics and Gynecology*–Volume 12, issue 3, December 2002, 175–179 I. Goldstein, A. Tamir, E.Z. Zimmer, J. Itskowitz-Eldor
GA gestational age, *CI* confidence interval

Table 7.4 Diameter of the fetal lense (mm)

GA (weeks)	N	Mean	95 % CI	Centiles 10	25	50	75	90
14	10	2.5	2.3–2.7	2.1	2.4	2.5	2.7	2.9
15	26	2.9	2.9–3.0	2.7	2.8	2.9	3.1	3.2
16	25	2.9	2.8–3.0	2.7	2.8	2.9	3.1	3.2
17–18	19	3.3	3.0–3.6	2.8	2.9	3.0	3.3	5.0
19–20	23	4.1	4.0–4.3	3.6	4.0	4.0	4.3	5.0
21	19	4.4	4.1–4.6	3.7	3.9	4.0	5.0	5.0
22	26	4.4	4.2–4.7	3.9	4.0	4.3	5.0	5.0
23	21	4.6	4.3–4.8	3.8	4.0	5.0	5.0	5.0
24	19	4.6	4.4–4.8	4.0	4.3	4.6	5.0	5.0
25	13	4.8	4.6–5.0	4.2	4.6	5.0	5.1	5.2
26	16	5.0	4.8–5.2	4.4	4.8	5.1	5.2	5.5
27	14	5.0	4.8–5.2	4.4	4.8	5.1	5.2	5.5
28	21	5.1	5.0–5.2	4.5	5.0	5.2	5.2	5.5
29	23	5.3	5.1–5.5	4.6	5.2	5.2	5.5	5.9
30–31	24	5.3	5.2–5.5	4.8	5.1	5.5	5.5	5.7
32–33	24	5.6	5.4–5.8	4.8	5.2	5.5	5.9	6.2
34–36	26	5.8	5.6–6.0	5.4	5.5	5.7	6.0	6.5

Growth of the fetal orbit and lens in normal pregnancies. *Ultrasound in Obstetrics and Gynecology*–Volume 12, issue 3, December 2002, 175–179 I. Goldstein, A. Tamir, E.Z. Zimmer, J. Itskowitz-Eldor
GA gestational age, *CI* confidence interval

Table 7.5 Area of the fetal lens (mm²)

GA (weeks)	N	Mean	95 % CI	Centiles 10	25	50	75	90
14	10	5.1	4.3–6.9	3.5	4.4	4.9	5.6	6.6
15	26	6.8	6.4–7.2	5.9	6.2	6.6	7.6	8.0
16	25	6.7	6.3–7.0	5.6	6.2	6.6	7.3	8.0
17–18	19	8.8	6.9–10.8	7.9	9.0	9.0	10.6	13.8
19–20	23	13.6	12.4–14.8	10.2	12.0	12.6	14.5	19.6
21	19	15.1	13.3–17.0	10.8	12.0	12.6	19.6	19.6
22	26	15.8	14.2–17.3	12.2	12.0	14.2	19.6	19.6
23	21	16.6	15.0–18.2	11.6	12.0	19.6	19.6	19.6
24	19	16.9	15.5–18.4	12.6	14.0	16.6	19.6	19.6
25	13	18.3	16.7–19.9	13.6	16.0	19.6	20.4	21.2
26	16	19.7	18.2–21.3	15.2	18.0	19.6	21.2	23.8
27	14	19.8	18.1–21.4	15.2	18.0	19.6	21.2	23.4
28	21	20.5	19.4–21.7	15.9	19.0	21.2	21.2	23.8
29	23	22.3	20.6–24.0	16.4	21.0	21.2	23.8	27.4
30–31	24	22.4	21.1–23.8	18.1	20.0	23.8	23.8	25.9
32–33	24	24.5	22.8–26.3	18.1	21.0	23.8	27.4	30.0
34–36	26	26.6	25.0–28.1	23.0	23.0	25.6	28.3	33.5

Growth of the fetal orbit and lens in normal pregnancies. *Ultrasound in Obstetrics and Gynecology*–Volume 12, issue 3, December 2002, 175–179 I. Goldstein, A. Tamir, E.Z. Zimmer, J. Itskowitz-Eldor
GA gestational age, *CI* confidence interval

Table 7.6 Circumference of the fetal lens (mm)

				Centiles				
GA (weeks)	N	Mean	95 % CI	10	25	50	75	90
14	10	8.0	7.3–8.6	6.7	7.5	7.9	8.4	9.1
15	26	9.2	9.0–9.5	8.6	8.8	9.1	9.7	10.1
16	25	9.1	8.9–9.4	8.4	8.8	9.1	9.6	10.0
17–18	19	11.0	9.3–11.4	9.5	10.0	11.0	11.5	12.5
19–20	23	13.0	12.4–13.6	11.3	12.0	12.6	13.5	15.7
21	19	13.7	12.8–14.5	11.6	12.0	12.6	15.7	15.7
22	26	14.0	13.3–14.7	12.4	12.0	13.4	15.7	15.7
23	21	14.4	13.7–15.1	12.1	12.0	15.7	15.7	15.7
24	19	14.5	13.9–15.2	12.6	130	14.5	15.7	15.7
25	13	15.1	14.5–15.8	13.1	14.0	15.7	16.0	16.3
26	16	15.7	15.1–16.3	13.8	15.0	15.7	16.3	17.3
27	14	15.7	15.1–16.4	13.8	15.0	15.9	16.3	17.1
28	21	16.3	15.6–16.5	14.1	15.0	16.3	16.3	17.3
29	23	16.7	16.1–17.3	14.3	16.0	16.3	17.3	18.5
30–31	24	16.7	16.2–17.2	15.1	16.0	17.3	17.3	18.0
32–33	24	17.5	16.9–18.1	15.1	16.0	17.3	18.5	19.4
34–36	26	18.2	17.7–18.8	17.0	17.0	17.9	18.9	20.5

Growth of the fetal orbit and lens in normal pregnancies. *Ultrasound in Obstetrics and Gynecology*–Volume 12, issue 3, December 2002, 175–179 I. Goldstein, A. Tamir, E.Z. Zimmer, J. Itskowitz-Eldor
GA gestational age, *CI* confidence interval

Table 7.7 The outer orbital diameter (OOD) and inner orbital diameter (IOD)

	OOD (mm)			IOD (mm)		
	Centiles			Centiles		
GA (weeks)	5	50	95	5	50	95
12	8	15	23	4	9	13
13	10	18	25	5	9	14
14	13	20	28	5	10	14
15	15	22	30	6	10	14
16	17	25	32	6	10	15
17	19	27	34	6	11	15
18	22	29	37	7	11	16
19	24	31	39	7	12	16
20	26	33	41	8	12	17
21	28	35	43	8	13	17
22	30	37	44	9	13	18
23	31	39	46	9	14	18
24	33	41	48	10	14	19
25	35	42	50	10	15	19
26	36	44	51	11	15	20
27	38	45	53	11	16	20
28	39	47	54	12	16	21
29	41	48	56	12	17	21
30	42	50	57	13	17	22
31	43	51	56	13	18	22
32	45	52	60	14	18	23
33	46	53	61	14	19	23
34	47	54	62	15	19	24
35	48	55	63	15	20	24
36	49	56	64	16	20	25
37	50	57	65	16	21	25
38	50	58	65	17	21	21
39	51	58	66	17	22	26
40	52	59	67	18	22	26

GA gestational age (Jeanty et al. 1984)

7.2 The Lens

Table 7.8 Lens diameters

12 GW	3 mm
17 GW	4 mm
Term	5–6 mm
Never more than 6 mm	

Pratique de l'échographie en gynécologie obstétrique (JM Bouton, M. Denhez, F. Eboué, Vigot, 1990)
GW gestational weeks

7.3 Vitreous Circumference

Measurement protocol

Coronal view: the transverse diameter and the supero-inferior diameter of the vitreous are measured from one internal edge to the other. The circumference is calculated automatically with a formula (D1 + D1) ×1.57

Table 7.9 Vitreous Circumference

	Centiles		
GA (weeks)	5	50	95
12	4	10.9	17.7
13	7.7	14.6	21.4
14	10.9	17.8	24.6
15	13.8	20.6	27.5
16	16.3	23.2	30.0
17	18.6	25.4	32.3
18	207	27.5	34.3
19	22.5	29.4	36.2
20	24.2	31.1	37.9
21	25.8	32.6	39.4
22	27.2	34.0	40.9
23	28.5	35.4	42.2
24	29.8	36.6	43.4
25	30.9	37.7	44.6
26	31.9	38.8	45.6
27	32.9	39.8	46.6
28	33.9	40.7	47.5
29	34.7	41.6	48.4
30	35.6	42.4	49.2
31	36.3	43.2	50.0
32	37.1	43.9	50.8
33	37.8	44.6	51.4
34	38.8	45.3	52.1
35	39.0	45.9	52.7
36	39.6	46.5	53.3
37	40.2	47.0	53.9

The development of the fetal eye: ultrasonographic measurements of the vitreous and lens in utero. E. Achiron, Z. Gottlieb, Y. Yaron, M Gabbay, S Lipitz, S Mashiach. *Prenat Diagn* 1995, 15, 155–160

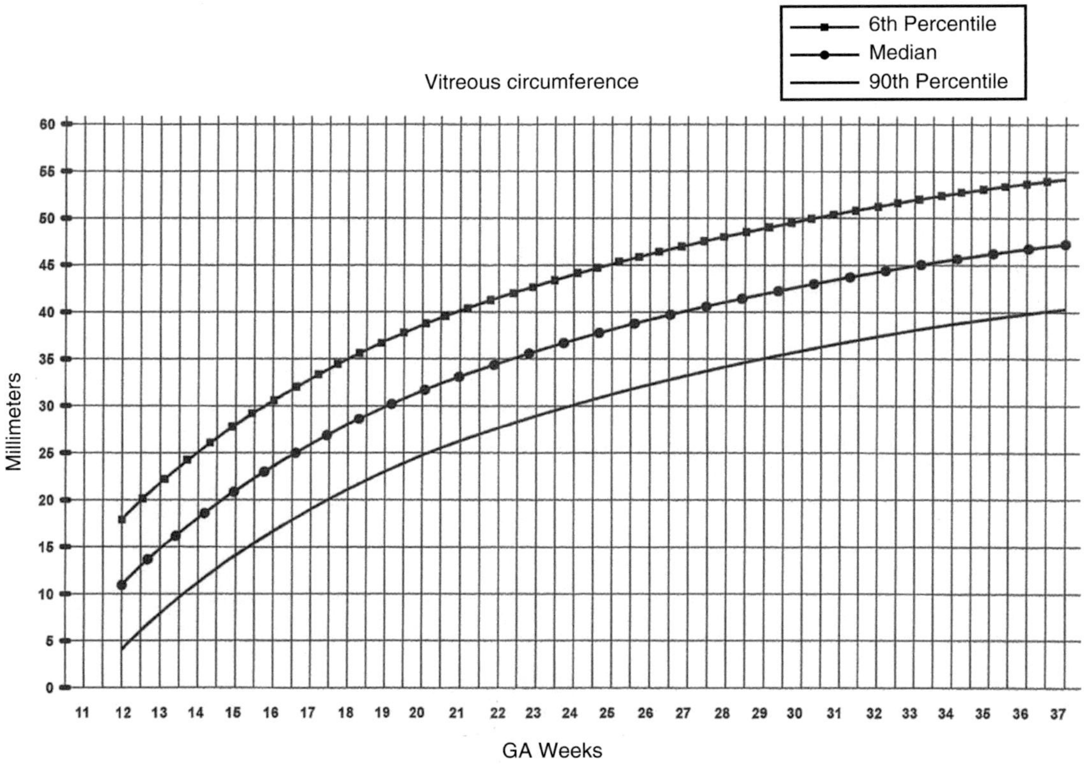

Fig. 7.1 Vitreous circumference

7.4 Ocular Diameter

Measurement protocol

The orbital diameter is measured from the internal bone edge on one side to the external bone edge on the same side.

Table 7.10 Ocular diameter

	Centiles		
GA (weeks)	5	50	95
12	1	3	6
13	2	4	7
14	3	5	8
15	4	6	9
16	5	7	9
17	5	8	10
18	6	9	11
19	7	9	12
20	8	10	13
21	8	11	13
22	9	12	14
23	10	12	15
24	10	13	15
25	11	13	16
26	12	14	16
27	12	14	17
28	13	15	17
29	13	15	18
30	14	16	18
31	14	16	19
32	14	17	19
33	15	17	19
34	15	17	20
35	15	18	20
36	16	18	20
37	16	18	21
38	16	18	21
39	16	19	21
40	16	19	21

Growth of the ocular parameters. Prenatal diagnosis of congenital anomalies, 1988, table 2–1, 83 R. Romero, G Pilu, P. Jeanty, A. Ghidini, JC Hobbins

7.5 Lens Circumference

Table 7.11 Lens circumference

	Centiles		
GA (weeks)	5	50	95
12	1.9	5.7	9.6
13	3.5	7.4	11.3
14	4.9	8.8	12.7
15	6.2	10.0	13.9
16	7.3	11.1	15
17	8.2	12.1	15.9
18	9.1	12.9	16.8
19	9.8	13.7	17.5
20	10.5	14.3	18.2
21	11.1	15.0	18.8
22	11.7	15.5	19.4
23	12.2	16.0	19.9
24	12.6	15.5	20.4
25	13.1	16.9	20.8
26	13.5	17.3	21.2
27	13.8	17.7	21.6
28	14.2	18.0	21.9
29	14.5	18.3	22.2
30	14.8	18.6	22.5
31	15.1	18.9	22.8
32	15.3	19.2	23.0
33	15.6	19.4	23.3
34	15.8	19.7	23.5
35	16.0	19.9	23.7
36	16.2	20.1	23.9
37	16.4	20.3	24.1

The development of the fetal eye: ultrasonographic measurements of the vitreous and lens in utero. E. Achiron, Z. Gottlieb, Y. Yaron, M Gabbay, S Lipitz, S Mashiach. *Prenat Diagn* 1995, 15, 155–160

7.6 The Ear

Table 7.12 The fetal ear

GA (weeks)	Cases (numbers)	Mean	Scale
18	7	12.42	2.01
19	11	14.35	1.02
20	44	14.91	1.12
21	135	16.35	1.20
22	200	17.22	1.11
23	97	18.38	1.34
24	28	19.32	1.66
25	16	19.62	1.60
26	6	21.51	2.08
27	9	23.43	1.66
28	16	24.90	2.44
29	18	25.29	2.03
30	12	26.09	2.88
31	27	28.13	1.74
32	49	28.77	1.46
33	19	29.52	1.67
34	8	30.90	1.55

(J. Kerleroux, S. Kerleroux, E. Masse, Médecine fœtale, 1997 no. 29)

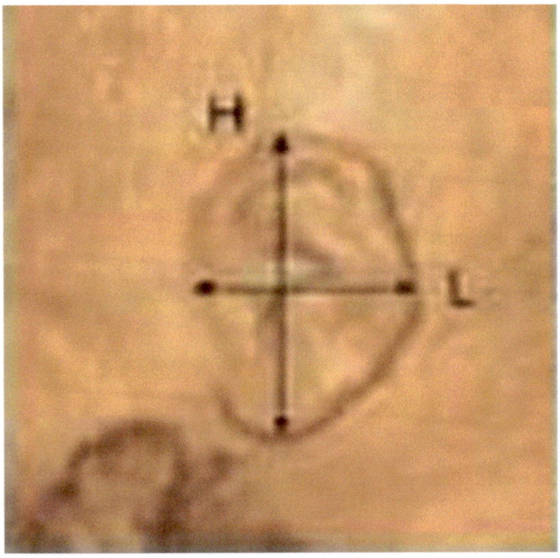

Fig. 7.2 Diameter of the lens

Fig. 7.3 Ear length

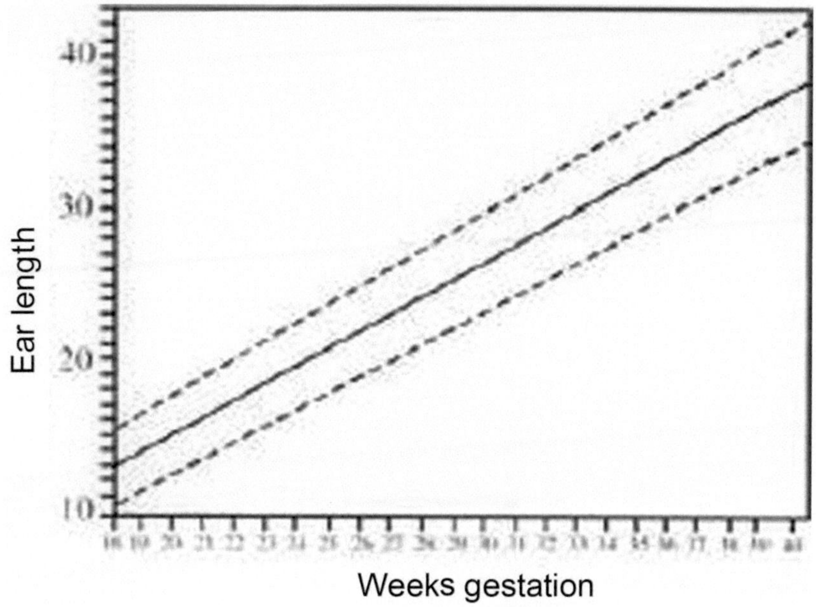

Table 7.13 CV indicates coefficient of variation

GA (weeks)	n	Ear length (mm)		
		Mean	SD	CV%
14	13	8	0.7	9.1
15	14	9	1.8	19.6
16	28	10	0.9	8.8
17	23	11	1.0	8.8
18	26	13	0.7	5.8
19	18	14	1.1	7.9
20	29	15	1.1	7.4
21	18	17	1.0	6.1
22	26	18	1.5	8.3
23	20	19	1.4	7.2
24	13	20	1.1	5.6
25	18	22	1.7	7.9
26	8	23	2.1	8.9
27	17	25	1.7	6.6
28	13	26	1.8	6.8
29	15	26	1.6	6.1
30	9	27	2.3	8.8
31	22	29	2.6	8.9
32	16	29	1.9	6.4
33	16	30	1.8	5.9
34	14	31	1.6	5.3
35	13	31	2.1	6.9
36	11	33	2.6	7.8
37	19	33	2.0	6.3
38	7	33	4.1	12.3
39	7	34	4.3	12.5
40	7	37	2.1	5.7
41	7	36	1.9	5.3

Fetal ear length statistical characteristics according to GA. Prenatal detection of fetal aneuploidy by sonographic ear length. *J. Ultrasound Med*, 2003–22:565. L Yeo, ER Guzman, CV Ananth, C Walters, D Day-Salvatore, AM Vintzileos

Table 7.14 Normogram of fetal pinna length (mm) according to percentile distribution

		Centiles						
GA (weeks)	n	5	10	25	50	75	90	95
15	10	8.18	8.36	9.00	9.20	10.08	11.47	12.24
16	12	8.83	9.54	9.98	10.60	13.03	13.97	14.00
17	12	12.91	13.02	13.43	14.20	14.73	15.25	15.35
18	11	12.55	12.70	13.80	15.00	16.20	16.70	17.05
19	23	14.00	14.00	15.55	16.00	17.25	17.94	18.00
20	60	15.00	15.40	16.00	17.00	18.00	19.60	20.00
21	88	15.80	16.00	17.00	17.90	19.00	20.20	20.90
22	92	16.20	17.00	18.00	19.00	20.00	21.40	22.50
23	63	17.00	17.60	1840	20.00	21.40	22.60	22.80
24	52	18.50	19.10	19.90	20.90	22.40	23.60	24.00
25	35	18.90	19.10	20.00	21.90	23.00	24.20	24.60
26	39	21.00	21.20	22.40	23.70	25.50	26.80	29.00
27	21	21.60	22.20	23.00	25.00	27.30	28.60	28.60
28	20	24.40	24.90	25.50	26.40	27.10	28.10	28.50
29	18	23.90	24.40	25.00	28.10	29.90	31.60	33.60
30	21	25.00	25.00	27.00	28.00	29.20	31.10	31.30
31	17	26.20	26.30	27.00	28.30	29.90	30.20	31.00
32	29	26.40	27.30	28.00	30.00	31.20	32.20	33.60
33	24	14.70	25.70	28.00	31.40	34.00	34.60	35.00
34	30	27.10	27.70	29.60	32.20	34.30	36.10	37.40
35	33	29.30	30.20	31.10	32.60	34.00	35.80	38.10
36	24	30.00	30.50	31.90	34.00	35.60	36.80	34.40
37	18	29.90	30.40	33.70	35.10	36.40	37.90	38.20
38	13	30.90	31.30	32.20	34.00	37.70	38.20	38.50
39	12	29.60	29.90	32.10	34.90	36.40	37.90	38.90
40	10	29.10	29.20	36.20	37.30	37.90	38.20	39.10

Sonography measurement of fetal pinna length in normal pregnancies. *Kathmandu Univ Med J (KUMJ)*. 2011 Apr-Jun;9 (34):49–53. KS Joshi, CD Chawla, S Karki, NC Shrestha

7.7 Nasal Bones

Table 7.15 Length of the nasal bone

GA (weeks)	Mean (mm)	SD	−2SD	+2SD
14	4.183	0.431	3.321	5.045
16	5.213	1.062	3.089	7.337
18	6.308	0.654	5.000	7.616
20	7.621	0.953	5.715	9.527
22	8.239	1.102	6.035	10.443
24	9.362	1.300	6.762	11.962
26	9.744	1.277	7.190	12.298
28	10.72	1.459	7.803	13.639
30	11.348	1.513	8.322	14.374
32	11.580	1.795	7.990	15.170
34	12.285	2.372	7.541	17.029

Mean, standard deviation (*SD*), mean +2SD and mean −2SD for length of the nasal bones (mm) throughout gestation (Guis et al. 1995)

Table 7.16 Fetal nasal bone length (mm), 11–20 weeks' gestation (Cuick et al. 2004)

GA (weeks)	Mean (mm)	SD (mm)
11–11.9	1.7	0.5
12–12.9	2.0	0.5
13–13.9	2.3	0.5
14–14.9	3.4	0.7
15–15.9	3.3	0.8
16–16.9	4.4	0.7
17–17.9	5.0	0.7
18–18.9	5.5	0.9
19–19.9	5.7	0.1
20–20.9	6.2	0.1

Table 7.17 Nomogram of fetal nasal bone length at 11–13 gestational weeks in fetuses (Sivri et al. 2006)

GA (weeks)	Mean (mm)	SD (mm)
11–11+6	1.69	0.26
12–12+6	2.11	0.37
13–13+6	2.34	0.39
14–14+6	2.94	0.48

7.8 Nose Width

Table 7.18 The fetal nose width (mm) (Goldstein et al. 1997)

	Centiles				
GA (weeks)	10	25	50	75	90
14–15	5.5	7.2	7.6	8.3	10.2
16–17	6.5	7.3	7.9	8.5	10.5
18–19	8.5	8.9	10.0	10.5	11.0
20–21	10.2	11.0	12.0	12.0	13.0
22	13.0	13.0	14.0	15.0	15.0
23	13.0	13.0	14.0	15.0	15.0
24	13.0	14.1	15.0	16.0	16.0
25	14.2	15.0	16.3	17.0	17.0
26	14.1	15.0	16.3	17.4	18.4
27	13.4	15.4	17.2	18.4	19.0
28	15.1	16.9	17.6	18.2	20.2
29–30	16.5	17.4	18.1	19.2	20.6
31–32	16.6	17.9	19.6	20.7	21.4
33–34	17.4	19.1	20.5	21.4	23.1
35–37	17.6	20.0	20.5	22.0	23.3
38–40	17.4	17.9	18.9	20.5	23.4

7.9 Nostrils

Table 7.19 The fetal nostril distance (mm) (Goldstein et al. 1997)

	Centiles				
GA (weeks)	10	25	50	75	90
14–15	3.3	3.6	4.2	4.7	5.4
16–17	3.5	3.9	4.4	4.8	5.9
18–19	4.0	4.4	4.6	5.0	5.8
20–21	4.2	5.0	5.0	5.7	6.0
22	5.0	5.0	6.0	6.4	7.0
23	5.0	5.6	6.0	7.0	7.0
24	5.8	6.0	6.2	7.3	7.9
25	5.9	6.0	6.4	7.0	7.7
26	5.1	6.2	7.7	8.0	9.0
27	6.4	6.8	7.8	8.4	9.4
28	6.4	7.0	7.9	8.6	9.4
29–30	5.4	7.0	7.0	8.2	9.6
31–32	4.6	7.4	7.9	9.2	10.7
33–34	5.4	6.4	8.1	9.0	9.7
35–37	5.8	6.6	8.5	9.6	10.2
38–40	6.0	6.8	8.5	9.5	10.5

Table 7.20 Measurement of the meant ±2SD of the frontal lobe distance and thalamic frontal lobe distance versus gestational age (Goldstein et al. 1988)

GA (weeks)	FLD (cm)	Mean ±2SD	TFLD (cm)	Mean ±2SD
15	1.4	0.4	3.2	0.4
16	1.4	0.4	3.2	0.4
17	1.6	0.2	3.6	0.6
18	1.6	0,2	3.7	0.6
19	1.7	0.2	3.8	0.4
20	1.7	0.2	4.1	0.4
21	1.8	0.4	4.1	0.4
22	1.8	0.4	4.6	0.4
23	1.8	0.4	4.6	0.4
24	1.9	0.2	4.7	0.4
25	2.2	0.4	5.1	0.6
26	2.3	0.4	5.2	0.6
27	2.5	0.6	5.6	0.8
28	2.8	0.2	5.7	0.4
29	2.7	0.2	6.1	0.4
30	2.8	0.6	6.2	1.2
31	2.9	0.4	6.2	0.8
32	3.0	0.6	6.4	0.8
33	3.1	0.6	6.5	0.6
34	3.2	0.2	6.7	0.6
35	3.2	0.4	6.9	0.6
36	3.2	0.4	7.0	0.4
37	3.4	0.4	7.2	0.6
38	3.5	0.4	7.3	0.8
39	3.7	0.6	7.5	0.8
40	4.0	0.6	7.7	0.8

GA gestational age, *FLD* frontal lobe distance, *TFLD* thalamic frontal lobe distance

7.10 Maxillary Bone Length

Table 7.21 Maxillary bone length across gestationnal age

				Xentiles		
GA (weeks)	n	Mean	SD	10	50	90
14	17	9.97	1.12	8.32	10.0	11.52
15	59	10.64	1.07	9.4	10.6	11.8
16	13	10.6	1.73	7.6	10.4	12.98
17–19	7	10.07	2.75	7.0	10.9	13.0
20–22	14	11.48	3.42	7.0	11.0	17.35
23–24	15	13.19	3.34	8.6	13.0	16.76
25–26	11	12.85	1.74	10.2	13.0	15.92
27–28	15	12.61	2.11	10.0	12.0	16.2
29–30	16	13.63	1.67	11.67	13.5	16.23
31	15	13.16	1.25	11.0	13.0	15.48
32	18	13.49	1.25	11.9	13.45	15.0
33	20	13.7	1.37	11.11	14.0	15.95
34	25	13.87	1.72	11.96	14.0	16.15
35	28	14.15	1.27	12.54	14.0	16.0
36	28	14.31	1.4	12.63	14.35	16.15
37	12	14.08	1.26	12.93	14.0	16.73
38–39	12	14.84	1.77	11.74	14.8	17.47

Nomogram of maxillary bone length in normal pregnancy. *Journal of ultrasound in medicine*, September 2005 – 24(9): 1229–1233, I Goldstein, A Reiss, BS Rajamim, A Tamir

7.11 Alveaolar Ridge Width

Table 7.22 Normal values of the fetal alveolar ridge width

GA (weeks)	Mean (mm)	+/–SO (mm)
14–15	10.5	1.3
16	11.7	1.1
17	16.6	2.5
18	17.5	1.1
19	18.0	1.1
20	18.5	1.1
21	18.5	2.1
22	19.9	1.7
23	20.5	1.9
24	21.3	2.7
25	22.8	1.9
26	23.6	2.6
27–28	23.6	2.1
29	25.5	2.2
30	26.3	2.5
31	26.5	2.1
32	26.7	2.0

Nomogram of the fetal alveolar ridge: a possible screening tool for the detection of primary cleft palate. *Ultrasound obstetric gynecology* – November 1999, 14 (5), 333–337

7.12 Tongue

Search of macroglossie (D. Nyberg)

Tongue circumference: age (GW) × 2.75

Ultrasound obstetric gynecology – January 1997 – 39–41. R Achiron, A Ben Arie, U Gabbay, S Mashiach, Z Rotstein, S Lipitz

Table 7.23 Tongue circumference (mm) by gestational age and the 95 % confidence interval

GA (weeks)	Lower 95 % Cl	Mean	Upper 95 % Cl
14	24	28	31
15	26	33	36
16	33	36	38
17	37	37	38
18	40	43	46
19	47	48	51
20	47	51	56
21	51	55	61
22	52	58	62
23	58	62	68
24	60	64	67
25	68	70	73
26	71	73	76

Development of the fetal tongue between 14 and 25 weeks of gestation: in utero ultrasonographic measurements

7.12.1 Skin Thickness Facing the Forehead

22 GW: 3.8 mm ± 0.6 mm

Table 7.24 Skin thickness

GA (weeks)	Skin thickness	SD
21	3.44	0.65
22	3.8	0.63
23	3.84	0.61
24	4.22	0.89
25–26	3.91	0.7
27–28	4.83	0.8
29–30	5.16	0.91
31–32	5.78	1.1
33	5.93	1.1

JM Levaillant
N = 638

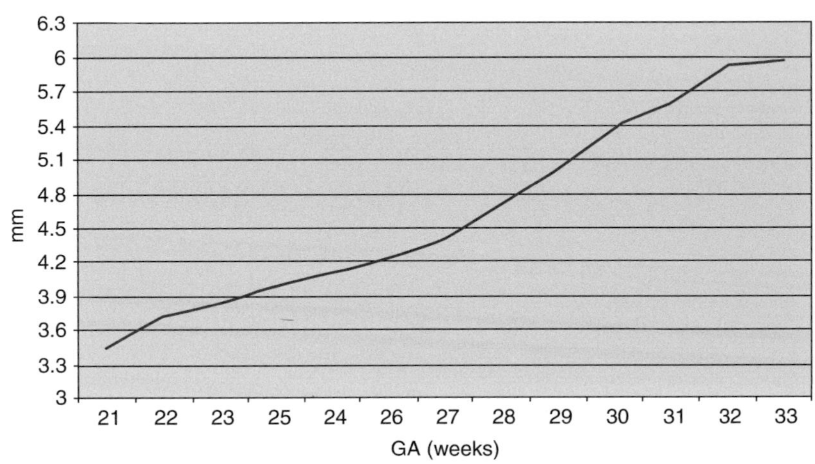

Fig. 7.4 Skin thickness

7.13 Length of Philtrum

Table 7.25 Length of Philtrum

GA (weeks)	Mean	SD
21	6.06	0.99
22	6.15	0.79
23	6.44	1.03
24	6.47	0.99
25–26	6.7	1.02
27–28	7.34	1
29–30	7.02	1.36
31–32	7.4	1.31
33–34	7.65	1.29

J.M. Levaillant. N = 651
22 GW: 6.15 mm ± 0.7
32 GW: 7.4 mm ± 1.37

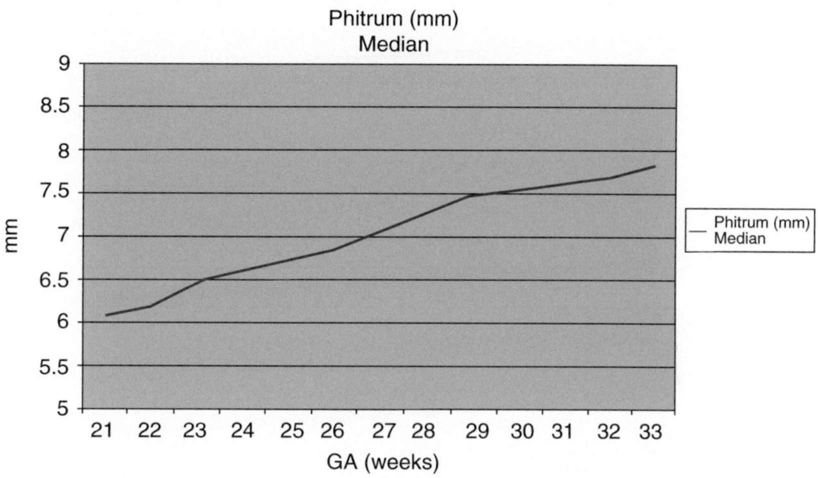

Fig. 7.5 Length of Philtrum

7.14 The Ears

Table 7.26 Ear length

GA (weeks)	Ear	SD
21	16.35	0.96
22	17.52	0.86
23	18.67	1.47
24	20.03	2.27

JM Levaillant
N= 135

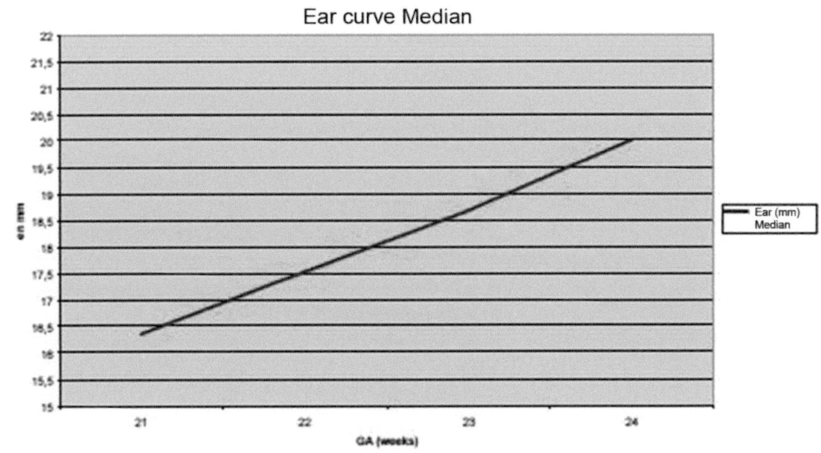

Fig. 7.6 Ear length

7.15 Lower Facial Angle

Table 7.27 Lower facial angle

GA (weeks)	Lower facial angle mean	SD
21–24	65.1	7.79
25–27	67.2	8.43
28–34	69.06	7.96

JM Levaillant
N= 422

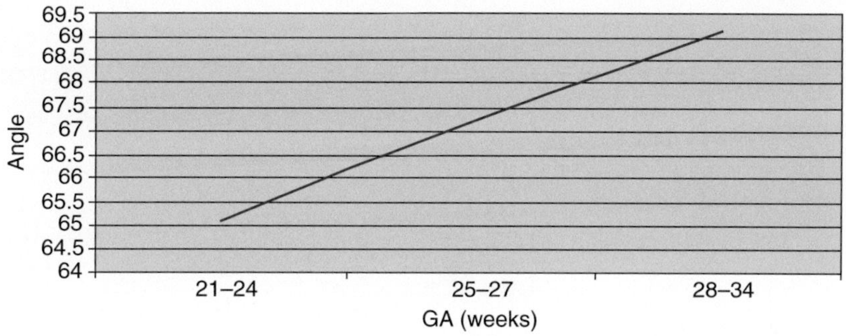

Fig. 7.7 Lower facial angle

7.16 Upper Facial Angle

Table 7.28 Upper facial angle

GA (weeks)	Upper facial angle mean	SD
21–24	128.12	10.99
25–28	129.36	13.42
29–34	131.14	12.25

JM Levaillant
N= 454

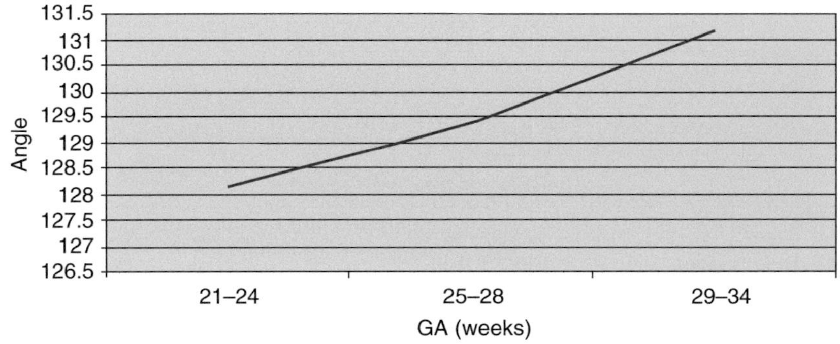

Fig. 7.8 Upper facial angle

7.17 Maxillary Width

Table 7.29 Maxillary width

GA (weeks)	Maxillary width mean	SD
22	24.8	3.67
23	26.19	4.05
24	27	4.14

JM Levaillant
N = 79

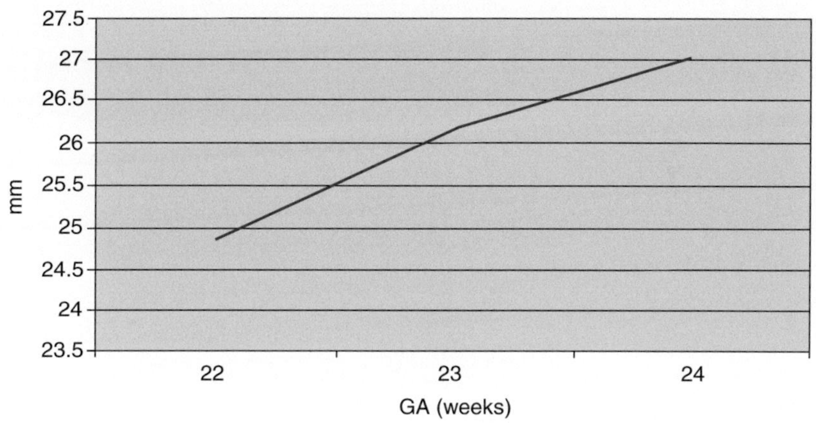

Fig. 7.9 Maxillary width

7.18 Mandibular Width

Table 7.30 Mandibular width

GA (weeks)	Mandibular width mean	SD
22	24.15	2.64
23	25.29	2.53
24	25.4	2.16

JM Levaillant
N = 79

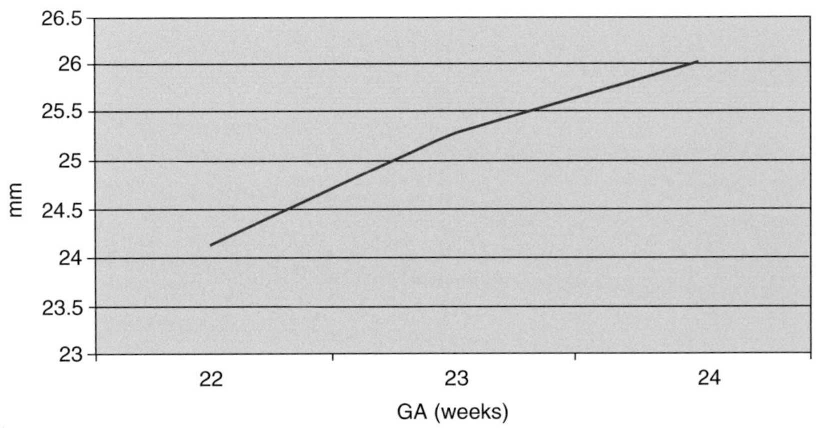

Fig. 7.10 Mandibular width

7.19 IOD

A few points of reference
Median IOD

22 GW: 25.2 mm ± 1.3 mm
32 SA: 34.8 mm ± 2.5 mm
IOD/BPD
22 GW: 0.47 ± 0.5–32 SA : 0.42 ± 0.5

Table 7.31 Median IOD

GA (weeks)	Median IOD mean	SD
21	24.24	1.61
22	25.26	1.25
23	25.97	2.32
24	27.04	1.59
25–27	28.86	1.99
28–29	30.91	2.43
31–32	34.15	2.35
33–34	34.49	2.14

JM Levaillant
N = 325

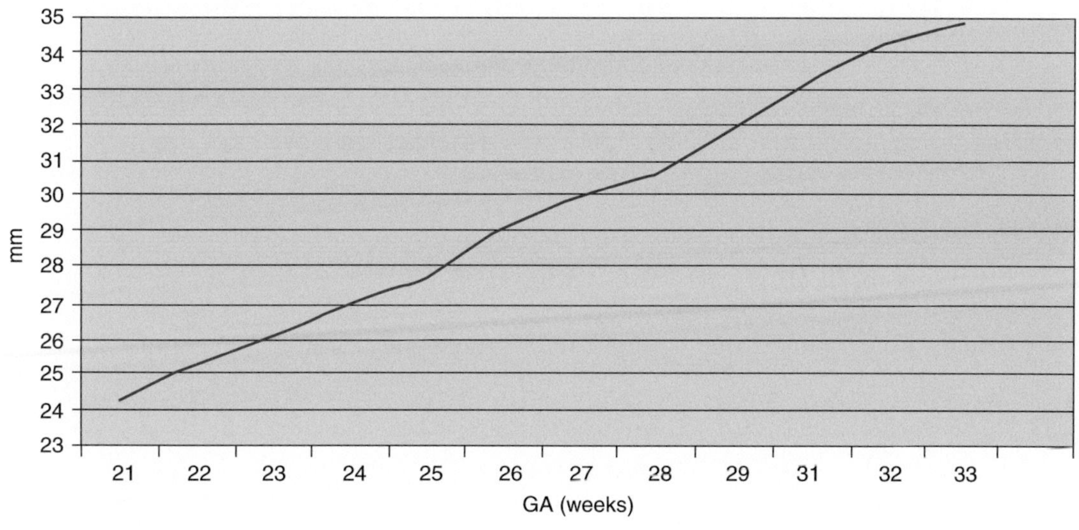

Fig. 7.11 Median IOD

Table 7.32 Internal IOD

GA (weeks)	Internal IOD mean	SD
21	13.66	1.49
22	14.09	1.42
23	14.22	1.57
24	14.57	1.89
25–26	15.27	1.81
27–28	16.47	1.56
29–30	16.5	1.63
31–32	17.65	2.19
33–34	18.14	2.26

JM Levaillant
N = 660

Table 7.33 External IOD

GA (weeks)	External IOD mean	SD
21	34.4	2.56
22	36.05	2.22
23	37.55	2.71
24	39.35	2.1
25–26	40.98	2.72
27–28	43.35	2.48
29–30	48.71	2.75
31–32	50.05	3
33–34	50.84	3.07

JM Levaillant
N = 618

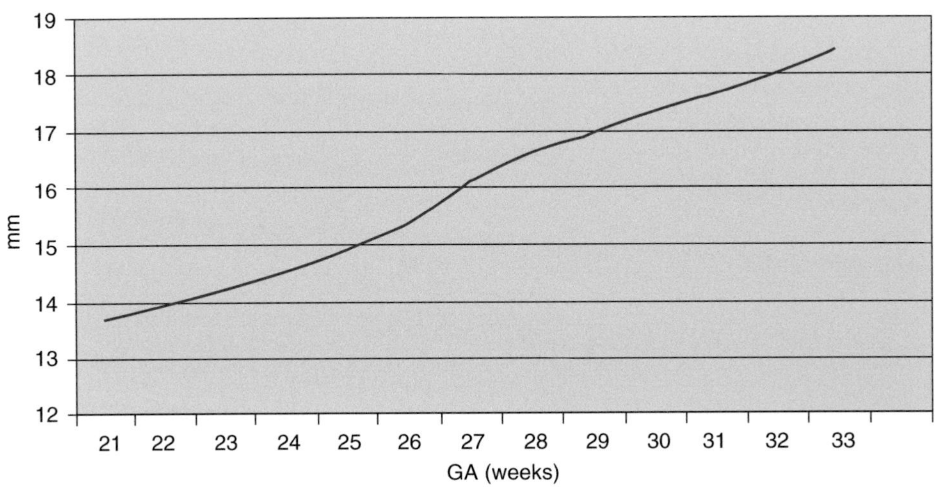

Fig. 7.12 Internal IOD

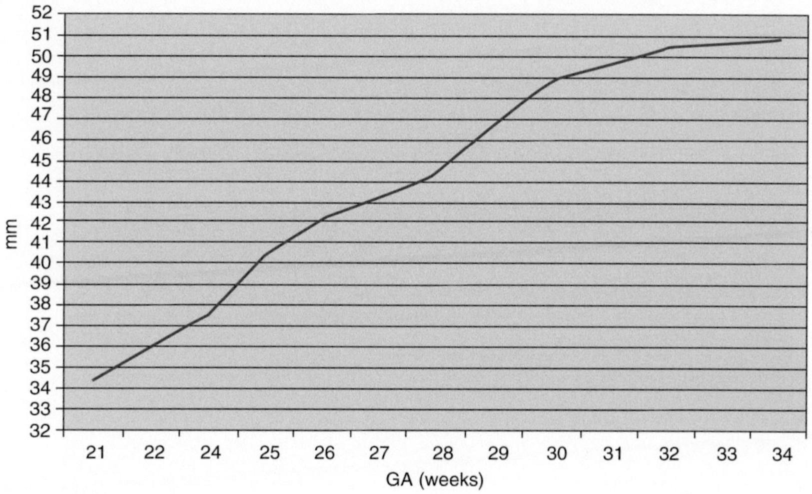

Fig. 7.13 External IOD

MIX
Papier aus verantwortungsvollen Quellen
Paper from responsible sources
FSC® C105338

If you have any concerns about our products,
you can contact us on
ProductSafety@springernature.com

In case Publisher is established outside the EU,
the EU authorized representative is:
Springer Nature Customer Service Center GmbH
Europaplatz 3, 69115 Heidelberg, Germany

Printed by Libri Plureos GmbH
in Hamburg, Germany